TRANSATLANTIC ENVIRONMENT AND ENERGY POLITICS

Global Environmental Governance

Series Editors: John J. Kirton, Munk Centre for International Studies, Trinity College, Canada and Miranda A. Schreurs, Freie Universität Berlin, Germany

Global Environmental Governance addresses the new generation of twenty-first century environmental problems and the challenges they pose for management and governance at the local, national, and global levels. Centred on the relationships among environmental change, economic forces, and political governance, the series explores the role of international institutions and instruments, national and sub-federal governments, private sector firms, scientists, and civil society, and provides a comprehensive body of progressive analyses on one of the world's most contentious international issues.

The Legitimacy of International Regimes
Helmut Breitmeier
ISBN 978-0-7546-4411-8

Governing Agrobiodiversity
Plant Genetics and Developing Countries
Regine Andersen
ISBN 978-0-7546-4741-6

The Social Construction of Climate Change
Power, Knowledge, Norms, Discourses
Edited by
Mary E. Pettenger
ISBN 978-0-7546-4802-4

Governing Global Health
Challenge, Response, Innovation
Edited by
Andrew F. Cooper, John J. Kirton and Ted Schrecker
ISBN 978-0-7546-4873-4

Participation for Sustainability in Trade
Edited by
Sophie Thoyer and Benoît Martimort-Asso
ISBN 978-0-7546-4679-2

Bilateral Ecopolitics
Continuity and Change in Canadian-American Environmental Relations
Edited by
Phillipe Le Prestre and Peter Stoett
ISBN 978-0-7546-4177-3

Transatlantic Environment and Energy Politics
Comparative and International Perspectives

Edited by

MIRANDA A. SCHREURS
Free University of Berlin, Germany

HENRIK SELIN
Boston University, USA

STACY D. VANDEVEER
University of New Hampshire, USA

Routledge
Taylor & Francis Group

LONDON AND NEW YORK

First published 2009 by Ashgate Publishing

2 Park Square, Milton Park, Abingdon, Oxfordshire OX14 4RN
711 Third Avenue, New York, NY 10017

Routledge is an imprint of the Taylor & Francis Group, an informa business

First issued in paperback 2018

British Library Cataloguing in Publication Data
Transatlantic environment and energy politics : comparative
  and international perspectives. - (Global environmental
  governance series)
  1. Environmental policy - European Union countries -
  Congresses 2. Environmental policy - United States -
  Congresses 3. Energy policy - European Union countries -
  Congresses 4. Energy policy - United States - Congresses
  I. Schreurs, Miranda A. (Miranda Alice), 1963- II. Selin,
  Henrik III. VanDeveer, Stacy D.
  333.7

Library of Congress Cataloging-in-Publication Data
Schreurs, Miranda A. (Miranda Alice), 1963-
  Transatlantic environment and energy politics : comparative and international
perspectives / by Miranda A. Schreurs, Henrik Selin, and Stacy D. VanDeveer.
     p. cm. -- (Global environmental governance)
  ISBN 978-0-7546-7597-6
  1. Sustainable development--United States. 2. Sustainable development--European
Union countries. 3. Environmental policy--United States. 4. Environmental policy-
-European Union countries. 5. International cooperation. I. Selin, Henrik, 1971- II.
VanDeveer, Stacy D. III. Title.

  HC110.E5S395 2009
  333.7094--dc22

                                                                    2008051704
  ISBN 978-0-7546-7597-6 (hbk)
  ISBN 978-1-138-37653-3 (pbk)

Contents

List of Figures and Tables

Figures

Tables

Contributors

Phillip Aerni is a Senior Researcher at the World Trade Institute in Bern, Switzerland.

Graeme Auld is a PhD candidate in the School of Forestry and Environmental Studies at Yale University.

Thomas Bernauer is Professor of Political Science at ETH Zurich.

Elizabeth Bomberg is Senior Lecturer in the School of Social and Political Studies at The University of Edinburgh.

Marcus Carson is a Researcher and Senior Lecturer in the Department of Sociology, Stockholm University.

Benjamin Cashore is Professor for Environmental Governance and Sustainable Forest Policy in the School of Forestry and Environmental Studies at Yale University, Director, Program on Forest Policy and Governance, and Director, Project on Forest Certification.

Elizabeth Egan is ecosystems services specialist in the Office of the Chief, US Forest Service, Washington DC.

Alastair Iles is Assistant Professor of Science, Technology and Environment in the Division of Society and Environment, Department of Environmental Science, Policy and Management, at the University of California, Berkeley.

Patricia M. Keilbach is Assistant Professor of Political Science at the University of Colorado at Colorado Springs.

Kate O'Neill is Associate Professor in the Division of Society and Environment, Department of Environmental Science, Policy and Management at the University of California, Berkeley.

Deanna Newsom is Researcher in the Rainforest Alliance's Evaluation and Research Program.

Ian H. Rowlands is Professor in the Faculty of Environment at the University of Waterloo.

Marcus Schaper is Visiting Assistant Professor in the Political Science Department at Reed College, Oregon.

Miranda A. Schreurs is Professor of Comparative Politics and Director of the Environmental Policy Research Center at the Free University of Berlin.

Henrik Selin is Assistant Professor in the Department of International Relations at Boston University.

Stacy D. VanDeveer is Associate Professor in the Department of Political Science at the University of New Hampshire.

Sonja Wälti is Assistant Professor at the School of Public Affairs, American University, Washington DC, and is also affiliated with the Hertie School of Governance in Berlin.

Acknowledgments

This project has been several years in the making. The editors have organized several conference panels and workshops in order to bring contributors together to discuss the status of the transatlantic environmental relationship and to work on this book.

Many of the ideas for this project emerged out of a panel we held at the International Studies Association annual convention in Montreal, Canada, March 17–20, 2004 entitled "Divergent Environmental Policy Styles Across the Atlantic: Implications for Theory and Practice." This was followed by an authors' workshop, which was held at the Massachusetts Institute of Technology, Cambridge, Massachusetts, July 15–17, 2004. As the book project began to take shape, we organized several panels to get feedback on contributors' chapters. These included a panel on "Environmental Relations in the US and an Expanded EU," at the 2004 Berlin Conference on the Human Dimensions of Climate Change, Berlin, December 3–4, 2004; a panel on "Transatlantic Environmental Relations" at the International Studies Association annual convention in Honolulu, Hawaii, March 1–5, 2005; and a conference on "Enlarging TransAtlantic Relations: Environment, Agriculture, and Trade Politics across the Atlantic" hosted by the American Consortium for European Studies, Johns Hopkins University, Washington DC, November 18, 2005. Chapter drafts were also presented at the Sixth International CISS Millennium Conference in The Hague, the Netherlands, July 2–4, 2006. Additional authors were pulled in to fill in gaps and based on reviewers' comments changes were made to strengthen the manuscript. We thank all those people who provided many helpful comments and suggestions along the way.

The project has received financial support from multiple sources. The workshop at the Massachusetts Institute of Technology in 2004 was funded by the Knut and Alice Wallenberg Foundation. Several of the authors (Benjamin Cashore, Alastair Iles, Patricia Keilbach, Kate O'Neill, Miranda Schreurs, Henrik Selin, and Stacy D. VanDeveer) benefited from seed grants for their work that were provided by the American Consortium for European Studies (ACES). Miranda Schreurs appreciates the research support provided by the Tamaki Foundation, the Department of Government and Politics, University of Maryland and the Environmental Policy Research Center, Freie Universität Berlin. Henrik Selin would like to thank the Knut and Alice Wallenberg Foundation, the European Commission's Jean Monnet Program, and the Department of International Relations at Boston University for supporting his research and this project. Stacy D. VanDeveer would also like to thank the University of New Hampshire's College of Liberal Arts and the Ronald

H. O'Neal Professorship, and the European Commission's Jean Monnet Program for supporting the project.

Finally, a very special thanks from all project participants is due to Marcus Schaper who has done much of the technical work associated with editing this manuscript. His help has been extensive and invaluable.

Miranda Schreurs, Henrik Selin and Stacy D. VanDeveer
September 2008

Abbreviations

AATF	African Agricultural Technology Foundation
AB	World Trade Organization Appellate Body
AI	Asbestos Institute
AI	Avian Influenza (bird flu)
AIA	Advanced Informed Agreement
AID	Agency for International Development
Amcham	American Chamber of Commerce
APHIS	Animal and Plant Health Inspection Service
BMP	best management practice
BSE	bovine spongiform encephalopathy
CAFÉ	corporate average fuel economy
CBD	Convention on Biological Diversity
CBP	Cartagena Biosafety Protocol
CCP	Cities for Climate Protection Program
CDC	Centers for Disease Control and Prevention
CDU	Christian Democratic Union
CEFIC	European Chemical Industry Council
CGIAR	Consultative Group of International Agricultural Research
CJD	Creutzfeldt-Jakob disease
CLRTAP	Convention on Long-Range Transboundary Air Pollution
CODEX	Codex Alimentarius Commission
COP	Conference of the Parties
CPGR	Commission on Plant Genetic Resources
CPSC	Consumer Product Safety Commission
CRSP	Collaborative Research Support Program
CSA	Canadian Standards Association
CSA	Climate and Stewardship Act
CSR	corporate social responsibility
CSTEE	Commission's Scientific Committee on Toxicity, Ecotoxicity and the Environment
DG	European Commission Directorate-General
DG SANCO	Directorate-General of Health and Consumer Protection
DNA	deoxyribonucleic acid
DSB	World Trade Organization Dispute Settlement Body
DSD	Duales System Deutschland
EAP	(European) Environmental Action Programme
EC	European Community
ECA	Export Credit Agency

ECDC	European Center for Disease Prevention and Control
ECG	OECD Export Credit Group
ECOS	Environmental Council of the States
ECSC	European Coal and Steel Community
EEA	European Environment Agency
EEB	European Environmental Bureau
EFSA	European Food Safety Authority
EIA	Energy Information Administration
EIA	Environmental Impact Assessment
EIS	Environmental Impact Statement
ELV	end-of-life vehicle
EMEP	Geneva Protocol on Long-term Financing of the Cooperative Programme for Monitoring Evaluation of the Long-Range Transmission of Air Pollutants in Europe
EP	European Parliament
EPA	Environmental Protection Agency
EPR	extended producer responsibility
ESA	Endangered Species Act
ETS	Emission Trading System
ETUC	European Trade Union Confederation
EU	European Union
Ex-Im	Export-Import Bank
FAO	Food and Agricultural Organization
FAS	Foreign Agricultural Service
FDA	Food and Drug Administration
FDP	Free Democratic Party
FSC	Forest Stewardship Council
FTAA	Free Trade Area of the Americas
GATT	General Agreement on Tariffs and Trade
GDP	gross domestic product
GE	genetic engineering
GEN	Global Ecolabelling Network
GHG	greenhouse gas
GKKE	Gemeinsame Konferenz Kirche und Entwicklung
GM	genetically modified
GMO	genetically modified organism
HPAI	Highly Pathogenic Avian Influenza
HSE	Health and Safety Executive
IBAS	International Ban Asbestos Secretariat
ICCP	Intergovernmental Committee for the Cartagena Protocol
ICLEI	International Council for Local Environmental Initiatives
IEA	International Energy Agency
IFIC	International Food Information Council
ILO	International Labor Organization

IMA	Interministerieller Ausschuß (interministerial committee)
IPCC	Intergovernmental Panel on Climate Change
IPCS	International Program on Chemical Safety
IPPC	International Plant Protection Convention
IPR	intellectual property rights
ISP	Independent Science Panel on genetically modified food products
JREC	Johannesburg Renewable Energy Coalition
LPAI	Low Pathogenic Avian Influenza
MBM	meat-and-bone meal
MEP	Member of the European Parliament
NATO	North Atlantic Treaty Organization
NAFTA	North American Free Trade Agreement
NEPA	National Environmental Protection Act
NGO	non-governmental organization
NIH	National Institutes of Health
NIOSH	National Institute for Occupational Safety and Health
NPR-PPMs	non-product-related processes and production methods
NRC	Natural Resources Canada
NSMD	non-state market driven
NTB	non-tariff barriers (to trade)
OECD	Organization for Economic Cooperation and Development
OIE	Office International des Epizooties (International Organization for Animal Health)
PBT	persistent, bioaccumulative and toxic
PC	Personal computer
PCB	Polychlorinated biphenyl
PCSD	(US) President's Council on Sustainable Development
PEFC	Pan European Forest Certification
PIC	prior informed consent
POP	persistent organic pollutant
PPP	purchasing power parity
PURPA	Public Utility Regulatory Policies Act
PV	photovoltaic
R&D	research and development
RAC	Recombinant Advisory Committee
REACH	European Union Registration, Evaluation and Authorization of Chemicals Regulation
RGGI	Regional Greenhouse Gas Initiative
ROHS	Restriction of Hazardous Substances
RPS	renewable portfolio standard
SAICM	Strategic Approach to International Chemicals Management
SDS	Sustainable Development Strategy
SFI	Sustainable Forestry Initiative

SFM	sustainable forest management
SPD	Social Democratic Party
SPS	sanitary and phytosanitary measures
SRM	specific risk materials
SSC	Scientific Steering Committee
SVTC	Silicon Valley Toxics Coalition
TAED	Transatlantic Environment Dialogue
TBT	technical barriers to trade
TSCA	Toxic Substances Control Act
TSE	transmissible spongiform encephalopathy
UN	United Nations
UNEP	United Nations Environment Programme
UNFF	United Nations Forum on Forests
UNICE	Union of Industrial and Employers' Confederations of Europe
UPOV	International Union for the Protection of New Varieties of Plants
US	United States of America
USDA	United States Department of Agriculture
vCJD	variant Creutzfeldt-Jakob disease
WEED	world economy, ecology and development
WEEE	waste electrical and electronic equipment
WHO	World Health Organization
WSSD	World Summit on Sustainable Development
WTO	World Trade Organization
WWF	World Wide Fund for Nature

Chapter 1
Expanding Transatlantic Relations: Implications for Environment and Energy Politics

Miranda A. Schreurs, Henrik Selin and Stacy D. VanDeveer

The transatlantic relationship is one of the strongest and most densely institutionalized transnational relationships in the world. The strength of transatlantic relations is based on many shared security interests, common historical experiences, shared values associated with free and open societies and markets, and deep economic interdependence through extensive trade and foreign direct investment. The European Union (EU) and the United States (US) are also the world's largest trading and investment partners. Yet, even the most casual observer knows that transatlantic relations were off to a rather rough start in the early twenty-first century. Contemporary transatlantic relations and their possible future trajectories are the subject of considerable popular and scholarly attention (e.g. Cohen-Tanugi 2003; Gordon 2003; Kagan 2003; Peterson and Pollack 2003; Garton Ash 2004; Gordon and Shapiro 2004; Hamilton 2004; Pond 2004; Reid 2004; Rifkin 2004; De Grazia 2005; Hodge 2005; Jasanoff 2005; Levy et al. 2005; Lindberg 2005; Motolla 2006; Martinelli 2007; Mahoney 2008; Svensson 2008).

During the Cold War, intense East-West conflict placed security issues at the forefront of much of the transatlantic relationship, and pushed the Western European states, the US and Canada to emphasize the cooperative nature of their relationship. In contrast, contemporary transatlantic relations cover a much broader range of environmental and energy issues pertaining to trade, agriculture and food safety, public health, biotechnology, and renewable energy. While there is much transatlantic cooperation in these issue domains, EU and US officials have not hesitated to bring their differences into the open. In fact, interconnected trade, agricultural, and consumer safety issues were at the heart of substantial transatlantic tension during much of the 1990s and 2000s.

The EU and the US have developed distinctly different approaches to a range of domestic and foreign policy issues. In addition to well-known differences related to security issues in the Middle East and the International Criminal Court, EU-US differences extend to a wide range of issues with major environmental and trade implications. These include the regulation of greenhouse gas (GHG) emissions, use of genetically modified organisms

(GMOs), the role of the state in the promotion of renewable energies, and control of hazardous chemicals (Barschdorff 2001; Busby 2003; Jasanoff 2005; Schaper 2005; Schreurs 2005b; Selin and VanDeveer 2006b). The EU and the US have also clashed on matters related to agricultural subsidies, environmental regulatory policies, and a host of product and accounting standards (Vogel 1995, 1997b; Lafferty and Meadowcroft 2000b; Princen 2002; Davis 2003; Levin and Shapiro 2004; Vig and Faure 2004b; Ansell and Vogel 2006).

As a result of greater economic interdependence and the internationalization of trade in all kinds of products and services, differences in transatlantic regulatory standards and consumer expectations have caused frictions that were less visible in the past. The many environment-related issues on which the EU and the US have taken different paths are a matter of both academic interest and policy concern. EU-US tendencies towards convergence or divergence, competition or cooperation have significant implications for global environmental politics, international problem solving, and transatlantic trade. The EU and the US are the world's two largest economies. As a result, they have significant influence on international decision making in economic, social, energy, resource and environmental outcomes around the globe. The policy positions adopted by the EU and the US affect policy opportunities and choices in other parts of the world. With approximately 12 percent of global population in 2008, the US and the EU together account for nearly half of global economic activity. When they choose to cooperate, they have great potential to address global issues, including natural resource degradation, climate change, food safety, sustainable energy, povery, and disease. In contrast, when the EU and US clash, global problem solving becomes more difficult and trade relations can chill.

In addition, due to their political and economic power and size, the EU and the US have the potential to accelerate global social and environmental problems when they assume policies that are aimed at protecting or furthering domestic interests irrespective of the environmental decline this may cause or accelerate in other parts of the world. EU-US attitudes toward agricultural policy and domestic farm subsidies under the World Trade Organization (WTO) are a prominent example of how EU-US agreement may not always benefit other countries and regions. Their support of biofuels—while well intentioned in terms of transitioning towards renewable energies—have also had unintended consequences in terms of food security and deforestation in developing countries. Clearly, transatlantic environmental and energy policies have significant meaning not only for the transatlantic relationship but also from a global perspective.

This book brings together long-term observers of environmental and energy politics and policy in the US and the EU to examine why, in so many prominent cases, such visible divides in regulatory approaches and outcomes have emerged. It focuses on policy areas particularly significant not only on environmental, energy, and health grounds, but also in terms of transatlantic trade relations.

The policy issues covered are sustainable development, safety of foods, genetic modified foods, asbestos, chemical management, the production and disposal of products, the production of energy, climate change, renewable energies, the standards placed on export credit agencies, and forest certification. The book is an exercise in inductive policy research. Each author was asked to compare the policy styles in the EU and the US in their area of special expertise, consider the extent to which a policy divide exists, explain the causes behind policy differences when they are there, and reflect on whether in the future policy rifts might be overcome or decline in significance with time. The conclusion looks across the individual case studies for patterns and trends in transatlantic environment and energy relations.

Many of the case studies confirm what has become something of a cliché—that the EU has become the global environment leader, a position the US could once claim but began to lose in the 1990s. The EU is pursuing policies that go well beyond those being adopted in the US—the case for example with "sustainable development," renewable energies, climate change mitigation, regulation of chemicals, and product standards. The asbestos case is another where the EU's complete ban of this carcinogen contrasts with the US, which still allows for some limited applications. The EU and US positions are closer to each other, however, than either are with the Canadian position, where asbestos mining is still permitted.

There are, however, exceptions to what has become the conventional wisdom about an EU in the lead. In relation to environmental regulations tied to export credit agencies, the US was the agenda setter and early mover although the EU eventually caught up with US restrictions. The case of food safety suggests that the EU and the US share many precautionary norms. Where there has been policy diffusion across the Atlantic, it is from the US to the EU in terms of the development of food safety institutions. The case of forest certification suggests that in some issue areas it is still difficult to talk of a US versus European position as policy convergence and divergence are not always defined at the federal level. Furthermore, as discussed in more detail below, at the sub-federal level many activities suggest considerable transatlantic similarity in norms and interests even in areas where politics are clashing at the federal levels.

Enlarging Transatlantic Relations

There are several reasons to emphasize the importance of EU-US environment and energy relations. Traditionally, transatlantic relations were primarily conducted between individual countries or groups of countries. As EU institutions (e.g. the European Commission (Commission), the Council of Ministers (Council), the European Parliament (EP), and the European Court of Justice) and actors (e.g. European non-governmental organizations (NGOs), lobby groups, and transnational corporations) have grown in strength and influence, transatlantic interactions increasingly occur at multiple levels of governance—between Washington DC and Brussels, among and between national capitals, as well as across many sub-national actors, both public and private.

Since the founding of the European Economic Community in the 1950s, what is now the EU has steadily grown in terms of the number of Member States, the size of its population, and the global importance of its economy. Especially significant was the accession of 10 new Member States in 2004 and Romania and Bulgaria in 2007, bringing total membership to 27. In addition, Croatia, Turkey and the Former Yugoslav Republic of Macedonia are recognized candidate countries (see Table 1.1). This enlarged EU has a population of almost 500 million, substantially larger than the approximately 300 million in the US (Table 1.2). When calculated based on purchasing power parity, the gross domestic products of the EU 27 and the US in 2008 were roughly equal, at $14.3 trillion and $13.84 trillion, respectively (Table 1.2). In many ways the EU now rivals the US in terms of economic power and its regional political influence continues to grow.

With each round of EU accessions, new members have had to transpose into national law the entire body of EU law (*acquis communautaire*). This

Table 1.1   Growth in EU membership

1951:	Belgium, France, Germany, Italy, Luxembourg, the Netherlands (ECE 6)
1973:	Denmark, Ireland, and the United Kingdom (ECE 10)
1981:	Greece (EC 11)
1986:	Portugal and Spain (EC 12)
1995:	Austria, Finland, and Sweden (EU 15)
2004:	Cyprus, the Czech Republic, Estonia, Hungary, Latvia, Lithuania, Malta, Poland, the Slovak Republic, and Slovenia (EU 25)
2007:	Bulgaria and Romania (EU 27)
	Official candidate countries: Croatia, Turkey, and Macedonia

Table 1.2   EU 27, US, NAFTA: population and GDP

	Population	GDP trillion US$ (PPP)
EU27	491,018,667	14.3.00
US	295,734,134	13.84
NAFTA	438,902,230	16.46

Source: Central Intelligence Agency, *The World Factbook* (CIA 2008), as of 19 June 2008.

includes all treaties, regulations, and directives passed within the EU as well as all judgments handed down by the European Court of Justice. The EU has had a powerful transformative influence on the political and legal institutions and bodies of law of its older Member States, its new Central and Eastern European Member States, and the would-be accession states currently aligning their domestic policies with European laws in an effort to increase the likelihood of eventual EU membership or because they are already deeply integrated into the EU market (Carmin and VanDeveer 2004). Beyond those negotiating for membership, other countries, ranging from Iceland, Ukraine, and Georgia to Israel and Morocco have at various times at least toyed with the idea of applying for possible membership. And, while Norwegians twice used referenda to reject membership, Norway has already adopted most EU legislation (Hovden 2004). This has put the EU in a powerful position to use its political and economic soft power to influence the behavior and policy of states beyond its outer borders much as the US has done for decades.

The EU has institutionalized close relations with many of its neighbors in other ways as well (Bretherton and Vogler 1999; Marsh and Mackenstein 2005). The European Economic Area promotes the free movement of goods, people, capital and services among the EU, Norway, Iceland and Liechtenstein, while Switzerland has more agreements with the EU than any other country. In addition, EU-sponsored partnership programs and association agreements, such as the European Neighborhood Policy, offer economic integration and political cooperation with EU-neighbors. The Euro-Mediterranean Association Agreement, which calls for cooperation on economic, political, social and migration issues, expands the EU's influence to North Africa, Central Asia and a host of former colonies. Some EU organizations including the European Environment Agency have also non-EU members.

The US is widely recognized as a global superpower. Due to its economic size and military and political influence, the US still exerts great influence internationally. Yet, in a growing number of cases, the EU has challenged US political dominance. This has been especially true in questions related to the environment. It is also important to note that the US has been less successful than Europe in promoting regional integration and harmonization. Public and private sector actors in North America have reacted to developments in Europe by developing their own free trade areas and common markets. The North America Free Trade Agreement (NAFTA) removed trade barriers between Canada, Mexico and the US. The construction and operation of NAFTA engendered much environmental debate and the creation of transnational and inter-state environmental organizations (Audley 1997; Deere and Esty 2002; Markell and Knox 2003; Gallagher 2004). In only a few cases, however, have these organizations facilitated limited harmonization of North American environmental regulations.

While NAFTA rivals the EU in economic importance and population size, its degree of institutionalized political and regulatory integration remains far lower. NAFTA does not promote the development of common policies and standards, except as these affect trade. This makes it difficult for NAFTA to ratchet regional environmental and product standards upward as the EU has done. Moreover, US officials have had difficulty expanding NAFTA to include other Latin America states, as exemplified by the 2005 collapse of the efforts to establish a Free Trade Area for the Americas (FTAA). Prior to their collapse, FTAA negotiations included almost no explicit attention to environmental issues or their links to trade. What this means is that while political and economic regionalism and integration has strengthened significantly the voice of EU Member States in international environmental debates, NAFTA has not had the same effect for North American states.

Many European leaders see the EU, at least in part, as a counterweight to US global power. Through enlargement and engagement, the EU has worked to spread political stability and expand capitalist markets across Europe. The EU, when successful, also offers the opportunity for Europe to speak and act as a large and strong actor in international politics. For decades, the US leveraged its market size to set global standards (Vogel 1995, 1997b; DeSombre 2000; Selin and VanDeveer 2006b). Now the EU increasingly plays this role. An illustrative example can be found in the 2004 international agreement on a common standard for barcodes on goods (Lohr 2004). The US standard had 12 digits, while the EU standard had 13. The expected benefits of having a single global standard, and the wider international acceptance of the European standard, forced North American producers to harmonize their bar code standard with the European one—despite the fact that bar codes were first developed in the US (Brown 1997; Haberman 2001). As noted in the *New York Times*, "the globalization of the bar code represents a small erosion of American industrial hegemony" (Lohr 2004). Similarly, Americans no longer buy a fifth of a gallon of Kentucky Bourbon (Reid 2004). Rather, whiskey comes in 75 centiliter bottles because that size meets the European standard.

While barcode and bourbon bottle standards may not command widespread attention, they reflect an important change in 21st century relations. Growing EU market power, driven by the size of the EU market and the EU's ability to impose common standards, results in increased opportunities for Europe to challenge the US in setting de facto global product and regulatory standards (Vogel 1995, 1997b). While changing standards for chemicals management, food safely, mandatory recycling, or energy efficiency may not grab headlines like UN Security Council debates, they have significant influence on how US companies invest and produce. Because many recent EU environmental and consumer safety standards frequently surpass older US standards, many US and international firms are finding that they must adopt EU standards if they are to sell their products in Europe and elsewhere.

The Emergence of a Policy Divide in Transatlantic Environmental Relations

European states were heavily influenced by US environmental policy developments in the 1960s and 1970s. Many environmental policy ideas and programs diffused across the Atlantic, particularly from the US to the EU (Vogel 1995; Jänicke and Weidner 1997; Vogel 1997b; Lafferty and Meadowcroft 2000b; Schreurs 2002). The EU and the US cooperated closely in the establishment of numerous multilateral environmental agreements, including the 1971 Ramsar Convention on Wetlands of International Importance; the 1973 Convention on International Trade in Endangered Species of Wild Flora and Fauna; the 1979 Convention on Long-Range Transboundary Air Pollution (CLRTAP); the 1985 Vienna Convention for the Protection of the Ozone Layer and the 1987 Montreal Protocol on Substances that Deplete the Ozone Layer (including subsequent amendments); the 1992 Framework Convention on Climate Change; the 1994 Convention to Combat Desertification; and the 1994 International Tropical Timber Agreement (Table 1.3).

This historical pattern of close EU-US cooperation however has changed as different policy styles and approaches came to dominate on the two sides of the Atlantic. Whereas in Europe an ongoing regulatory role for the state in environmental protection remains generally well accepted, in the US, strong regulatory intervention by the state for conservation and pollution control has been increasingly challenged (Kraft and Kamieniecki 2007; Klyza and Sousa 2008). In addition, while the EU has moved towards greater multilateralism, the US has moved towards unilateralism. As a result, since at least the early 1990s, there has been a growing rift between the EU and the US in relation to numerous multilateral environmental agreements (Vogel 2003a; Vig and Faure 2004b; Schreurs 2005b). The best known case is the Kyoto Protocol, designed to reduce the GHG emissions of the world's industrialized countries (Bodansky 2003; Busby 2003; Hovi et al. 2004; Schreurs 2004c). US withdrawal from Kyoto in 2001 deeply strained the transatlantic relationship.

Beyond this, the US signed but never ratified the 1992 Convention on Biological Diversity (CBD) and did not sign the related Cartagena Protocol on Biosafety, which aims to establish safety standards related to the development, use, and transfer of GMOs. Similarly, the US has not become a party to the 1989 Convention on Transboundary Hazardous Waste Movements and their Disposal; the 1988 Aarhus Protocol on Persistent Organic Pollutants to the CLRTAP; the 1991 Convention on Environmental Impact Assessment in a Transboundary Context; the 1998 Convention on Access to Environmental Information, Public Participation in Decision-making and Access to Justice in Environmental Matters, and; the 2001 Stockholm Convention on Persistent Organic Pollutants (Table 1.3). In contrast, there are no major environmental agreements pioneered by the US that have been rejected by the EU.

Differences in environmental policy approaches across the Atlantic also contribute to trade disputes. Between 1995 (the formation of the WTO) and

Table 1.3   European Community (EC) and US ratification of major multilateral environmental agreements (as of July 2008)

	EC	US
1959 Antarctic Treaty	Ratified by 18 EU members	18 August 1960
1971 Convention on Wetlands of International Importance (Ramsar Convention)**	Ratified by all 27 EU members	18 April 1987
1972 Convention for the Conservation of Arctic Seals*	Ratified by 6 EU members	28 December 1976
1973 Convention on International Trade in Endangered Species of Wild Fauna and Flora**	Ratified by all 27 EU members	14 January 1974
1979 Convention on Long-Range Transboundary Air Pollution (CLRTAP)	15 July 1982	30 November 1981
1982 Convention for the Conservation of Antarctic Marine Living Resources*	21 April 1982	18 February 1982
1984 Geneva Protocol on Long-term Financing of the Cooperative Programme for Monitoring Evaluation of the Long-Range Transmission of Air Pollutants in Europe (EMEP) (CLRTAP)	29 October 1984	17 July 1986
1985 Protocol on the Reduction of Sulphur Emissions or their Transboundary Fluxes by at Least 30 per cent (CLRTAP)	Ratified by 18 EU members	Neither signed nor ratified
1985 Vienna Convention for the Protection of the Ozone Layer	17 October 1988	27 August 1986
1987 Montreal Protocol on Substances that Deplete the Ozone Layer	16 December 1988	21 April 1988
1988 Protocol Concerning the Control of Emissions of Nitrogen Oxides (CLRTAP)	17 December 1993	13 July 1989
1989 Convention on the Transboundary Movement of Hazardous Wastes and their Disposal (Basel Convention)	07 February 1994	Signed, not ratified
1990 London Amendment (to the 1987 Montreal Protocol)	20 December 1991	18 December 1991
1991 Protocol on Environmental Protection to the Antarctic Treaty*	Ratified by 11 EU members; signed but not ratified by 7	17 April 1997
1991 Geneva Protocol Concerning the Control of Emissions of Volatile Organic Compounds and their Transboundary Fluxes (CLRTAP)	Ratified by 18 EU members	Signed, not ratified
1991 Convention on Environmental Impact Assessment in a Transboundary Context (Espoo EIA Convention)	24 June 1997	Signed, not ratified
1992 Convention on Biological Diversity (CBD)	21 December 1993	Signed, not ratified

Table 1.3 cont'd

	EC	US
1992 United Nations Framework Convention on Climate Change (UNFCCC)	21 December 1993	15 October 1992
1992 Copenhagen Amendment (Montreal Protocol)	20 November 1995	02 March 1994
1994 Convention to Combat Desertification	26 March 1998	17 November 2000
1994 Protocol on Further Reduction of Sulphur Emissions (LRTAP)	24 April 1998	Neither signed nor ratified
1994 International Tropical Timber Agreement	29 March 1996	14 November 1996
1996 Comprehensive Nuclear Test Ban Treaty	Ratified by all 27 EU members	Signed, not ratified
1997 Montreal Amendment (to the 1987 Montreal Protocol)	17 November 2000	01 October 2003
1997 Kyoto Protocol to UNFCCC	31 May 2002	Signed, not ratified
1998 Protocol on Heavy Metals (CLRTAP)	03 May 2001	10 January 2001
1998 Protocol on Persistent Organic Pollutants (CLRTAP)	20 April 2004	Signed, not ratified
1998 Convention on Prior Informed Consent Procedure for Certain Hazardous Chemicals and Pesticides in International Trade UNEP/ FAO (Rotterdam Convention)	19 December 2002	Signed, not ratified
1998 Convention on Access to Environmental Information, Public Participation in Decision-making and Access to Justice in Environmental Matters (Aarhus Convention)	17 February 2005	Neither signed nor ratified
1999 Beijing Amendment (to the 1987 Montreal Protocol)	25 March 2002	01 October 2003
1999 Protocol to Abate Acidification, Eutrophication, and Ground-level Ozone (CLRTAP)	23 June 2003	22 November 2004
2000 Cartagena Protocol on Biosafety (Convention on Biological Diversity)	27 August 2002	Neither signed nor ratified
2001 Stockholm Convention on Persistent Organic Pollutants	16 November 2004	Signed, not ratified

*   This treaty does not apply to all EU Member States and thus, only a sub-set have ratified.

** These treaties only apply to individual EU Member States; the EC does not have the authority to become party to these agreements.

2006, the EU (technically, the European Communities) lodged 29 complaints against the US and the US filed 16 against the EU. While most EU complaints against the US were related to non-agricultural products or trade laws, nine of the 16 complaints the US made against the EU dealt with agricultural trade or biotech matters. While some of these disputes dealt with concerns related to trademarks, subsidies, and tariffs, others were linked to differences in policies protecting human, animal and plant health. Most important was the EU decision to ban hormone-treated beef in the mid-1990s. The US also brought the EU before the WTO for its practice of labeling genetically modified (GM) foods, a practice which the EU argues provides consumers with necessary information.

While renewable energy is still a small share of total energy in both the EU and the US and both sides of the Atlantic introduced energy conservation measures in response to the 1973 and 1979 oil shocks, Europe persisted with such efforts at the supranational level far more aggressively than did the US. In 2001, the EU introduced a directive with the goal of meeting 12 percent of energy consumption from renewable sources by 2010 and in December 2008, the European Council confirmed the Communities' commitment to increase the share of renewables to 20 percent of total energy consumption by 2020. In the US, several states promote renewable energy, but national programs remain limited (Rabe 2004; Selin and VanDeveer 2005, 2006a, 2009a). At the 2002 World Summit on Sustainable Development, EU officials failed to win US support for agreed national targets for renewable energy development.

These trends raise the question of why the EU and the US have moved in such different directions, leading to considerable discordance across the Atlantic? With the growing influence of global environmental norms, international expert communities, multinational corporations, international organizations, and international agreements, should not environmental policy differences between the EU and the US be narrowing, not expanding (Holzinger et al. 2008)? Given that the forces of globalization are suppose to be strong and bringing countries closer together, why in the past decade or more have the EU and the US diverged on so many important policy issues (Andrews 2005)? Why, when transatlantic economies are increasingly integrated, are the accompanying politics so often discordant?

Several plausible explanations exist. One is that differences indicate that societal and cultural norms are in fact diverging (Martinelli 2007; Guehlstorf and Hallstrom 2008). According to this line of reasoning, Europeans have come to embrace more deeply than have Americans such concepts as the precautionary principle and sustainable development (Grant et al. 2000; Vig and Faure 2004b; Sadeleer 2007). Sustainable development may simply be more suited to the kinds of social democratic political systems found in Europe than to the more conservative economic and political milieu of the US. Moreover, green parties and environmental movements in Europe have been more influential in changing political and societal norms than has the environmental movement in the US (Bomberg 1998; Burchell 2002; Müller-Rommel and Poguntke 2002). Conversely, Americans have more firmly adopted the goals of liberal economic competitiveness and small government than their European counterparts.

A second line of argumentation is that the policy differences seen across the Atlantic are primarily a political artifact tied to the rise in influence of the Republican Party in the different branches of US politics during much of the 1990s and the first half of the 2000s (the US Congress from 1994–2006 and the Presidency from 2001–2008). The Republican Party has traditionally been a stronger supporter of business interests than environmental ones—which recently have been more the domain of the Democratic Party. The Republican Congress, with the backing of the George W. Bush administration favored industrial, mining, land use, and energy interests (Kraft and Kamieniecki 2007; Klyza and Sousa 2008). In the case of Europe, the rise of Green parties influencing the stands of Social Democratic, Liberal and Christian Democratic parties—partly a response to major environmental crises that have confronted Europe—helps explain the greater focus placed on environmental protection in Europe than the US.

A third perspective points less towards the emergence of new cultural and normative divides or shorter term political differences and more towards the impact that institutional changes in US and European politics have had on the ability of different groups to influence political outcomes. This approach suggests that the neoconservative revolution that began under the Reagan administration, was strengthened with the 1994 appointment of Newt Gingrich as Speaker of the House of Representatives, and solidified with the two-term George W. Bush administration has led to relatively deep institutional changes and a shift in the balance of the strength of different actor groups (the weakening of the Environmental Protection Administration, the empowerment of conservative think tanks, the decline of the activist court). These changes have put environmentalists on the defensive and encouraged a search for alternative—non-regulatory approaches to pollution control and environmental protection (e.g. voluntary agreements, emissions trading, public-private partnerships) that are less likely to result in an all out assault from entrenched interests (Morgenstern and Pizer 2007; Klyza and Sousa 2008).

Conversely the development and greatly expanded authority of EU organizations have provided new avenues for environmental and other societal interests to influence EU policy outcomes. The Commission and EP have been strengthened over time. New environmental NGOs have also formed all over Europe, shaping local and regional policy development. Furthermore, Green parties became members of national parliaments and governments, and many European parties across the political spectrum have been more inclined to push green ideas and pursue sustainable development than their US counterparts. In addition, concepts of sustainable development and the precautionary principle became deeply embedded in member state and EU laws and treaties (e.g. Baker 1997; Hunter and Smith 2005; Baker 2006).

As discussed more fully in the conclusion, this volume suggests that each of these explanations holds some explanatory power. Culture does matter. Liberal (i.e. free market) economic ideas and policies have a stronger hold in the US than the EU. An energy conservation culture has taken a firmer hold in Europe than the US. The political make-up of governments also clearly can influence environmental policy outcomes. This has been very visible in the US with the dominance of the Republican Party in US politics in the latter 1990s and first half of the 2000s. It is also visible in the EU member states. Spain's policies towards climate change and renewable energy changed quite dramatically when the Socialists regained power in 2004. Yet, the persistence of the differences in regulatory appoaches that have emerged between the EU and the US across a rather wide swath of environmental issues suggests that these differences have become more than short-term political differences. They are also quite deeply institutionalized.

So then, what is the likelihood that domestic political forces or environmental understandings will change enough on one or the other side of the Atlantic to result in greater similarity in EU and US environmental policies and programs and a return to a more cooperative transatlantic environmental relationship in the future?

Expanding Transatlantic Relations: Politics and Governance at Multiple Levels

Typically, when the EU and the US are compared this is done at the national/ supranational level and in relation to the politics of state actors (Desai 2002; Scruggs 2003; Harrington and Morgenstern 2004; Vig and Faure 2004b; de Bruijn and Norberg-Bohm 2005). Indeed, many of the divides that exist across the Atlantic in areas like climate change, product standards, and regulation of hazardous chemicals are most evident at the federal level, when the policies of Brussels and Washington DC are compared. Yet, the EU-US relationship needs to be understood as more than just a relationship at the federal level.

Individuals, organizations, and governments interact across the Atlantic at all levels, from the sub-national to the supranational, publically and privately. US public and civil society advocates for more stringent US policies to combat growing US GHG emissions or to manage chemical risks have become increasingly engaged with their European counterparts in attempts to import information, discourses and political lessons into North America. Many officials in US states, such as Massachusetts and California pursuing active climate change mitigation policies, have gone out of their way to meet and exchange information with colleagues who work in EU institutions, Danish and British national ministries, and/or German Länder (states). Municipal level officials and civil servants on both sides of the Atlantic belong to transnational sustainability networks (Slaughter 2004).

Transatlantic politics includes a diverse array of actors, including national and EU-level officials from an expanding set of different ministries and agencies, sub-national public officials and organizations, a host of intergovernmental organizations, NGOs of many types, corporations and national and international trade and industry associations. In fact, it can be argued that one of the most significant changes in the dynamics of transatlantic relations over the last generation is associated with the dramatic growth in formal and informal connections across the Atlantic. In more theoretical terms, the evidence suggests that agency is diffusing from a small number of powerful state actors to a larger and more diverse set of agents operating at local, national and transnational levels (O'Neill et al. 2004). Environmental and consumer NGOs, industry and trade associations, public officials and professionals of all stripes are increasingly embedded in transatlantic environmental, food safety, health, and consumer networks and organizations.

The growing number of sub-national and civil society actors engaged in transatlantic environmental and energy relations has important implications for transatlantic politics of related policy issues. While tensions have been prevalent in the relations between the EU and the US at the federal level, there has been a noticeable degree of policy convergence and a more cooperative transatlantic environmental relationship developing at the sub-national level over the past decade (Lopes and Durfee 1999; Tews et al. 2003; Levi-Faur and Jordana 2005). These multiple pathways of transatlantic politics can serve as both important channels of norm diffusion and learning, and influential avenues for strategic action (O'Neill et al. 2004; Slaughter 2004; Vogel 2005; Selin and VanDeveer 2007). For example, in 2006 Tony Blair and Arnold Schwarzenegger discussed potential means for transatlantic cooperation in climate mitigation, including with carbon emissions trading.

It should also be emphasized that while the differences that have emerged between European and US approaches to international environmental regulation are quite dramatic, only comparing these two economic powers at the federal level masks the many differences that exist among the states that comprise the EU and the US. There is considerable difference among European states in the extent to which domestic and international environmental protection policies and programs are developed and implemented (e.g. Hanf and Jansen 1998; Börzel 2002; Jordan and Liefferink 2004; Harris 2007). In general, despite many noteworthy exceptions, the richer states of northern Europe have been stronger supporters of international environmental agreements and better at domestic implementation than have their still developing southern neighbors. Similarly, there is a wide range of opinions on international environmental matters across the states of the US. In relation to a number of the environmental matters considered in this book, we will see that California and New England have tended to assume environmental positions closer to those embraced by the European Union than by Washington DC. This suggests the need for more nuanced and multi-level comparative approaches to the study of transatlantic relations.

An Overview of the Book

This book examines transatlantic relations around sustainable development, GMOs, chemical management, public health issues, Export Credit Agency (ECA) standards, forest certification, interstate environmental competition, sustainable development initiatives, climate change action, and WTO cases involving environmental regulations. The chapters demonstrate that transatlantic tensions, most of them deeply embedded in cooperative institutions, are commonplace in areas where environmental protection and trade policies intersect. Yet, while the volume as a whole suggests substantial transatlantic discord, several cases illustrate the deepening integration and institutionalized cooperation that is emerging due to market forces and transnational linkages that are at times forged at the local level.

The book is divided into four parts. Part I addresses issues of comparative environmental governance. In Chapter 2, Elizabeth Bomberg explores the question of why, when the US was the source of many of the ideas associated with the concept of sustainable development, it is the EU that has done more to formally embrace the notion in policy making. While President Clinton established a President's Council on Sustainable Development, it was shut down in 1999 and official reference to sustainable development since then has been difficult to find. Bomberg argues that the EU sustainable development steering network that included the Commission, the EP, and a community of scientific experts, NGOs, think tanks, and industry, have promoted the inclusion of sustainable development in European regulations and programs. In the US, actors have failed to forge such strategic alliances.

In Chapter 3, Sonja Wälti addresses how EU and US versions of federalism influence the way that businesses and environmental interests shape environmental policy making. Wälti focuses on key stakeholders and their use of institutional venues and opportunity structures. Businesses in Europe have comparatively better access to Member State governments because of the corporatist traditions and third-party accommodation found in many European countries. Businesses that lobby in Brussels tend to engage regulatory debates and discussions, rather than seek to oppose virtually all attempts to increase regulatory standards. This is in sharp contrast with the US where many industrial associations lobby Washington to prevent the adoption of environmental regulations. The US Congress has an institutional bias toward the particular interests of constituents as opposed to the diffuse interests of environmental NGOs. In contrast, the Commission has been receptive to NGO demands, and the EP has been an ardent defender of NGO interests.

Part II addresses specific environment cases: chemical regulations, asbestos bans, product standards, and GMOs. In these areas, the EU has outpaced the rest of the world in developing precautionary controls restricting the use of known and potentially hazardous chemicals, banning the use of asbestos, promoting product take back and recycling, and restricting the entry of GM products into the European market place. These are issues where regulatory differences have not only important environmental and health implications, but major trade and economic ones. They have put the EU and the US into direct competition with each other.

In Chapter 4, Henrik Selin examines EU-US cooperation and competition on chemicals management. While North American and European concerns about hazardous chemicals have contributed to the formation of several international organizations and multilateral treaties and programs for their management, the EU and the US often disagree over specific regulatory issues and approaches. In particular, European efforts to revise and expand Community chemicals assessment and controls in the form of the new Registration, Evaluation and Authorization of Chemicals (REACH) regulation are resulting in growing EU-US controversy over the future direction of chemicals management. This is likely to have major consequences for transatlantic relations and international policy making in areas of risk management.

In Chapter 5, Marcus Carson considers the different approaches of the EU, Canada, and the US to asbestos use. The EU has banned the use of asbestos in large part due to actions by the French and UK governments. In contrast, neither Canada nor the US has banned all uses of asbestos, and the Canadian government still permits mining. The Canadian government has operated as the coordinator for international lobbying and public relations efforts to protect the global asbestos market. In the US new uses of asbestos have been banned, but existing ones have not. Carson shows why in Europe scientific data indicating asbestos as a risk led to a complete phase out of asbestos use in Europe, but not in the US and Canada.

Alastair Iles in Chapter 6 compares US and EU approaches to the establishment of product standards for automobiles and electronics. Iles outlines the environmental and health problems posed by cars and electronics and the different approaches of the EU and the US toward their regulation. EU policies surpass those in the US in requiring industry to change manufacturing processes and procedures to reduce environmental risks and facilitate mandatory recycling requirements. The EU has taken the lead in this area because of changing consumer views of product risks, the agenda-setting role of European institutions, and industry willingness in the end to acquiesce to growing environmental pressures. Desires to harmonize product standards across the EU and to influence global standards are significant.

In Chapter 7, Patricia Keilbach asks why transatlantic tensions have mounted over the trade in food products. EU-US disputes over GM food reveal the growing complexity of international trade conflicts. Many US actors advocate the spread of GM food, arguing that population growth means that our future may be dependent on the success of the promise of GM food to deliver plentiful, more nutritious food. EU actors argue that information about the impact of GM food on human health and the environment is relatively scarce, and their promise is uncertain. The divergent regulatory approaches to GM food across the Atlantic stem from ideological differences rather than from economic considerations, making harmonization of policies difficult.

In Chapter 8, Thomas Bernauer and Phillip Aerni show that developing countries have become an important target of the transatlantic agri-biotechnology debate. The EU and the US have sought to influence the position of developing countries in the context of the Cartagena Protocol on Biosafety, which governs transboundary movements, handling, transit, and use of living GMOs and is supported by the EU and opposed by the US. EU and other GMO-adverse stakeholders have been more successful in exporting their preferences and regulatory approaches to developing countries in the past decade. The tide appears to be turning, however, because a more pragmatic approach to GMOs is emerging in many developing countries.

Part III focuses on issues of renewable energy and climate change. Ian Rowlands compares EU and US policy positions and performance in relation to the promotion of renewable energies in Chapter 9. The development and operation of electricity systems have major economic, social and environmental implications. Interactions between the EU and the US on renewable energy could promote sustainability. Yet, European and North American attitudes are quite different, at least at the federal level, towards the promotion of renewables. Rowland finds that regulatory action in the EU has been encouraged by climate change, the environmental effects of conventional electricity generation, energy security concerns, and energy costs. The US lags behind the EU on issues of renewable electricity although policy progress is being made in a growing number of US states.

Chapter 10 by Miranda Schreurs, Henrik Selin, and Stacy VanDeveer addresses the case of climate change. In the EU there were multiple leaders, including some Member States, the Commission, and the EP that made possible the formation of a relatively ambitious EU climate policy. In the US, advocates of climate change action were not able to form a sufficiently strong lobby to counter the opposition to climate change action that came from powerful industrial opponents. Yet, a growing number of US states and municipalities are adopting more progressive climate change policy. Expanding policy initiatives in California and the east coast in particular, suggest greater potential for transatlantic cooperation in climate change mitigation in the future.

Part IV considers issues of standard setting as they apply to export credit agencies and the forestry sector. These cases add a healthy dose of caution into generalizations that the EU always leads. In Chapter 11, Marcus Schaper examines a case where the US led: the establishment of environmental standards for export credit agencies. Export credits provide companies with financial assistance and insurance when investing in projects that are perceived as risky. There is a thin line, however, between trade distortion and government support of a company's exports. Schaper focuses on US and German responses to international negotiations within the OECD on environmental standard setting for export credit agencies as well as the international negotiations themselves.

Within the OECD, the US pushed for its higher standards to become the standards required for all OECD Member States. Schaper examines why the positions of the US, Germany, and other OECD Member States differed significantly going into the negotiations, and how those differences were overcome in the 2003 OECD agreement.

Chapter 12 by Benjamin Cashore, Graeme Auld, Deanna Newsom, and Elizabeth Egan questions the common characterization that the EU is the champion of innovative environmental policy development while the US is lagging behind. There is increasing use of non-state market driven (NSMD) governance systems in Europe and North America. Yet, the kind of NSMD system that is chosen varies depending on the place of the country/region in the global economy, the structure of the domestic forest sector, and the history of forestry on the pubic policy agenda. This can be seen in relation to forest certification politics in British Columbia, Canada, the US, Germany, Sweden, the United Kingdom and Finland. There is tension between supporters of global, prescriptive standards as represented initially by the Forest Stewardship Council and supporters of domestic initiated, controlled, and discretionary approaches. In this issue area, there is no clear divide between the EU and the US.

In Chapter 13, Kate O'Neill looks at the transatlantic dimensions of outbreaks of mad cow disease and avian influenza. Both the EU and the US are quite precautionary in their responses, not only at the initial, outbreak stage of a disease but also over the longer term. Yet, there are differences in institutional responses. One of the prime motivations for EU activities in this area has been to build and expand its authority as a new, supranational form of governance. The US, in contrast, has not responded to these diseases with institutional change and reform, relying on its existing configuration of agencies. O'Neill documents policy diffusion from the US to the EU in development of two new agencies: the European Food Safety Authority and the European Center for Disease Prevention and Control. The EU modeled these agencies' structures and functions on their US counterparts (the Food and Drug Administration and the US Centers for Disease Control and Prevention), adapting these models to fit the realities of EU politics.

Finally, the concluding chapter by Miranda Schreurs, Henrik Selin and Stacy VanDeveer returns to the questions raised in the introduction. Looking across the case study chapters, the conclusion finds that many of the divergent policy positions on environmental, health, energy and agricultural trade issues between the EU and the US can be explained by institutional structures and the political opportunities that they provide to actors. Important to this has been the expansion of the powers of EU institutions. As the EU has broadened its policy competencies, it has sought to strengthen its power and influence both within Europe and internationally. In the US, leading efforts in contrast have attempted to decrease the scope and reach of the federal government with substantial implications for US environmental policymaking including a shift in regulatory leadership from federal authorities to states and municipalities. As such, some of the differences between the EU and the US are being mitigated by a range of forces, including transnational actors and sub-national policy diffusion, and international legal developments.

PART I
Governing Within and Beyond the State: Comparative Environmental Governance and its Implications for Policy Development

Chapter 2
Governance for Sustainable Development: The United States and the European Union Compared

Elizabeth Bomberg

How should societies best meet present and future generations' basic economic and social needs without undermining their environmental quality of life? This question lies at the heart of sustainable development. But beyond this, the concept of sustainable development is interpreted and embraced differently across the globe and across the Atlantic. A common perception is that the EU is the champion of sustainable development, the US an irredeemable laggard. Sustainable development is seen as an area where the US and EU are most "clearly diverging" (Vig and Faure 2004c, 6).

This depiction is not wholly lacking in foundation. The EU's engagement is far more advanced rhetorically, legally and constitutionally; the EU's overall embrace of key sustainable development policies (such as climate change) appears far more serious and elaborate than that in the US, and the EU ranks well above the US in many global environmental performance measures.[1] Explanations for this discrepancy usually mention institutional factors such as a hostile US Republican administration and Congress beholden to industrial interests, or the highly fragmented American policy making system. Explanations of difference also emphasize more general issues related to political culture, including a greater environmental awareness and concern in the EU than in the US; the EU's deeper sense of moral responsibility and collective responsibility; or fundamental American values—individualism, property rights, anti-statism—inimical to sustainable development.

Yet this portrayal bears closer scrutiny. It can not explain how the US was a key initiator of principles and practices central to sustainable development, and why it still has some of the most ambitious environmental and sustainable development-related laws, institutions and mechanisms. Many of these originated under the conservative Republican administration of Richard Nixon. Moreover, key sustainable development principles (precaution, policy integration) and mechanisms (economic tools, policy learning, corporate social responsibility) either originated in the US, or developed faster there than in the EU. Further, public concern and awareness of environmental issues is not

---

1   Yale's 2006 Environmental Performance Index, http://www.yale.edu/epi/.

dramatically different across the two polities (Dryzek et al. 2002, 679). Finally, powerful producer groups are very well organized and represented within EU policymaking, which is itself highly fragmented and chaotic.

This chapter argues that differences between EU and US engagement with sustainable development are real, but not as profound as often depicted, and located primarily in different mobilization patterns. A particularly important factor is the existence or lack of networks able to steer polities towards particular goals.

This chapter first introduces the subject of comparison by outlining the core features of the governance concept in general and governance for sustainable development in particular. Subsequent sections compare the extent to which the US and EU officially and constitutionally have recognized and embraced sustainable development; explore institutionalization (the extent to which key institutions and actors have adopted sustainable development principles and practices); and examine mobilization—especially the presence or lack of networks able to steer policies and practices in a sustainable direction. Finally, the last section summarizes the study's preliminary finding and identifies possible trends and further areas of research.

Comparing Sustainable Development Governance

Why Compare?

Understanding the differences and similarities between the EU and US matters practically, empirically and conceptually. The US and EU are powerful polities with tremendous potential or real impacts on global environment, energy, trade and development politics. The US and EU's global share of GDP, trade, energy consumption and pollution mean their actions shape fundamentally the pace and form of sustainable development at the domestic and global levels.

Conceptually, we can learn a lot about the nature of sustainable development, particularly its potential implementation and governance, by examining these two major polities. Despite their similarities as high consumption, liberal democracies, the structural and constitutional differences between these two polities are significant. One is a sovereign federal state, the other a unique quasi-federal polity made up of 27 sovereign states. Their key policymaking institutions differ in power, authority and role, even if many of their central functions are similar. Political, historical and cultural differences make comparison challenging. But in terms of sustainable development governance the two have much in common; both have committed themselves to the broad sustainable development principles agreed at the international level; both are signatories to the Rio Declaration and its Agenda 21, a massive document setting out a detailed plan of action for implementing agreed principles (UNCED 1992). Both have engaged—albeit to different degrees and in different ways—with key principles, practices and mechanisms of sustainable development. This study compares that overall engagement (formal and informal), with a particular focus on the domestic level.

Governance, Steering and Sustainable Development

A richer comparison of the US and EU can be achieved by broadening our focus from policies and outputs to the wider notion of governance. Governance refers here to established patterns of rules and norms *steering* a polity in a stipulated direction. It implies the incorporation of principles, practices and mechanisms which enable a community to be governed even without a government or ruler. It may well include declarations, laws and policies mandated by government or from the center, but it is much broader, including soft law, non-regulatory tools and policy learning (Kjær 2004, 3–4).

Governance for *sustainable development* requires further explanation. The woolliness of the term sustainable development causes some to despair (it is meaningless fudge! (Victor 2006)), but others see its semantic openness as key to its importance: "Sustainable development is now like democracy or freedom: it is universally desired, diversely understood, extremely difficult to achieve, and won't go away" (Lafferty 2004, 26). This chapter does not offer a precise definition because it explores the varying ways two polities view and embrace the concept. Yet, we can identify the core of sustainable development as the idea of integrating or balancing environmental, economic and social objectives within a framework of generational and global equity. Or as described by the OECD (2001b): sustainable development involves the "integration of economic, social and environmental objectives across sectors, territories and generations." This broad understanding, as formulated in the Brundtland Commission (World Commission on Environment and Development 1987) has achieved authoritative status (Baker 2006) and is commonly used as a base line.

Some scholars have provided a continuum of sustainable development types, ranging from strong (a robust embrace of the full spectrum of principles) to a weak or more selective embrace. The latter form is often referred to as ecological modernization and is primarily focused on reconciling the supposed trade-off between environmental sustainability and economic growth (Mol and Sonnenfeld 2000). Its advocates suggest that both are possible; pollution prevention pays and businesses can profit by protecting the environment. The ecological modernization form of sustainable development thus advocates a new way of thinking but stops short of radical green demands for a fundamental restructuring of the market economy and liberal democratic state (Carter 2007, 211).

Beyond broad definitions and typologies, other scholars (Lafferty and Meadowcroft 2000c) and organizations such as the UN, World Bank and OECD have fleshed out the concept of sustainable development (in both weak and strong forms) by identifying integral principles and practices. These include:

- substantive principles (policy integration, eco-efficiency, equity);
- procedural principles (precaution, broad participation, transparency);
- mechanisms (new policy instruments, policy learning, multi-level coordination).

We see from these principles that sustainable development is incredibly ambitious and clearly aspirational. It seeks to reconcile ecological, social and economic dimensions of development, now and into the future, locally and globally. It is not an end state to be reached but a good (such as freedom, peace, equality, democracy) to be pursued, however imperfectly. This procedural and aspirational character of sustainable development makes governance a useful focus; governance as steering focuses our attention on process over policies, direction over destination.

This analysis suggests governance for sustainable development involves not just a change in policies, but the incorporation of more demanding components. These include:

- formal *recognition/awareness* of sustainable development issues and strategies to address them;
- *institutionalization* of operating principles underpinning sustainable development;
- *mobilization* of key stakeholders able to steer polities and policies in sustainable direction.

The chapter now examines each of these components in turn.

Official Recognition of Sustainable Development

This section examines how/if the idea of sustainable development is recognized and embraced in the EU and the US by tracking how sustainable development appears in their official documents, laws, treaties and agreements.

European Union[2]

Sustainable development has not been defined consistently in EU treaties or documents, but it is still possible to trace a growing recognition of sustainable development goals, and the explication of strategies to achieve them. The 1992 EU 5th Environmental Action Programme (EAP), borrowing from Brundtland, defined sustainable development as "continued economic and social development without detriment to the environment and natural resources" (European Commission 1992). This recognition in soft law was given formal

---

2   This section borrows from Bomberg (2004).

treaty status in Article 6 of the 1997 Amsterdam Treaty which stipulated that "environmental protection requirements must be integrated into the definition and implementation of Community policies ... in particular with a view to promoting sustainable development."

In 1998, European leaders launched what became known as the Cardiff Process, which requires the Council of Ministers in all its formations (for instance, fisheries, transport, agriculture) to integrate environmental and sustainable development objectives into their respective policy areas. The Commission (the EU's executive/civil service) proposed a European Sustainable Development Strategy (SDS) which outlined some overwhelming challenges to sustainability; set out key principles for sustainable development, established priorities, and offered concrete objectives and targets. Several of these were also evident in the Commission's 2001 6th Environmental Action Programme due to run to 2010.

A particularly robust demonstration of EU leaders' recognition and development of strategic goals occurred at the Gothenburg Summit in June 2001. The summit's conclusions stressed that the "economic, social and environmental effects of all policies should be examined in a coordinated way and taken into account in decision-making" (European Council 2001). Pursuing this principle, leaders at Gothenburg agreed to widen the EU's existing commitment to promote socio-economic goals (the Lisbon Process)[3] to one promoting sustainable development.

The external dimension of the EU's sustainable development role was highlighted in the EU's preparation for the World Summit on Sustainable Development (WSSD) (European Commission 2002d). The EU's large WSSD delegation included two Commissioners and the Commission President, ample staff, plus nearly 100 Members of the European Parliament (MEPs). No radical initiatives were agreed at Johannesburg, and the Commission regretted a lack of progress. But it applauded (and took credit for) the agreed commitments to increased development assistance, good governance and a better protection of the environment.[4] Since WSSD the European Council (2005) has issued further Guiding Principles underlining the EU's commitment to "safeguard the earth's capacity to support life in all its diversity," and the Commission has further revised its Sustainable Development Strategy (European Commission 2007d). Finally, the EU's 2008 ambitious climate action and renewable energy package sought to make Europe "the first economy for the low-carbon age" and a "world leader" in combating climate change and furthering sustainable growth. (European Commission 2008b).

---

3   At the Lisbon Summit in March 2000 the European Council agreed to commence a ten year intergovernmental process aimed at making the EU "the most competitive and dynamic knowledge-based economy in the world capable of sustainable economic growth with more and better jobs and greater social cohesion" (European Council 2000).

4   http://europa.eu.int/comm/environment/wssustainable development).

Yet the EU's engagement is clearly of a shallow, ecological modernization sort. First, the documents discussed above feature much about integrating environmental objectives into economic and social concerns, but less about *prioritizing* them which, according to Lafferty (2004), constitutes a core assumption of strong sustainable development. Moreover, and to be expected from an institution whose founding purpose was the creation of a common market, the strategies outlined in the documents above often reflect EU's desire to address environmental degradation through economic means and, often, economic growth. Typical is the Commission's "Towards a Global Partnership," which emphasized that "market forces can be harnessed to maintain and increase growth and to create jobs, while preserving the environment for future generations and strengthening social cohesion" (European Commission 2002d, 5). Like most ecological modernization approaches, there is little recognition here of the potential conflict between increased trade and environmental sustainability.

Moreover, official EU reference to sustainable development since 2002 has become less conspicuous. In the Lisbon Treaty signed in 2007) sustainable development was not given the prominent position (or separate Article) it had enjoyed in the Amsterdam Treaty (indeed, it did not appear at all in earlier drafts). The 2004 accession of ten new states, several of whom had far less interest in sustainable development (Bomberg 2007) distracted attention in 2004. The Commission review of the EU's SDS revealed several shortcomings, environmental groups bemoaned lack of progress and huge implementation gaps (EEB and G-10 2006), and more neutral commentators noted that the SDS had "subsided from view, overshadowed by the Lisbon Agenda for economic competitiveness" (*European Voice* 23 March 2006, 18). Finally, the EU's declaratory commitment in more recent trade and aid talks has been less obvious. In sum, EU's official recognition of sustainable development is notable, though it takes the weaker form of ecological modernization, and has become less robust in recent years.

United States

Formal official recognition of sustainable development as a policy goal in the US is far less prevalent. Yet, to focus solely on the explicit term "sustainable development" neglects important episodes of official recognition of key sustainable development principles, practices and politics. For instance, several of sustainable development's basic principles, couched in terms such as "wise use" or "harmonious and coordinated management of resources," have early origins (Baker and McCormick 2004, 290), pre-dating even the emergence of the EU. Moreover, several key sustainable development-related policies, embodying principles such as precaution and policy integration, emerged first in the US. In the 1970s the US witnessed a "massive burst of environmental policy innovation" (Dryzek et al. 2002, 665), and was the trailblazer in setting up a federal Environmental Protection Agency and passing acts which, inter alia, required government agencies to complete environmental impact assessments of all federal projects (Kraft and Vig 2006). And the notion of corporate social responsibility (CSR)—which embodies the sustainable development notion that business should make decisions based not just on the economic bottom line but also on social and environmental considerations—was developed and took hold first in the US, not Europe (Vogel 2005). By the 1970s, the federal government's Council on Environmental Quality had already begun to integrate sustainable development aims into its documents, advocating a "more holistic approach to environmental problems, simplification of regulations, more flexible problem solving and a more interactive approach with stakeholders and the community at large" (quoted in Bryner 2000, 277).

The peak of official explicit recognition of sustainable development came with the Clinton administration's creation of the President's Council on Sustainable Development (PCSD) in 1993. The Council's remit was to bring together representatives from government (including several cabinet-level departments) industry, NGOs and labor groups to develop a sustainable development strategy. The resulting PCSD's report featured rhetoric rivalling the EU's ambitious pronouncements. It advocated a sustainable US with "a growing economy that provides equitable opportunities for satisfying livelihoods and a safe, healthy, high quality of life for current and future generations." The stated national goals included a "healthy environment, economic prosperity, justice for all, the conservation of nature, sustainable communities, civic engagement and a leadership role for the US in the development and implementation of global sustainable policies" (PCSD 1999, 9ff). The PCSD identified specific areas in need of change (for instance, making environmental regulation more effective and efficient, developing an ethic of stewardship to guide human interaction with the environment, and fostering US leadership in international efforts), and it set up various tasks forces to work towards these goals (Bryner 2000, 278). Under Clinton's PCSD, the official rhetoric on sustainable development was as close as it had ever been to that set out in Agenda 21: an aspirational set of ambitious goals linking social, and environmental and economic development aims, and a set of strategies aimed to achieve them.

Yet unlike in the EU, where several different institutions (Council, Commission and Parliament) played an important role, US official recognition and strategy in the 1990s was almost exclusively located in the executive branch, and it was thwarted by opposition in Congress (Bryner 2000). Moreover, executive attention quickly evaporated under George W. Bush's administration; the PCSD's work wound down in 1999 and was not renewed, and official recognition of sustainable development as a domestic issue since then has been difficult to find (Krämer 2004). Similarly, on the global level, despite PCSD's call for international leadership, the US commitment has not been evident. Most sustainable development pronouncements and strategies emanated not from the White House but from the US Agency for International Development

(AID), State Department or the Environmental Protection Agency (EPA), and these have tended to emphasize economic growth and good governance rather than environmental sustainability, social equity or the need to reconcile these objectives (Purvis 2003). In the run-up to the 2002 WSSD, for instance, the US delegation was muted in comparison with the EU's. The Secretary of State's speech, while full of optimism and references to God, failed to mention environmental protection, climate change, sustainability, or equity (Powell 2002). Since the WSSD, formal recognition—or even mention—of sustainable development by Bush or his cabinet or congressional leaders has been strikingly absent.

This short overview suggests clearly different degrees of official recognition and awareness of sustainable development. The EU's rhetorical embrace since the 1980s, although it may have flagged recently, is remarkable. It reflects awareness across levels of government and across institutions. The US official embrace is appears weak by comparison, especially since 2000. That the EU has taken a legal, declaratory, semantic leadership role is not in doubt. But the explanations for this difference vary. Several analysts have suggested the discrepancy is based on fundamentally different cultural values and norms (Wallace 2001; Smith and Steffenson 2005). For instance, because sustainable development is "seeped in the rhetoric of compelling urgency and long term commitment" (Lafferty 2004, 20), it is acceptable to Europeans but frightening to a US society wary of intrusive government action (Victor 2006). Baker and McCormick (2004, 288) go further, suggesting the EU's more enthusiastic embrace of sustainable development is a result of the EU's "deep seated belief in the ethos of collective societal responsibility for the welfare of the community as a whole." Linked to the above is the rejection by many Americans of the compatibility of economic and environmental goals: "In the United States environment and economy remain cast in zero-sum conflict" (Dryzek et al. 2002, 666–7). Moreover, Krämer (2004, 62) suggests that (in implicit contrast to the US) the EU enjoys a sort of consensus that environmental protection cannot be left to market forces.

No doubt values play an important—if sometimes ill defined—role in shaping the two polities' divergent overall embrace. In particular, they are useful background variables—political culture and values predispose societies to be more or less supportive of sustainable development. But a focus on values tends to overstate differences between the polities, and provides an insufficient explanation for that difference. First, Americans' individualist, anti-government, anti-regulatory national values may limit the embrace of sustainable development policies, but it can not explain variation across time nor the variation found below the federal level (Rabe 2004). Secondly, some so-called American values mesh with sustainable development, at least in its ecological modernization form. Faith in market tools, an embrace of wide participation, and the advantages of decentralized policymaking are all important underlying principles of sustainable development (OECD 2001b). Finally, a focus on values underplays the importance of interests which are crucial in explaining the EU's enthusiastic recognition of sustainable development. The EU's embrace of ecological modernization enables it to enjoy a leadership role globally (especially when the US forfeits that role), allowing it to serve important economic interests in the area of green technology where it competes favorably (Weale et al. 2000, 270). These alternative explanations for the EU's rhetorical lead are discussed below.

Institutionalization of Sustainable Development

Governance denotes an institutionlized set of principles, rules and norms within which actors function (Weale et al. 2000, 1). The key norms and principles of sustainable development encompass substantive or policy-specific principles such as policy integration, decoupling of economic growth from environmental degradation, cost effectiveness and environmental effectiveness, as well as procedural principles referring to how policies should be made and delivered. These procedural principles include long-term planning horizons, precaution, and accountability (OECD 2001b, 6). Also inherent in sustainable development are normative principles such as social and inter-generational justice. The extent to which these principles are embraced by EU and US institutions varies across and within institutions (Bomberg 2004). This section examines a few key institutions of both the EU and US, highlighting their varying and often selective acceptance of key sustainable development principles and practices.

European Council

The European Council (Council)—made up of heads of government or state—is a major agenda-setting body in the EU. Goals of sustainable development have been addressed and publicized at this highest tier of EU decision making. While lax in upholding procedural norms of transparency or accountability, the Council has paid significant attention to sustainable development principles such as integration and international cooperation. Embracing sustainable development serves the Council, providing a highly visible and salient issue demanding common action that is abstract enough to be amenable to intergovernmental agreement. Sustainable development allows leaders to offer political leadership on an issue that, in its abstract form, few oppose.

The Council is an increasingly overburdened body which meets only a few times a year and pays little attention to the institutional or operational details of its commitments or strategies (de Schoutheete 2006). Nor is it significantly involved in implementation of the goals and strategies it pronounces in Summit declarations. Actual commitment to principles embraced at Summits is particularly contingent on the priorities of the Member State holding the Presidency of the Council at any given time. The Council's lack of follow-up helps explain the gap between the EU's declarations—where it clearly outshines the US—and implementation.

(Sectoral) Council of Ministers

The Council of Ministers, in its various formations, has had uneven success internalizing or even embracing some sustainable development principles and norms. It brings together national ministers to agree on decisions, and thus incorporate the positions of 27 countries with often sharply contrasting views on the value and meaning of sustainable development and its principles. This diversity includes contrasts between so-called pioneer or laggard states, but the starker variation is sectoral rather than national. Environment ministers are far more likely to endorse and institutionalize sustainable development principles such as integration, environmental effectiveness, precaution and accountability than their counterparts in other sectoral Councils, especially trade, economic, transport and agriculture.

European Commission

The Commission—or at least several of its constituent departments (called Directorates General (DGs))—is probably the most advanced institution in terms of internalizing many of the key norms of sustainable development. While often acting on the lead of the Council, the flexibility of the Commission to shape the tone and details of proposals is immense. It has seized on sustainable development (especially its relation to governance) as an area in which it can play an active role, steer policy and expand its institutional remit (Zito 1999). This combination of factors has allowed several key principles to become firmly established in parts of the Commission. DG Environment has been at the front of efforts to push integration as a key substantive principle of sustainable development, with most of its strategic documents and legislative proposals highlighting the need to integrate environmental concerns into other areas. The Commission has pushed the use of sustainable development mechanisms designed to ensure that the polluter pays, such as environmental liability and, less successfully, energy or carbon taxes (Jordan et al. 2003; Bomberg 2007).

Another principle, that of precaution, has been pursued with gusto, often to the dismay of industry groups and the EU's trading partners. The principle, which first developed in the US, implies a willingness to take action in advance of formal or certain scientific proof. It features in the EU's Environmental Action Programmes and Treaties (Article 174), and now constitutes a principle of customary international law. But the precise meaning of the principle remains unclear, and its application by the Commission—it has been accused of using this sustainable development principle as a disguised trade barrier—is contentious both within and outside the EU's institutions (Majone 2002).

In short, the Commission's institutional embrace is mixed. Its concern with good governance is clear (European Commission 2001a) and its leadership role, especially in international global affairs is unmistakable if also self-interested. Moreover, DG Environment has moved to institutionalize both substantive and procedural principles. Yet, DG Environment, like environmental ministries or departments at the domestic level, are often losers in inter-institutional battles within the Commission.

European Parliament

Historically the Parliament has championed sustainable development and principles of integration, transparency and accountability. The EP has used sustainable development to play its favoured role as the environmental watchdog of other institutions. It heavily criticized the Commission's original SDS, for instance, for failing to take integration seriously enough in its own procedures. It is worth noting, however, that as the Parliament's powers increase, and as it is lobbied ever more intensely by business groups, its traditional green reputation may well be fading (Watson and Shackleton 2003). But playing watchdog can serve its interests and promote sustainable development. Institutional competition can weaken initiatives, such as when DG Environment is out-manoeuvred. But, it can also work for sustainable development: the EP's past attempts to ratchet up sustainable development commitments and check other institutions' progress arguably created more robust internalization of principles at the EU level.

In sum, a powerful explanation for inter-institutional variation is found in new institutionalists' insights into institutional behaviour. How and why principles of sustainable development are internalized has much to do with how they fit with existing institutional norms and patterns (March and Olsen 1989; Armstrong and Bulmer 1998). Sustainable development principles of environmental efficiency, integration, and transparency fit comfortably in the existing remit of Environment ministers or DG Environment. Their internalization is thus far less disruptive to these actors than they are to, say, the Council of Economic and Finance ministers steeped in norms of economic efficiency, stability and secrecy. The overall result has been the absorption of key sustainable development norms within several EU institutions, though that absorption remains uneven and unable to match the EU's legal or constitutional rhetoric.

United States

US institutionalization of sustainable development principles and practices appears more uneven and patchy than in the EU, but the explanatory role of institutional norms and interests remains important.

Executive   Presidential leadership can be crucial for the institutionalization of sustainable development. An overview of sustainable development engagement reminds us that the federal government's current reputation as sustainable development laggard is relatively recent. In the 1970s the Nixon administration used its tenure to institutionalize an array of legislation and programs heavily imbued with sustainable development principles, especially integration, precaution and transparency. The 1969 National Environmental Protection Act (NEPA) was highly precautionary and among the first-ever national efforts to integrate environmental concerns into other policy areas. The Endangered Species Act (ESA), by requiring the US government to protect hundreds of endangered species regardless of the economic effect on the surrounding region, anticipated (and exceeded) subsequent global efforts embodied in Agenda 21. And the EPA—one of the world's first—was established to monitor and oversee these policies and practices transparently. These innovative measures were in addition to far reaching pollution legislation such as the Clean Air Act with its rigorous targets that challenged assumptions about the proper balance between reducing risks to environment or health, and harming economic growth. None of these ambitious policies was labelled as sustainable development, but their underlying principles, and the practices they created, could have been.

Institutional competition and interests can help explain these 1970s measures. Dryzek et al. (2002, 665) convincingly present these environmental initiatives as part of the executive's attempt for institutional legitimation, specifically the need to stave off protests and regain lost legitimacy in an unstable time. Others have pointed to the competition between branches (usually seen as a barrier to concerted federal action) as spur for ever more ambitious policies (Foley and Owens 1996; Bryner 2008).

Presidential leadership since the 1970s has been noticeable by its absence. The Clinton/Gore administration encouraged institutionalization of key sustainable development principles within the PCSD, especially those of integration, eco-efficiency and cost effectiveness. The latter were less difficult for the administration to endorse; they dovetailed with its quest for "modern, efficient, effective governance" underlying the "Reinventing Government" initiatives (Vig and Faure 2004a, 354). This attempt aside, recent administrations in the era preceding Barack Obama's presidency made little effort to internalize even less radical principles of eco-efficiency, to say nothing of more ambitious norms of equity and integration.

Executive agencies   The nature of sustainable development means the number of executive agencies potentially needing to institutionalize principles is staggering. Federal responsibilities for sustainable development cross many departments: Energy, Transportation, Interior, and indirectly, Commerce, Agriculture, and Health. The US Agency for International Development is also responsible for promoting sustainable development as a guiding principle in foreign aid, though environmental criteria do not today feature prominently in aid projects (Purvis 2003). No one environment department coordinates and consolidates duties or serves (as does the Commission's DG Environment) as central clearing house for sustainable development initiatives.

Most responsibilities fall on EPA, but its engagement is mixed at best. On one hand, it has the most potential to integrate, enforce, monitor, and steer. It has been a key actor pushing mechanisms such as tradable pollution permits, and it has promoted a range of partnership efforts with industry "to encourage efficiency and conservation" and reduce emissions (Bryner 2000, 284). Further, it has emphasized transparency and policy learning both within the US and internationally as means to improve sustainable development measures. But its task is daunting and it is often criticized for an inability to carry out its duties or stand up to other more powerful actors.[5] Nor has it demonstrated much inclination to engage in the wider sustainable development agenda (Lafferty and Meadowcroft 2000a, 350).

Congress　Since the 1970s Congress has not internalized to any great extent key principles or practices of sustainable development (Shabecoff 2000). A few principles, such as eco-efficiency and precaution (Wiener 2004), have been promoted to a limited degree, but the more challenging demands of sustainable development such as integration, accountability and any principle requiring long term planning or international commitment are lacking. As Bryner (2000, 278) concludes: for Congress, sustainable development is simply someone else's problem.

Limited institutionalization of sustainable development principles in Congress is not surprising. Case work and pork barrelling—the supposed mainstays of congressional politics- are not easily amenable to sustainable development , and the holistic treatment required by sustainable development is not easily achieved in the decentralized Congress without effective mobilization or entrepreneurship. Yet the latter is possible. Congress played crucial roles in the initiation and formulation of key sustainable development policies in earlier eras. Several earlier landmark initiatives embedding sustainable development principles—such as NEPA, the Clean Air Act or the ESA—were launched by Congress. Bailey's institutionalist study on Congress and air pollution (1998) suggests environmental and sustainable development policy can flourish in Congress when it allows members to make a personal mark on policy or gain power and prestige within the legislature. Since 2003 Congressional debates related to climate change suggest awareness and receptivity within Congress may be increasing (Schreurs, Selin and VanDeveer, this volume; Selin and VanDeveer 2007).

5　See claims made by several EPA scientists that the EPA is bowing to industry pressure and ignoring sound science in permitting several toxic chemicals to be used in pesticides (*New York Times* 2 August 2006, A13).

Thus, there is nothing inherently sustainable development-inhospitable about Congress' operation or structure, however fragmented it may appear. After all, EU policymaking machinery suffers from some similar pathologies (Peterson and Bomberg 1999). What this does suggest is that effective mobilization across institutions and actors is crucial for any shift towards sustainable development.

Mobilization and Steering: The Role of Networks

If governance is polity steering, or steering without government, special attention needs to be paid to steerers—those pushing or nudging a polity in a particular direction. In fact, EU-US differences in the take-up of sustainable development principles and practices are a reflection of a different constellation of actors, and the bargaining power they hold. Policy network analysis, which focuses on stakeholders and exchange of resources amongst them, is a useful way to capture this mobilization dynamic.

Policy networks have been defined as strategic constellations of actors "forged around a common agenda (however contested, however dynamic) of mutual advantage through collective action" (Hay 1998, 38). Networks are products of mutual advantage and exchange of resources. Although usually used to explain policy outcomes (Marsh 1998; Peterson 2004), this chapter adapts the idea of networks to help explain polity steering rather than specific outputs.

Sustainable Development Mobilization in the EU

An examination of EU's engagement reveals a specific sustainable development steering network—a confluence of shared interests, a willingness to negotiate and a fruitful exchange of resources among key stakeholders. First, as analyzed above, each of the EU's central institutions played a central role promoting sustainable development governance, albeit to a different extent and for different reasons. In addition, during this period several so-called leader or pioneer Member States with advanced sustainable development measures on the domestic level (Sweden, Finland, Germany) held the Presidency of the Council, affording them key agenda setting power (Hayes-Renshaw 2006). The ability of these Member States' representatives to push national policy to the EU level is well documented (Andersen and Liefferink 1997).

In addition the sustainable development steering network involved non-institutional actors including scientific experts, think tanks, non-governmental organizations (NGOs) and industry groups. In contrast to the US, the network's privileging of business actors encouraged engagement of actors who might otherwise block moves toward sustainable development policies and practices. Industry federations in the EU such as the Union of Industrial and Employers'

Confederations of Europe (UNICE) or the American Chamber of Commerce (Amcham) were closely involved in the elaboration and practice of sustainable development principles. They tended to emphasize the need to balance environmental and economic concerns, ensure that principles such as cost effectiveness be given ample consideration in sustainable development strategies, and make clear their desire that sustainable development policies (especially market-based ones such as taxes or incentives) be spread evenly. The loose or weak form of sustainable development adopted (ecological modernization) very much reflects the influence of industries and firms.

Similarly, incorporation of environmental NGO representatives into the network provided expertise not readily available within the Commission or governments, and ensured a level of legitimacy and public support. These Brussels-based environmental NGOs, especially the European Environmental Bureau (EEB), a federation of over 100 national environmental groups from countries within and outside of the EU, were among the most resolute proponents of sustainable development in the EU. The EEB grasped sustainable development as a "defining issue of our time" and lobbied the EU's institutions and Member States intensely (EEB 2002b). Its influence within the network lay in links to Member States representatives, access to the Commission, long established credentials as reasonable voices, and cooperation with labor groups.[6]

Add to this mix of stakeholders certain external stimuli, both in the form of global environmental conferences (which allow the EU to act collectively in one of the few areas where a majority of citizens desire EU action above national action), and a leadership vacuum left by the US administration. Sustainable development promotion represented a chance for the EU, if it could act together, to compete with the US, and wrest from it the mantle of global leader. The result of these institutional, mobilization and structural forces was the congealing of a loose network able to steer sustainable development policies and practices through EU decision-making machinery and negotiations. It facilitated the open and swift spread of ideas, both horizontally amongst institutions and stakeholders, and vertically across levels of government.

Mobilization for sustainable development remains visible in the EU, but it is hardly permanent or unassailable. Generally the looser the network the less stable the results (Marsh 1998). The contingent character of the European sustainable development steering network means its embrace of sustainable development may be subject to change. The weakening of sustainable development as an EU priority issue in recent years underlines this dynamic. But in comparative terms, sustainable development mobilization in the EU appears robust.

---

6   See joint statement by EEB, the European Trade Union Confederation and the European Social Platform, "Make Lisbon work for sustainable development", http://www.etuc.org/a/982).

Sustainable Development Mobilization in the United States

If networks comprise strategic alliances forged around a common agenda, no such sustainable development network is readily visible at the US national level. The potential key actors and stakeholders—officials, politicians, experts, industry, environmental groups—are active in the US, and there's no one veto stakeholder, such as President or industry, blocking formation of the network. Rather, stakeholders in this distinctly adversarial system have not forged meaningful strategic alliances. Particularly noticeable by their disengagement are national environmental groups. Whereas European NGOs have individually and collectively pursued adoption of sustainable development initiatives at the EU level, US groups have remained unfocused nationally, and comparatively uninterested in the concept of sustainable development (Schlosberg and Bomberg 2008). The term sustainable development is not widely used by the groups, and professional US NGOs have resisted its integrative holistic character, preferring tried and tested strategies built around discrete court battles and legislative campaigns.[7] Similarly, whereas organized labor unions in the EU have shown broad (if shallow) support for the EU's sustainable development efforts, their US counterparts remain generally opposed to sustainable development policies fearing job losses (Bryner 2000, 279). The result has meant a lack of meaningful negotiation and bargaining requisite of networks, and instead a reliance on "well-established patterns of regulation ... confrontation and litigation" (Lafferty and Meadowcroft 2000a, 379). Yet the US political system, however fragmented, does not render mobilization impossible.

Despite (or because of) a lack of national mobilization, policies and initiatives at the sub-national level reveal that the most interesting and far-reaching efforts to steer the US towards sustainable development, albeit perhaps under different labels, are developing at local, state and regional levels. Many projects underway at the local and regional levels suggest the eventual incorporation of sustainable development as a defining goal and organizing idea. An array of community, university and local projects—on transport, smart growth or sustainable cities—embrace the fundamental requirement of sustainable development by explicitly integrating social, economic and environmental goals (Mazmanian and Kraft 1999; Vig and Kraft 2006, 381). On the issue of climate change, several hundred city mayors, representing nearly 50 million Americans, have signed up to cut their cities' green gas emissions, and they've done so as part of the Kyoto requirements to which their federal government has not agreed (*Financial Times* 12/13 August 2006). Of course these measures are by definition small scale, and cities have little power to order firms to reduce emissions or

---

7   Note M. Schellenberger and T. Nordhaus' much debated warning of "The Death of Environmentalism," should the movement not adopt new strategies and vision: 13 January 2005, *Grist Magazine*, http://www.grist.org/news/maindish/2005/01/13/doe-reprint/ and *Environmental Politics* 17:2 (2008).

make infrastructure changes, but they can shape policy on streetlights, housing or congestion charges to cut city traffic.

Action and initiatives on the state level also suggests the considerable potential within the US to develop a wider range of policies, instruments and practices that respond to the challenge of sustainable development. Most important of these have been state-level efforts to reduce greenhouse gas emissions (GHG). These measures range from GHG inventories, to more ambitious action plans, reporting and registries, through to normally contentious emission caps for power stations. Particularly striking is the extent to which these measures are a result of stakeholders bargaining and working together. In contrast to federal level stagnation, the state experience in climate change initiatives has generally been bipartisan and consensual (Rabe 2004; Selin and VanDeveer 2009a). Affected businesses have not found policies particularly costly and have welcomed chances to take credit for reductions (Rabe 2002, 2). Similarly, environmental groups are better mobilized at this level (Mazmanian and Kraft 1999). The result is emerging state-level networks made up of local and state leaders, agencies, business groups, universities and environmental groups.

A further development is creation of a number of regional initiatives. These include the Western Governors Association's resolution to expand significantly the use of clean energy in their states, or the agreement by the governors of California, Oregon and Washington to coordinate state polices to combat global warming, or the Climate Change Action Plan between several New England states and Canadian provinces (Knigge and Bausche 2006, 21). Unlike the national level, where cooperation between business, officials and environmental groups is more difficult, these sub-national stakeholders have, under certain circumstances, been able to work together to build more sustainable communities and tackle the central sustainable development issue of climate change. In so doing these networks may push or steer their wider polities towards more sustainable society. To illustrate: the OECD's sustainability report card, reviewing period 1996–2004 and published in 2005 (OECD 2005a), applauded the many initiatives by states, municipalities, and corporations to address climate change and related measures, but omitted the US federal government from its list of kudos. Of course sustainable development is broader than climate change, and the holistic, international character of sustainable development limits the ability of local or state governments to address issues in a comprehensive way. But these mobilization initiatives are important because of their potential steering role in a fragmented federal system. A widely recognized benefit of federalism is the extent to which it allows and encourages state or local innovations to flourish and influence policy at the federal level (Scheberle 1997; Nicolaïdis and Howse 2001). Obviously, spill over or spill up of innovation to the federal level faces many hurdles, legal, financial and political (Rabe 2004; Knigge and Bausche 2006). And as Victor et al. (2005) suggest, sub-state ventures still remain too atomized to exert much leverage on the federal government. But many of these hurdles also confront EU

Member States which is why the success of 'pioneer states' pushing sustainable development initiatives up to the EU level is so instructive. More generally, in terms of governance and steering, it is the direction that matters. Clearly, state and local actors have demonstrated a significant ability to experiment, innovate and steer state, regional and perhaps federal policy in particular directions.

More speculatively, there are other signs of changes in mobilization and advocacy at the national level including corporate groups' growing receptiveness to ecological modernization ideas that embrace environmental sustainability as good for business. Witness the July 2006 *Newsweek* cover story ("New Greening of America") documenting Wal-Mart's shift towards sustainability, Gore's popular warnings on climate change or his burnished claim that eco-efficiency is an integral part of a firm's capability to create value and profits. These prominent business or political leaders, reluctant to embrace sustainable development in the 1990s, have greeted this ecological modernization re-branding in ways culminate in a sort of a sort of "Ecological Modernization, American Style" (Schlosberg and Rinfret 2008). Finally, witness the mobilization of some leaders of the Christian Right and their discussion of environmental sustainability and climate change (Bomberg and Schlosberg 2008).

For many advocates of sustainable development (in both the US and EU) this putative shift of stakeholders towards ecological modernization means little; these converts aim neither to limit growth nor to change existing patterns of high consumption, and they ignore the developing world. But nonetheless if our focus is more modest—on a shift towards transformation, rather than that transformation itself, these wider trends are significant because they highlight the mobilization of previously uninterested parties—religious and business leaders—who, combined with the on-going advocacy of environmental justice campaigners, university academics and environmental activists, could form the basis of a nascent sustainable development steering network.

Conclusion

A focus on sustainable development governance as steering reveals interesting preliminary findings regarding the durability and direction of sustainable development governance in general, and the differences between US and EU sustainable development governance in particular. The findings summarized here relate to the three different components of sustainable development governance explored: recognition, institutionalization and mobilization.

On recognition, the differences in US and EU engagement are real. They are most noticeable in the declaratory, legal and official embrace of the term sustainable development, where EU engagement is more extensive. This chapter adds several qualifications to this finding. First it is important not to equate an absence of the term sustainable development in US discourse, with an absence of sustainable development principles. Though referred to by different labels in the US (sustainability, smart growth, and so on) key policies and practices reflect the core aim of sustainable development—a future-oriented attempt to balance economic, environmental and social needs. Second, what transatlantic difference does exist is due as much to varying interests as to varying values. Put another way, the EU's lead role is not just the result of moral superiority or greener norms, but interests related to economic gain, global leadership and legitimation. Secondly, formal recognition (in laws, treaties, and regulations) is important, but only part of the governance story. The latter must also include analysis of institutionalization and mobilization.

The short comparative overview offered here confirmed that institutionalization—that is, the incorporation of sustainable development principles and practices by main institutional actors—is more developed in the EU than in the US. Insights from new institutionalism regarding institutional interests and competition were used to explain institutions' selective embrace. But the chapter revealed that the US' perceived current laggardness in sustainable development is more complex than often depicted. US institutions have proven capable of institutionalizing, often before the EU, key principles and mechanisms. Moreover, networks rather than institutions are key actors steering polities towards sustainable development.

Finally, sustainable development governance requires the mobilization of ideas, interests, and institutions. Or to put it in structure/agency terms, the template of inertia or inaction in federal or quasi-federal systems is such that an agency of some kind is needed that transcends institutions, interest group and levels of government. The notion of steering networks—a constellation of actors forged around a common agenda and capable of pushing that agenda forward- is useful for exploring transatlantic differences in this mobilization dynamic. Sustainable development governance in the EU has advanced because of the steering nudge of a loose but effective constellation of actors. Its loose and inclusive character allowed for consensus on sustainable development around which most could mobilize. But these characteristics also mean the nature of the mobilization and commitment to sustainable development is itself loose and subject to weakening. In the US not even a weak national steering network was found, though significant developments at the sub-national level, and amongst some national groups, are apparent.

These patterns suggest intriguing prospects for EU-US cooperation and collaboration on sustainable development issues. They do not, on the face of it, signal a renewed or invigorated convergence between US and EU views, policies or approaches to sustainable development. Official promises emanating from Brussels or Washington, such as the announcement in June 2006 of a new US-EU High Level Dialogue on Climate Change, Clean Energy and Sustainable Development are seductive, but offer no guarantee of action beyond proclamation. Earlier statements of high level transatlantic collaboration, such as that on the UN Millennium Development Goals agreed in 2000, are far from realized.

Nor do national environmental NGOs based in Washington, or Brussels-based European NGOs appear more willing than before to collaborate or build bilateral advocacy networks. Previous formalized attempts to build bridges between NGO communities in Washington and Brussels, such as the Transatlantic Environmental Dialogue (TAED) never took off. Of course collaboration between many groups occurs through international umbrellas organizations, such as Friends of the Earth International, or informal campaigns. Moreover, this international collaboration has increased on issues of climate change. But on broader issues of sustainable development, a transatlantic divergence of strategies and priorities remains more conspicuous than any collaboration.

Perhaps the place to look for transatlantic collaboration is not at the national level but precisely where most advanced sustainable development initiatives are taking place: the sub-national level. Here we find an array of initiatives, incidences of policy learning and sharing of innovative policies and principles across the Atlantic. Urban planners in the US have taken from their European counterparts projects and ideas linked to smart-growth, justice campaigns, renewable energy and brownfield sites (Medearis and Swett 2003; Knigge and Collins 2005). In exchange, European cities, regions and states and leaders are increasingly looking to US cities and states for advice, experience and lessons and collaboration on sustainable market tools and climate change. These initiatives and lessons from them have yet to be widely disseminated. And in comparison to the high-level bilateral or international cooperation required to meet global sustainable development challenges, they appear pretty small beer. But if we look at sustainable governance as the creation of practices, rules and norms steering polity in a sustainable direction, these developments suggest the potential of sub-national partnerships to bridge the putative transatlantic gap on sustainable development issues.

Chapter 3
Intergovernmental Management of Environmental Policy in the United States and the EU

Sonja Wälti

On theoretical grounds, the literature on environmental federalism suggests that multi-tiered systems face particularly difficult challenges in addressing large-scale problems such as environmental degradation. They inherently have more difficulties adopting national standards and policies than do unitary systems; and when they do pursue national policies, they have a harder time implementing them. This is because multilevel governance structures multiply the opportunities for economic actors to subvert environmental goals. Capture of environmental progress by economic interests is believed to be especially prevalent at the local level (e.g. Ringquist 1993). In the absence of national environmental standards and programs, political systems composed of multiple jurisdictions have a tendency to under-regulate environmental problems because the jurisdictions can offload externalities to neighbors (e.g. Oates 1998). This tendency is aggravated by the fact that jurisdictions compete with one another over jobs and fiscal resources.

Empirical research shows that this daunting picture of environmental governance in multi-tiered polities is incorrect (e.g. Wälti 2004). Environmental policies have seen a continuous expansion in a number of federal countries. In the late 1960s and early 1970s, the US emerged as a worldwide leader in environmental matters. Since the 1980s, Germany, another federal country, has been a driving force in developing the EU's environmental policies. Starting in the late 1980s, the EU, a quasi-federal system, has become a leader in environmental performance, in some areas bypassing the US. Between 1990 and 2002, the EU was able to curb SO_x emissions by 65 percent while the US reduced its SO_x emissions by 34 percent. In the EU NO_x emissions went down by 30 percent as compared to 18 percent in the US. And CO_2 emissions went up by 18 percent in the US but only by 2 percent in the EU.[1]

However, multilevel structures may shape the way in which other determinants of environmental progress operate. Thus, recent comparative research stresses the importance of "corporatist arrangements"—meaning

---

[1]   Data reflect percentage changes in total levels of emissions, compiled from the OECD Environmental Compendium 2004 (OECD 2005b, 23–8).

well established links between businesses, unions, and government—to improve environmental outcomes (Crepaz 1995; Scruggs 2003, 152–61). Environmental progress is fostered by such tripartite arrangements likely because they also provide access to environmental groups. Wälti (2004) shows that this favorable link only holds true in multi-tiered polities, suggesting that environmental advocates are particularly dependent on the *multiple* access points existing at the local, state, and national level. Based on these findings, this chapter asks whether and how divergent multilevel structures in the US and the EU help to explain differences in environmental policies between the US and the EU. Both the EU and the US have developed comprehensive environmental programs and administrations, and both are international leaders in policy development. Yet, the two entities diverge over an increasing number of environmental issues. As this volume demonstrates, climate change policies and biotechnology are among the most prominent.

The ongoing European integration process has brought the EU to a level of political development basically on a par with the US. The EU has become sufficiently integrated and institutionalized, especially in specific policy areas like the environment, to be viewed as a fully-fledged political system and examined using tools of comparative politics and comparative policy analysis (e.g. Glim 1994; Braden et al. 1996; Burgess 1999; Hix 1999; Hoornbeek 2000; Kelemen 2000; Hooghe and Marks 2001; Delmas 2002; Börzel and Hösli 2003; Fabbrini 2004; Hoornbeek 2004; Vig and Faure 2004b).

Comparing the US and the EU requires a common analytic framework that enables us to discuss both systems in commensurable terms. For that reason, this chapter calls for an actor-centered perspective, which focuses on actors by considering in particular the institutional opportunities their respective multilevel political systems offer and the resulting policy options (Wälti 2004). In part the differences that have emerged between the US and the EU over approaches to environmental policy may be explained by the way in which business and environmental advocates act within the multilevel settings in which they operate. Thus, while the regulatory federal systems of the US and the EU have converged in many respects (e.g. Pfander 1996; Kelemen 2000), crucial differences persist in the access they offer environmental and business advocates.

Environmental Policy Expansion in the United States and the European Union

Comparing the multilevel dynamics in the US and the EU and their effects on policy processes and outcomes is made difficult by the fact that the two entities have long been seen in fundamentally different terms, the former as a federal system and the latter as an international or supranational regime (e.g. Hix 1999; Sbragia 2000; Fabbrini 2004).

The multi-tiered polity of the US is usually understood using a federalist lens. Federalism traditionally focuses on the allocation of powers between levels of government and on federal-state relations. American federalism has seen various developments over time (e.g. Conlan 1998; Wright 1998). It tends to be depicted in dual (competitive) terms, in which responsibilities are separated by policy area between levels of government (Watts 1999). This means that the federal government and the constituent units tend to be responsible for different sets of policies and administer them independently. Consistent with the dual federal tradition, the US federal government can implement its federal policies via regional offices. Thus, the US Environmental Protection Agency (EPA) maintains 10 regional offices that implement federal policies alongside state environmental agencies and it can even rely on its own environmental police force. The dual court system, which provides both individuals and environmental advocacy groups with numerous access points, is also characteristic of the American federal tradition (O'Leary 2003, 155).

Until the 1960s, the federal government left environmental policy to the states (Switzer 2001, 20; Sussman et al. 2002, 21–2). In the 1960s and early 1970s, as environmental degradation became more apparent and the environmental movement gained momentum, the US federal government— that is Congress—stepped in by adopting landmark environmental legislation, such as the National Environmental Policy Act in 1970, the Clean Water Act in 1972, and the Endangered Species Act in 1973 (Schreurs 2002, 32–5). The 1960s saw 11 major federal environmental laws and the 1970s another 17 being enacted (O'Leary 2003, 152). The federal government sought to regulate policy areas in which states were increasingly seen as inactive and inclined to offload externalities onto one another (Stewart 1977). The Commerce Clause in the US Federal Constitution, which entrusts the federal government with the regulatory levelling of market conditions across states, was an additional legitimizing element favouring federal oversight, as state environmental regulations threatened to introduce market barriers. These environmental policy advances, which tended to be regulatory in nature, departed from the dual tradition by pre-empting states' rights in environmental matters and enlisting them in the implementation of federal environmental policies (Sussman et al. 2002, 21–2; Rosenbaum 2005, 79–80 #2053). Consistent with the dual federal tradition, however, the US Supreme Court has repeatedly denied Congress the right to direct environmental policy implementation (Pfander 1996, 60).

During the 1980s states became vocal against so-called "mandated" policies, which imposed requirements on them without adequate funding. Posner (1998, 233–7) reports 31 major mandate legislations passed between 1984 and 1991, of which seven relate to environmental matters. The states' hesitation to further empower the federal government was supported by an increasing inclination of the US Supreme Court to protect states' rights during the 1980s. These trends coincided with the successful mobilization of industry against a continued expansion of regulatory environmental programs (Schreurs 2002; Sussman et al.

2002, 22) and the federal government's recognition that the implementation of its regulatory environmental programs was resource intensive and inefficient.

As a result of these trends, the 1990s were marked by significant changes in federal-state relations, which also affected environmental policy. Environmental policy was infused with federal-state partnerships and public-private cooperation throughout the 1990s (Scheberle 2004). US environmental federalism was subject to significant shifts, affecting not only practice but also theory. Where the emphasis used to be on traditional federal concerns of how powers are distributed or pre-empted, intergovernmental relations were seen increasingly in strategic and relational terms. The study of environmental policy diffusion and interstate environmental competition among American states can be seen as part of the same trend (e.g. Vogel 1995; Kern 2000; Esty and Geradin 2001; Fredriksson and Millimet 2002).

In contrast to the federalist tradition in the US, the EU has traditionally been studied as an international or supranational regime (Hix 1999, 14–6; Sbragia 2000; Fabbrini 2004), resulting in two main vantage points: Attention may be paid to the European integration process by examining the evolution of treaties and EU competencies; or one can study the strategic interplay between national governments from a comparative, intergovernmentalist, and state-centered stance. Both perspectives have been used to examine environmental policies in the EU.

From a "top-down" vantage point, it is striking to see that environmental policy is today one of the most integrated and centralized policy areas of the EU (McCormick 2001). Pfander (1996) points out that—paradoxically, given its limited statehood—the EU now seems to have more power over the Member States than the US Congress over the states. The expansion of EU environmental policies owes much to the growing awareness of spillovers between Member States during the 1980s, especially in combating acid rain and improving air quality. Thus, in 1984 and 1985 several important Directives were adopted to limit emissions from large industrial plants, set air quality standards for nitrogen dioxide, limit the lead content of petrol, and regulate the shipment of hazardous waste (McCormick 2001, 55).

However, more important in explaining the expansion of environmental policies in the EU is the fact that they became an integral part of economic integration. In order to reduce trade barriers, the Member States should neither push ahead nor fall behind in the adoption of environmental regulations. This commitment was anchored in the EU legislation by the 1987 Single European Act, whose main objective was to create a single European market that allowed for the free movement of capital, labour, goods, and services. The commitment is also backed by the European Court of Justice (Bzdera 1993; Koppen 2002), which has become an important proponent of environmental matters in the EU. Following the 1987 Single European Act, the Council of Ministers, composed of Member State ministers, can pass new environmental legislation by a majority (rather than unanimity) vote. The Council ultimately makes decisions. By means of Directives it can compel Member States to transpose EU environmental legislation into domestic law, and the Member States have to report regularly to the Commission, the EU's administrative body, on the measures adopted in response to EU environmental legislation. The use of Directives is consistent with the cooperative (Germanic) federalist model, with which the EU is sometimes compared (Börzel and Hösli 2003). Cooperative federal systems are characterized by a functional division of powers in which the federal government has the upper hand in legislation and oversight (Demmke 2004, 139), while the constituent units dominate the implementation process. In rare cases, the Commission can bring infringement proceedings against Member States that do not comply with EU regulations (McCormick 2001, 134–42). However, unlike the US EPA, the European environmental administration cannot deploy its own offices or oversight personnel into the Member States.

Environmental policy expansion in the EU can also be understood "bottom-up" by examining intergovernmental dynamics among Member States. Thus, the presence of a strong pro-environmental coalition among Germany, Denmark, and the Netherlands explains why environmental laggards could be forced into line by environmental leaders even before the unanimity rule was introduced in environmental matters (Liefferink and Andersen 2002). Likewise, state-centered arguments have been used to explain why the accession of the southern, environmentally less developed countries—Greece, in 1981; Portugal and Spain, in 1986—did not lead to an overall lessening of concern for the environment. It has been argued that these countries, driven by their desire to achieve economic integration, readily paid the price of environmental concessions as part of a package deal leading to their incorporation into the European Community's institutions (Liefferink and Andersen 2002). The accession of the environmentally progressive Finland, Sweden, and Austria in 1995 continued this trend. State-centered arguments are also used to explain why the eastern expansion of the EU turned out to be less of a threat to continued environmental expansion than could be expected based on their environmental track record (Schreurs 2004a). Indeed, many of these countries made tremendous environmental progress in the years prior to their accession precisely to pave the way for full economic integration. Finally, the growing attention paid to policy diffusion processes among European Member States helps to explain environmental policy expansion in the EU (Jörgens 2002).

While scholars have started to regard the EU as a multilevel or even federal polity (e.g. Pfander 1996; Kelemen 2000; Börzel and Hösli 2003; Woll 2006), they often continue to view Member State governments as the "most important pieces of the puzzle," but emphasize that the EU is an "interconnected" (not merely nested and superimposed) political arena composed of state and non-state actors (Hooghe and Marks 2001, 2–12). The multilevel governance approach seeks to locate explanatory factors of policy developments and outcomes in the strategic interplay that occurs between governmental and non-governmental actors at the various levels of governance (Grande 1996; Benz and Eberlein

1999; Weale et al. 2000). The multilevel governance framework has also paved the way for a growing number of studies about the implementation of EU policies (e.g. Grant et al. 2000; Knill and Lenschow 2000).

Explaining Differences between the EU and the United States

American federalism, European intergovernmentalism as well as supra-nationalism, and European multilevel governance (Bache and Flinders 2004; Enderlein et al. 2009) help to describe the development and expansion of environmental policy on both sides of the Atlantic. These frameworks are, however, of limited use when attempting to explain differences between the US and the EU more systematically in commensurable terms. An actor-centered perspective is better suited for comparing the EU and the US in policy-relevant terms. It places the locus of attention on key stakeholders and their use of institutional venues and opportunity structures (Scharpf 1997; Braun 2000; Wälti 2004; Woll 2006).

The main non-state stakeholder groups in environmental policy are industries (large firms and business associations) and environmental advocates (environmental groups and green parties). To explain the comparatively higher environmental performance of the EU in recent years, an actor-centered perspective calls for a study of ways the two main stakeholder groups press their causes at various levels of government. Differences in actor constellations that emerge as a result of the varying institutional settings in which they operate can help to explain divergent policies. In addition, the incentives and limits offered by institutional structures at the federal and state levels are likely to provide particular rationales to governments and public agencies to respond to this pressure in specific ways.

Comparing the Role of Leading Industries and Business Interests

Few comparative studies of lobbying in the US and EU exist because scholars have long regarded the EU as unique system and have made little reference to comparable work regarding the US. Recent comparative studies suggest that lobbyists in the US practice a more adversarial style than their EU counterparts (McGrath 2002; Woll 2006) and that they try to block unwanted legislation rather than amend it (Mahoney 2005). According to these studies, EU lobbying groups—true to the corporatist tradition in Europe—also appear less fragmented in pursuing amendment strategies (but see Mazey and Richardson 2002, 108). Bernauer and Meins (Bernauer 2003; 2003) explain why the EU has been more eager to regulate biotechnology than the US by the fact that it was faced with comparatively more unified business interests. The European biotech industry was brought together in the face of mounting public pressure against GM foods. In addition, France and the United Kingdom, traditionally not among the environmental leaders, had to address food scandals with new policies, the outcome of which they were eager to upload to the European level in a concerted effort to make them applicable EU-wide.

Besides differences in style, the authors point to a stronger impact of business lobbies in the US than in the EU. Mahoney (2005) stresses that lobbying in the EU is weakened by the fact that lobbyist have less leverage over Commission officials, Council ministers and members of the EP because the electoral accountability of governmental officials is much weaker than that of their counterparts in the US. American business lobbies enjoy great influence at the federal level via their considerable support of both parties in presidential and Congressional elections. In the EU, Mahoney suggests, by the time lobbies come in, so much work has already gone into designing a new policy that they are left with the sole option to amend a policy rather than to block it altogether. This is because the policy formulation cycle in the EU takes place in intergovernmental channels. The comparative strength, especially of export industries such as the automobile sector, at the national level in the US may also be explained by their involvement in international trade disputes during the 1970 and early 1980s (Shiroyama 2007).

The differences in the ways business lobbies affect environmental policies in the US and the EU arise from how they interplay with the respective multilevel institutions in which they must operate. The main targets of business lobbies in the EU remain the Member State governments (Grande 1996; Hix 1999, 188–208; Jones and Clark 2001; Mazey and Richardson 2002). This can be explained primarily by the fact that, despite the increasing role of the EP and the relative independence of the Commission and its directorate-generals, the Member States remain central in the policy formulation process, formally—via the Council—and informally. This difference is not merely due to the more limited degree of integration and statehood that the EU enjoys compared to the US. Rather, it is a fundamentally different understanding of the role of states within a quasi-federal setup, as is also evident in the Germanic federal tradition (Börzel and Hösli 2003; Woll 2006, 460). The Council, whose members are appointed by state governments, assures territorial and sectoral (as opposed to partisan) representation. Adding to the comparatively better access business interests enjoy at the EU Member State level is many European countries' longstanding tradition of corporatism and third-party accommodation. As a result, territorial (national) interests are generally better represented than functional (policy-relevant, sectoral) interests, and functional interests are defended via territorial channels (Hoornbeek 2004).

By contrast, business lobbies in the US have little reason to turn to the states when trying to influence federal legislation, as there are few institutional channels for American states to participate in federal policymaking. The Senate, although formally representing the states, operates largely on a partisan basis. US states tend to be most influential via the lobbying route, for example via the Council of the State Governments or the National Governors Association. In environmental matters, the states joined forces by creating the Environmental Council of the States (ECOS) in 1993.

This is not to say that business lobbying is altogether weak in EU politics. Rather, the argument here is that business interests enjoy comparatively greater influence via their access at the Member State level. This greater degree of influence of business interests at the national level, paradoxically, does not necessarily lead to capture and, therefore, decreasing levels of environmental protection but instead to environmental improvements. This must be understood in conjunction with the interest of key transnational industries in practicing environmental protectionism and uploading (even stringent) national standards to the European level in order to harmonize the conditions in which they do business. In short, industries that lobby in Brussels tend to do so for "pro-environmental" reasons. Select transnational industries do enjoy privileged access to certain policy processes and sectors of the European bureaucracy, which is quite heavily dependent on their support and input. This is not only because the EU bureaucracy is comparatively small, but also because, unlike its American counterpart, it cannot rely on its own offices to enforce policies.

Industry-driven ratcheting up of environmental policies at the intergovernmental level tends to occur when a powerful, environmentally proactive state is itself pushed to adopt higher than average standards (Vogel 1995, 1997b; DeSombre 2000; Bernauer 2003). These standards are then "exported" to other states via trans-border industries and their propensity to seek uniform regulatory conditions. In the US, California plays a leading role in setting environmental standards, especially in regard to clean air. For example, in 2005, Califorinia, followed by several other states, sought permission from the EPA to exceed federal standards in curbing greenhouse gas emissions from new motor vehicles. States are granted such "waivers" to address state pollution problems. Despite subsequent protracted lawsuits between state and federal agencies, California has helped bring global warming on the federal government's agenda, despite President Bush's refusal to develop a federal climate change policy.

California's role in the United States is mirrored in Europe by Germany and its supporters (Denmark, the Netherlands, and varying alliances of newer Member States). As in the US, European industries dislike Member States' unilateral adoption of environmental policies, preferring to wait for "Brussels" to take harmonizing action, even if that results in more stringent regulation (Glim 1994, 277; Andersen and Liefferink 1997, 7). Moreover, the EU is particularly responsive to pro-active trendsetters because Member States enjoy broader leeway to opt out of trade harmonizing policies on environmental grounds than US states.

The Proactive Role of Environmental Advocacy Groups

The picture drawn so far is incomplete without an equivalent analysis of the opportunities environmental advocacy groups face in both multilevel systems. These opportunities appear greater at all levels of the EU than in the US, and particularly more effective at the central level. This argument is supported by Daniel Bodansky (2003, 59), the climate change coordinator and attorney-advisor to the US Department of State under the Clinton administration, who asks, "Why do non-governmental groups seem so much more influential in Europe than in America?"

Environmental advocates encounter similar multilevel conditions in the American and the European (quasi-)federal systems, most notably in terms of their vertical cooperation, oversight, and enforcement combined with their horizontal division of powers between a strong executive and the legislature. On both sides of the Atlantic, environmental groups enjoy numerous venues at the local, state and central levels. In both the US and the EU, these venues are multiplied by the division between executive and legislative powers. In addition, both the US Supreme Court and the European Court of Justice constitute important veto points for environmental groups against environmental policy setbacks (Pollack 1997, 582; Sussman et al. 2002, 240; Kelemen 2004a, 114).

Despite these similarities, essential differences exist. In the US, environmental advocacy groups are particularly effective at the state level. States in which environmental groups are well engaged have more expansive environmental policies, whereas the presence of business interests does not appear to matter one way or another (Ringquist 1993). This has to do with the grassroots tradition of environmental activism in the US and with the comparatively early development of local mediation and environmental conflict resolution strategies in land use planning and conservation (Switzer 2001; Schreurs 2002; Rosenbaum 2005). Recent trends toward environmental devolution and the development of state-federal partnerships further strengthened the role of the states in environmental policy (Scheberle 2004).

Just as important, however, is the political opportunity structure that American environmental groups face. Their effectiveness at the federal level is weakened by the fact that environmental policy in the US has never become a (quasi-)constitutional principle, as it has in the EU (Hoornbeek 2004). Instead, environmental advocates have to convince Congress, step by step, to take action on various matters. Yet, their effectiveness in doing so is diminished by the limited accessibility of federal policy makers to environmental groups. Representatives in Congress are strongly committed to their constituencies and therefore more receptive to particularistic rather than "diffuse" policy concerns. As Rosenbaum (2005) illustrates by recounting the nomination hearings of Christine Todd Whitman as head of the EPA in January 2001, when environmental issues are at stake, those too tend to be expressed in particularistic terms. For Congress to act in environmental matters high public awareness and immediate electoral pressure is needed (Mahoney 2005).

The administration does not provide an effective alternative lobbying route for environmental advocates. While the US EPA and other environmental agencies are part of iron triangles and issue networks that characterize the American environmental policymaking process and while NGOs play an increasing role in implementing policies, there is a general reluctance to let private parties into the administrative process. To be sure, agencies give advocates opportunities to comment and hear them at meetings and workshops. They may also be called to participate in so-called "negotiated rulemaking." Yet, the neo-corporatist tradition of many European countries and the EU administrative culture may well allow greater informal access. Moreover, in contrast to the Commission's Environment Directorate-General, the US EPA is not a fully-fledged ministry. Environmental policy is therefore very dependent on (and at times tempered by) the White House.

In sum, environmental groups in the US face inconsistency and often considerable obstacles in trying to shape policies via the executive and the legislature at the federal level. In contrast, they enjoy considerable litigation power and have been highly successful in influencing state as well as federal environmental policies by means of lawsuits against industries and developers as well as governmental agencies. This constellation of opportunities and constraints may help explain why US environmental groups are found more often in relatively stable advocacy coalitions than their counterparts in the EU (Mahoney 2007).

In the EU, environmental groups also enjoy a grassroots tradition and could conceivably focus on the local and sub-national levels. Yet, due to their early alliance with the anti-nuclear movement, they have looked more quickly to the national legislator as well as to supranational organizations for influence. However, their comparatively greater influence in central policymaking is not alone the result of divergent pasts but likely due to more numerous access points.

European environmental advocacy groups, although likely less unified and centralized than their counterparts in the US, are very effective in directly influencing central EU policymaking (Pollack 1997). The Commission, the EU's administrative body is particularly receptive to non-governmental organizations' demands throughout the policy process, providing them with considerable assistance (McCormick 2001, 60). For example, it has initiated the General Consultative Forum on the Environment, which is designed to provide representatives of NGOs, business, and unions with a channel to advise the Commission on policy development (McCormick 2001, 60). Generally, NGO activities have expanded since the 1987 Single European Act.

The greater receptiveness of the Commission vis-à-vis NGOs may result from its greater independence and more extensive agenda setting privileges than those of the American bureaucracy. At the same time, the European environmental bureaucracy is more dependent on intermediaries to enforce its policies at the Member State level (McCormick 2008, 205). Unlike the US EPA, which counts 17,000 employees across the US, the EU Environment Directorate-General, which employs around 700 staff, cannot rely on its own resources to ensure compliance.[2] Rather, it is forced to entrust Member State agencies with the implementation of European directives. The role of NGOs is further legitimized by the lack of direct citizen control over much of the EU policy process at the executive level. Finally, the Commission's environmental policies are shaped significantly by the environment ministers of Member States, which have a more distinct sectoral and hence environmental orientation.

The EP, although more limited in its powers than the US Congress, has been shown to be a particularly ardent defender of so-called "diffuse interests," not least because of the notoriously strong presence of green parties (Pollack 1997, 580). In 2008, for example, 43 or 5.5 percent of the 785 seats are held by the Greens and European Free Alliance representatives[3]—more than in many Member State parliaments. The US Congress' responsiveness to environmental pressure fluctuates significantly depending on prevailing majorities and public opinion (e.g. Sussman et al. 2002). Green voters in the US do not have the benefit of a proportional representation system that helps European environmentalists elect distinct Green parties; nevertheless, they can rely on a similar number of environmentally oriented congressmen and congresswomen to push their cause.

Finally, European environmental advocacy groups, like business lobbies, can influence EU environmental policies via traditional access points at the Member State level. For example, Germany's leadership in pressing for limits on car exhaust emissions was not only due to the interests of the German car industry, but also considerable domestic green pressure (Mazey and Richardson 2002, 108–9). Likewise, the German waste packaging policy was pushed by green advocates via the German legislative channels and later influenced the EU Packaging Waste Directive (Mazey and Richardson 2002, 109). Such indirect influences are less developed in the US than in the EU because the US states are less well represented in central policymaking than the EU Member States. The fact that the Council, much more so than the US Senate, assures a territorial and sectoral representation of Member States without direct electoral accountability opens many more links between the policy venues environmental advocates enjoy at the Member State level and the EU policymaking process. As a result national "politico-administrative elites" have considerable weight (Mazey and Richardson 2002, 107). Therefore, "federal-state relations" in the EU therefore tend to be more policy than politics driven than in the US. In

---

2   http://ec.europa.eu/dgs/environment/index_en.htm; http://www.epa.gov/epahome/aboutepa.htm (accessed in June 2008).

3   http://www.greens-efa.org/cms/default/rubrik/6/6552.members@en.htm.

such a setting, complex policies whose distribution of costs and benefits are diffuse, have a higher chance of succeeding.

Substantial differences can also be found between the US and the EU in terms of the propensity of state politicians to respond to pro-environmental pressure for electoral reasons, when such pressure exists. European governments are generally subject to stronger constraints from the voter-taxpayer than their counterparts in the US for several reasons. *Exit* ("voting with one's feet") is not as realistic an option for Europeans as it is for Americans. Although EU citizens can locate freely within Member States, their rights are still more intimately linked with work permits and citizenship than that of Americans, not to mention language barriers to relocation. At the same time, *voice* (sanctioning state governments by means of elections) is a more powerful means of expressing voter discontent in the EU than in the US, as Member State governments take direct part in the policymaking process within the Council. Also, when aspiring to a federal career, US state politicians may occasionally compromise preferences of the state electorate in the name of the "greater (federal) good." Electoral politics in the EU is still more driven by national agendas and EU careers are too sparse and unattractive to be worth compromising national electoral concerns. Finally, European citizens may be more focused on sanctioning their governments because they cannot look to the "central government" for electoral sanctions the same way Americans can. Except for elections to the EP, whose powers are more limited than those of the US Congress, citizens cannot directly sanction the European leadership. In sum, European state governments when prompted by an environmentally sensitive electorate are more likely to choose a proactive course than are American state governments.

Conclusion

The comparative assessment of US and EU intergovernmental environmental policymaking demonstrates the utility of thinking of the US and the EU in commensurate terms. In studying similarities, differences and relations between the two entities, it is crucial to create a common framework that is informed by comparative politics and comparative policy analysis. To go beyond a mere comparison of institutions in order to better understand how different outcomes are achieved; namely by examining the constraints stakeholders face and the opportunities they are given in each of the multilevel systems.

In sum, the EU has taken a greener path than the US in recent years because its environmental decision process since the Single European Act came into effect offers more access points to environmental advocates at the domestic and EU levels. The incentives for governments to respond to pro-environmental pressure at the Member State level are greater in the EU than in the US. Moreover, there are more effective linkages between access points at the Member State level and those at the EU level because the Member States have a greater say in the agenda setting and policy formulation process of the EU than their American counterparts. In the US, environmental groups may enjoy great leverage at the state level, especially in some environmentally progressive states, but they have a harder time affecting federal policies or uploading environmental advances from the state to the federal level. They are able to do so only when the federal government is fully committed to environmental advances, as was the case in the early environmental expansion phase in the 1970s and early 1980s. As in the EU, they are also successful in uploading environmental advances when there are significant business interests in harmonizing market conditions across the country.

Business lobbies in the US, on the other hand, enjoy more extensive lobbying structures at the federal level than their EU counterparts. At the same time, the latter appear to be more successful in influencing the environmental agenda setting and policy formulation process than their American counterparts. An important pro-environmental dynamic stems from the fact that the better organized and more effective lobbies at the EU level are at the same time those particularly interested in market harmonizing measures. And such harmonizing measures often work in favour of increasing environmental standards.

Intergovernmental dynamics are by no means the only factors driving environmental policy on both sides of the Atlantic. Diverging environmental priorities are also due to other than institutional and actor-related factors. Most importantly, as Bernauer (2003) shows in his study on biotechnology, environmental and consumer protection values have diverged in the EU and the US in part because the events shaping these values have differed. Similarly, Vogel (2003b) argues that European environmental policy is surpassing American environmental policies due to higher political support for environmental regulation, not least because important regulatory failures have occurred, most notably in the food safety area, and because the EU has steadily increased and demonstrated its regulatory competence.

However, it is useful to isolate intergovernmental dynamics because they are likely to remain important driving forces of environmental policy. The vulnerability of the US to a reversal of its federal environmental achievements can be attributed in important ways to its multilevel dynamics. These achievements were made during times when public opinion drove environmental policy at both the state and federal level. In the absence of such a national consensus environmental progress in the US is very dependent on the actions of environmental leaders such as California and environmental followers such as the north-eastern states. Indeed, the most significant environmental advances the 1990s have been made at the state level. This tendency has become more pronounced under George W. Bush's leadership, with California pushing ahead on climate change policies (Schreurs, Selin and VanDeveer, this volume).

In contrast, the multilevel dynamics in the EU have a consolidating effect on environmental achievements. The enlargement process does not seem to have significantly altered the opportunity structures and actor dynamics in environmental matters. As we have seen, the propensity of Member States to adopt and stay a pro-environmental course as well as to upload environmental concerns to the EU level is intimately linked to the prevalence of environmental concerns among their electorate. The Member States driving EU enlargement have shared environmental priorities to buy into the "European dream." A rolling back of environmental achievements at the EU level seems unlikely for two main reasons: First, the EU's environmental policies are intimately tied to economic integration and have thus become a core element of the EU's very existence. Secondly, the EU enjoys high credibility in environmental matters, both with the public and with business and environmental advocates as well as on the international stage.

PART II
Governing Risk: Chemical Regulations, Asbestos Bans, Product Standards, and Genetically Modified Organisms

Chapter 4

Transatlantic Politics of Chemicals Management[1]

Henrik Selin

The European Union (EU) and the United States (US) have well-developed institutions for chemicals management, and EU and US authorities assume general responsibility for protecting human health and the environment from hazardous chemicals (Rosenbaum 2002, 229–68; Schörling and Lund 2004; Wiener 2004). Much European and North American chemicals legislation has developed in parallel since the 1960s, often as a result of transatlantic learning and policy diffusion. In addition, multilateral efforts since the early 1990s involving much North American and European participation have established several international organizations, treaties and programs on hazardous chemicals (Krueger and Selin 2002; Downie et al. 2004). These actions were motivated in part by a shared European and North American desire to have their relatively stringent protection standards adopted globally.

While the EU and the US have cooperated extensively on chemicals issues, and share a basic concern over hazardous chemicals, they frequently disagree over regulatory issues and approaches (Selin 2003; Selin and Eckley 2003). Transatlantic disagreements include controls on specific substances as well as the application of principles and procedures for chemicals assessment and regulation. Furthermore, EU efforts to substantially revise and expand its chemicals legislation and management through the implementation of the Registration, Evaluation and Authorization of Chemicals (REACH) regulation adopted in 2007 are causing growing EU-US controversy over the future direction of chemicals management, despite pressures from economic globalization for transnational regulatory harmonization.

Transatlantic politics of chemicals management is furthermore important beyond issues of improved environmental and human health protection. Chemicals regulations affect not only the production and use of chemicals, but also the trade in chemicals and goods that contain chemicals, including textiles, pharmaceutical, electronics and automobiles. Because of the domestic and international economic significance of the chemicals industry and the

1   The author thanks the following people for their assistance and helpful comments: Miranda Schreurs, Stacy VanDeveer, Noelle Eckley Selin, Alastair Iles, Joel Tickner, Rachel Massey, Robert Donkers, Chris Blunck, and Janice Jensen. The author is solely responsible for the arguments made in the text.

high value of trade in chemicals and goods between Europe and the US, the US government and industry organizations closely follow, and often try to influence, European policy developments on chemicals management including the creation and implementation of REACH. This shapes both transatlantic relations and global chemicals policy.

This chapter explores EU-US cooperation and competition in the area of chemicals management. It begins with a brief discussion of why chemicals are an important political, economic, environmental and human health issue. This is followed by an examination of EU-US collaboration on international chemicals management, including issues over which the EU and the US have disagreed. The next section looks at US and EU chemicals legislation and regulation, respectively, identifying similarities and differences between the two systems. The chapter continues with a summary of REACH including shortcomings in earlier EU chemicals management that the regulation is intended to address.

Next, it is argued that REACH is accentuating distinctions between EU and US approaches to chemicals management. The Bush administration and the US chemicals industry voiced opposition to REACH for regulatory, financial and market-based reasons. Regulatory opposition stems from differing transatlantic regulatory cultures. Financial arguments focus on the costs of REACH to the chemicals industry and regulatory agencies. Market-based opposition involves competition over international setting of product standards and transatlantic disagreement over trade issues. The chapter ends with a discussion of policy divergence and convergence issues between the EU and the US and their implications for transatlantic relations and global cooperation.

Chemicals and the Chemical Industry

Since the end of World War II, a rapidly growing number of chemicals have been used in industrial manufacturing, agriculture, consumer products and human health protection. Over 100,000 chemicals may be in world-wide use, but no exact number is available (European Commission 2001b). Global sales in chemicals (excluding pharmaceuticals) were worth €1.5 trillion in 2005 (CEFIC 2006, 2). 80 percent of global chemicals production takes place in sixteen countries including the US and eight EU members (OECD, 2001a).[2] The chemical industry is Europe's fourth largest industrial sector, accounting for over 11 percent of European manufacturing and employing 1.6 million workers (Geiser and Tickner 2003). The chemical industry is America's largest exporting sector employing one million people (American Chemistry Council 2007).

---

2   The countries are: the United States, Japan, Germany, China, France, the United Kingdom, Italy, Korea, Brazil, Belgium, Luxembourg, Spain, The Netherlands, Taiwan, Switzerland, and the Russian Federation.

Chemicals have been subject to continuous government controls since the 1960s. Yet, chemical regulation is often plagued by scientific uncertainty and competing interests, presenting policy makers "with some of the most intractable dilemmas of social regulation" (Brickman et al. 1985, 21). Societies see many commercial and social benefits from the growing production, use and trade in chemicals. However, some chemicals can have severe negative impacts on human health and the environment (European Environment Agency and United Nations Environment Programme 1999). Hazardous chemicals may be released during all stages of the chemical life-cycle: production, use, trade and disposal. In addition, combustion and production processes result in the unintentional release of hazardous by-products into the environment (Krueger and Selin 2002; Selin 2003).

Concentrations of chemicals can build up in individuals over time (e.g. bioaccumulation) and increase up through food webs (e.g. biomagnification). Chemicals can persist in the environment for long periods of time, ranging from years to decades (Arctic Monitoring and Assessment Programme 2002a). Chemical contamination in animals is linked with disruption of endocrine functions, impairments of immune system functions, and functional and physiological effects on reproductive capabilities. Carcinogenic and tumorigenic risks and developmental effects in small children because of exposure to chemicals are attracting growing scientific and public attention (Arctic Monitoring and Assessment Programme 2002b). Health authorities in many countries have issued dietary guidelines for pregnant women and small children to reduce exposure to hazardous chemicals.

International Chemicals Management

Hazardous chemicals permeate national borders. They are extensively traded, including a vast range of goods containing chemicals. Emissions of hazardous chemicals frequently are transported long distances, in some cases on a global scale. Over the past two decades, international cooperation and policy making aimed at improving the management of the full life-cycle of hazardous chemicals have intensified. On these issues, transatlantic collaboration and competition are significant.

At the 2002 World Summit on Sustainable Development (WSSD) in Johannesburg, governments agreed that by 2020 chemicals should be "used and produced in ways that lead to the minimization of significant adverse effects on human health and the environment" (United Nations 2002, paragraph 23). Because of the political importance of the EU and the US and their positions as the two largest producers, users and exporters of chemicals, their support and active engagement is necessary to realize this objective. The EU and the US play central roles in strengthening legal structures, generating scientific knowledge about chemicals, aiding in the identification and application of alternative techniques and substitutes to hazardous chemicals, and improving domestic management through capacity building.

The international community, with strong EU and US involvement, recently concluded three major treaties addressing different parts of the life-cycle of chemicals. The 1998 Protocol on Persistent Organic Pollutants (POPs) under the Convention on Long-Range Transboundary Air Pollution (CLRTAP) covers production, trade and disposal; the 1998 Rotterdam Convention on the Prior Informed Consent Procedure for Certain Hazardous Chemicals and Pesticides in International Trade focuses on import and export; and the 2001 Stockholm Convention on Persistent Organic Pollutants targets production, use, trade and disposal (Downie et al. 2004). These treaties initially regulated a small number of chemicals, but the international community is working to expand the number of chemicals that are covered under each of the treaties.

The EU and the US were also engaged in much early multilateral cooperation (Buccini 2004; Downie et al. 2004). The 1989 Basel Convention on the Control of Transboundary Movements of Hazardous Wastes and Their Disposal covers the disposal of chemicals that are categorized as hazardous wastes. Conventions under the International Labor Organization address chemicals affecting workers. Treaties under the International Maritime Organization cover marine chemical pollution. Supplementary regional marine chemical pollution agreements have been negotiated in both Europe and North America, including around the Baltic Sea, the Northeast Atlantic, the Mediterranean Sea, and the Great Lakes region.

In 2002, the EU and the US were part of a larger group of countries that adopted the Globally Harmonized System of Classification and Labeling of Chemicals, a standardized way of classifying chemicals according to their hazards and communicating information through labels and safety data sheets. Furthermore, the US and the EU work together on the Strategic Approach to International Chemicals Management (SAICM), which was adopted in 2006. SAICM outlines a global policy strategy and plan of action for improved governance towards the 2020 goal set at the WSSD, including measures to: support further risk reduction; improve knowledge and information sharing; strengthen institutions, law and policy; enhance capacity building; and address illegal traffic.

Even though EU-US collaboration is a significant driver behind much multilateral policy making on chemicals, the EU and the US regularly disagree. Important disagreements include: which chemicals should be regulated under specific treaties and how they should be regulated; how approaches to risk management should be applied; what the role of the precautionary principle should be in chemicals assessment and regulation; and how trade regulations should be used in environmental agreements.

During treaty negotiations, the Commission and Member States pushed for more substances to be regulated under the 1998 CLRTAP POPs Protocol than did the US (Selin 2003). In addition, the US has argued for more exempted uses of regulated substances under the CLRTAP POPs Protocol and the Stockholm POPs Convention than have European countries, which have instead pushed for a complete phase-out of the production and use of regulated chemicals (Downie 2003; Selin 2003). As efforts to expand the number of regulated substances have begun under the CLRTAP POPs Protocol and the Stockholm POPs Convention, EU-US differences over regulations and exemptions are continuing under these treaties. Similar controversies have also occurred under the Rotterdam Convention.

The EU is also a vocal supporter of more precautionary-based chemicals regulation while the US has been more reluctant to embrace the precautionary principle (Loewenberg 2003; Selin 2003; Selin and Eckley 2003; Vogel 2003b). The US advocates regulation-based on detailed risk assessments using set data requirements and fixed numerical criteria. In contrast, European officials argue that assessments should be flexible, and should allow the use of different combinations of scientific and socio-economic data. Whereas US policy makers wish to take regulatory action only in cases of clear evidence of harm, the EU rejects the notion of an unambiguous line between harmless and hazardous substances. Instead, EU policy makers want the precautionary principle to guide individual assessments and regulatory decisions.

In addition, the EU has been more willing to include trade-related provisions in multilateral environmental treaties than the US. For example, the Commission and several Member States advocated for export and import controls on regulated chemicals during the negotiations for the CLRTAP POPs Protocol (Selin 2003). This was rejected by US government officials, who argued that it would be inappropriate to include trade restrictions in a regional environmental agreement. Trade regulations are, however, included in the Rotterdam Convention and the Stockholm Convention. Nevertheless, the US has not yet ratified either of these agreements, or the Basel Convention managing the trade in hazardous wastes. In contrast, the EU and the vast majority of Member States are parties to all the major chemicals treaties.

US and EU Chemicals Policy

Not surprisingly, EU and US positions on chemicals management issues in multilateral forums are shaped by their internal legislation, norms and practices. To understand transatlantic relations on chemicals policy, it is necessary to look at the design of early and recent European and US chemicals legislation and regulation, factors driving EU efforts to revise and expand its chemicals management, and the influence of expanded EU policy making on transatlantic relations.

US Chemicals Legislation and Regulation

In the US, industrial chemicals and pesticides are regulated under separate sets of legislation that are noticeably different in how they control hazardous substances. The *Toxic Substances Control Act* (TSCA) of 1976, which is administered by the Environmental Protection Agency's (EPA) Office of Pollution Prevention and Toxics, regulates industrial chemicals. Regulatory options range from labeling standards to complete bans. TSCA introduced pre-manufacture notification only for industrial chemicals that were introduced after December 1979 (so-called "new" substances). Substances that were on the market prior to December 1979 (so-called "existing" substances) did not have to be retrospectively notified.

The Chemical Substance Inventory of "existing" chemicals established under TSCA originally covered approximately 61,000 substances (generally industrial chemicals). During the 25 years of TSCA's operation, over 20,000 "new" chemicals have been added. The inventory currently includes approximately 82,000 chemicals produced in or imported to the US (US Environmental Protection Agency 2003). For pre-manufacture notification, manufacturers and importers are required to notify the EPA 90 days in advance of the scheduled introduction of a new chemical. If the EPA takes no action to prevent commercialization, commercial production and sales may begin. TSCA does not, however, define a base set of data required for notification. Rather, determinations are made by the EPA on a case by case basis.

Several reviews by the US General Accounting Office suggest that the protection of human health and the environment should be improved (United States General Accounting Office 1994, 2005, 2006). The reviews note that TSCA rests on the assumption that companies have a right to produce and market industrial chemicals and that the EPA has to demonstrate that a chemical poses "unreasonable risk" to human health or the environment before the agency can take action. The EPA's Office of Pollution Prevention and Toxics thereby is caught in a catch-22 situation: it has to provide evidence of unreasonable risk or substantial exposure before it can request data from companies to evaluate whether there is an unreasonable risk (Powell 1999, 83). TSCA also does not define what constitutes unreasonable risk.

Pesticides are covered by several pieces of legislation. The main act is the *Federal Insecticide, Fungicide and Rodenticide Act (FIFRA)*, adopted in 1947, and substantially revised in 1972 and 1996. This law requires companies to register and label all pesticides sold in the US. The EPA processes approximately 5,000 new pesticide registrations every year (Powell 1999, 83). The act requires industry to apply for registration before a pesticide is marketed, and instructs the EPA to weigh the environmental and social costs and benefits before allowing registration. A company has to prove that a pesticide will not have "unreasonable adverse effects on the environment" to pass registration (Bodansky 1994).

The Pesticide Product Information System, available on-line, lists all pesticide products that are registered in the US.

The *Federal Food, Drug and Cosmetic Act (FFDCA)*, which was enacted in 1906 and is administered jointly by the EPA and the Food and Drug Administration, aims to ensure that food, drug, and cosmetic products put on the market are safe. The so-called "Delaney Clause" of 1958 introduced a zero-tolerance policy for carcinogenic food additives (Brickman et al. 1985, 34). This clause was repealed in 1997. Instead of a blanket ban of carcinogenic additives in food, the act sets maximum residue levels of pesticides in food and products and requires documentation of all hazardous pesticide residues, not just cancer risks (Andrews 2003, 238). The 1996 *Food Quality Protection Act* amended parts of FIFRA and FFDCA introducing stricter safety standards and new ways of determining pesticide tolerance levels.

Several other chemical-related pieces of legislation are also implemented by the EPA (Rosenbaum 2003, 178). For example, the *Emergency Planning and Community Right-to-Know Act* of 1986 requires companies to notify state authorities and communities about chemicals releases. The *Clean Air Act* sets emission standards for hazardous air pollutants, while the *Clean Water Act* sets standards for hazardous water pollutants. Similarly, the *Safe Drinking Water Act* sets standards regarding hazardous substances for drinking water. The *Resource Conservation and Recovery Act* regulates solid waste management practices, including all forms of hazardous substances. The *Comprehensive Environmental Response, Compensation, and Liability Act* (the so-called Superfund Act) authorizes the cleanup of many sites contaminated by hazardous substances.

EU Chemicals Legislation and Regulation Pre-REACH

EU chemicals legislation pre-REACH was structured around four major pieces of legislation: three directives and one regulation, all of which were subject to several amendments. These covered the classification, packaging and labeling of dangerous substances; "existing" substances; the classification of dangerous preparations; and the restriction of marketing and use. The four legislative items were administered jointly by two of the Commission's Directorates-General (DGs): DG Environment and DG Enterprise and Industry.

The 1967 *Directive on the Classification, Packaging and Labeling of Dangerous Substances* introduced uniform labeling and packaging requirements for "dangerous" chemicals, but mainly to facilitate trade (Montfort 2002; Schörling and Lund 2004, 52–3). The sixth amendment, passed in 1979, introduced environmental concerns (Schörling and Lund 2004, 53). In addition, this amendment made a distinction between "new" and "existing" chemicals (similar to TSCA). It required a pre-market notification for "new" chemicals that entered the market after 18 September 1979 (Montfort 2002, 271). This notification procedure involved requirements for testing by the applicant depending on the substance's marketing volume. All "new" chemicals sold in quantities over 10 kilograms annually had to be registered.

In contrast, the 1979 amendment did not introduce any regulatory requirements on chemicals already on the common market before 18 September 1979. It was not until the 1993 *Regulation on Existing Substances* that also "existing" chemicals came under EU regulatory control (Montfort 2002, 272). Yet, "existing" chemicals were not subject to the same regulations as "new" chemicals; manufacturers and importers needed only to provide competent authorities with basic data depending on the volumes in which these chemicals were produced or imported. This regulatory distinction between "existing" and "new" chemicals was made despite the fact that "existing" chemicals make up over 95 percent of the chemicals on the common market.

The 1993 regulation furthermore introduced a set of more uniform principles for risk assessment and increased testing and labeling requirements for "existing" substances. Significantly, it also determined that notification of a new chemical to the competent authority of one Member State equated to notification throughout the EU, creating a situation where it was difficult to get an EU-wide overview of notification. There have been approximately 2,700 notifications of new substances since 1981, with notifications of over 300 new substances each year since 1996. Of these, approximately 70 percent have been classified as dangerous.

The 1988 *Directive on the Classification of Dangerous Preparations*, which was updated in 1999, set out harmonized classification, packaging and labelling requirements for preparations, similar to those applying to dangerous substances under the *Directive on the Classification, Packaging and Labeling of Dangerous Substances*. A preparation is defined as a mixture or solution of two or more substances. The directive made no distinction between "new" and "existing" preparations, and those new substances included in mixtures or solutions were subject to notification requirements. Most chemicals on the European market— an estimated 90–95 percent—are preparations (Montfort 2002, 272).

The 1976 *Directive on the Restriction of Marketing and Use* regulated the sale of chemicals. Under "ban with exemptions," only explicitly approved uses were allowed. Under "controlled use," the most common type of restriction, marketing and use of a substance or preparation was allowed except in cases of specifically identified prohibitions. A 1994 amendment banned public sale of all substances and preparations that are carcinogens, mutagens, or reproductive toxicants. This directive covered approximately 900 chemicals, including 850 categorized as carcinogenic, mutagenic or reproductive toxicants (Geiser and Tickner 2003, 74).

In addition to these four main sets of legislation, a multitude of directives cover specific chemicals issues (Montfort 2002, 272; Royal Commission on Environmental Pollution 2003, 86–94). For example, the *Water Framework Directive* contains a long list of regulated chemicals pertaining to water quality. Similarly, air pollution directives set emission standards on by-products such as dioxins and furans. A long list of directives dating back to the 1970s regulates pesticide residue levels in different kinds of food. There are also directives pertaining to worker protection and waste management that address hazardous chemicals.

Comparing US and EU Chemicals Legislation and Regulation

Early US and EU chemicals legislation and regulation shared many similarities, and there has been much transatlantic policy interaction and information exchange over the years (Wyman 1980; Brickman et al. 1985). The US led on much early legislation, which was later copied by the EU. In both, legislation was complex and split across many separate large pieces of law (directives and regulations in the EU and acts in the US) and implemented by multiple public agencies. First developed in the US, the US and the EU also divided chemicals into "existing" and "new" substances for regulatory purposes. Yet, neither US nor EU authorities have assessed all "existing" substances, and there is little public risk assessment data on the vast majority of chemicals available on North American and European markets (largely the same set of substances).

Transatlantic differences are, however, growing as the EU moves ahead of the US on many legal and regulatory issues with REACH. Whereas REACH harmonizes procedures for assessing and regulating "existing" and "new" chemicals, TSCA continues to set a very high threshold for regulating "existing" chemicals and does not require industry to generate toxicological data for pre-manufacture notification (only to submit data that they have). The US EPA has also been conservative in its use of TSCA to control chemicals (Wyman 1980; Brickman et al. 1985), particularly after a court decision in 1993 that repealed (in part) an EPA ban on asbestos (Wyman 1980; Brickman et al. 1985; Lowell Center for Sustainable Production 2003). In contrast, much regulatory competence has been transferred from Member States to the European level since the 1980s, resulting in REACH.

EU Chemicals Management and REACH

The expansion of EU chemicals legislation and regulation is part of a larger effort to develop more stringent environmental and human health protection standards. These broader efforts are driven by failures of past risk management policies, growing European political and public support for more risk-adverse regulations, and expanded competence of EU organizations (Vogel 2003b). In response to the political and public demands for more effective and precaution-based chemicals policy, the EU strengthened its legal and institutional capacity for chemicals risk management through the adoption of REACH in 2007 following a decade-long political process (Selin 2007).

The implementation of REACH is linked to the EU's Sustainable Development Strategy, adopted by the Council in June 2001, which calls for better management of hazardous chemicals. Similar to the 2020 goal agreed to at the WSSD, the Sustainable Development Strategy sets the goal to "by 2020, ensure that chemicals are only produced and used in ways that do not pose significant threats to human health and the environment" (European Commission 2002a, 35). The implementation of REACH over an (at least) 11-year long period will involve a series of actions throughout the registration, evaluation, and authorization phases.

During *registration*, producers or users of a chemical have to compile a dossier containing basic risk assessment data as well as a provisional risk assessment based on intended uses. REACH requires approximately 30,000 "existing" chemicals to be registered between 2007 and 2018. This covers most chemicals that are produced or imported into the EU in quantities greater than one metric ton annually (exceptions are made for polymers and intermediaries as well as products such as pesticides, pharmaceuticals and cosmetics covered by separate pieces of legislation). "Existing" chemicals that are produced in quantities greater than 1,000 metric tons per year and those that are carcinogenic, mutagenic, and reproductive toxicants must be registered during the first three years. "New" chemicals are also subject to registration.

During *evaluation*, data on all registered substances submitted by industry will be collected in a central European database. Under the guidance of the European Chemicals Agency, located in Helsinki, designated authorities in Member States will evaluate those chemicals used in the greatest quantities or those of particular concern to develop appropriate risk reduction measures—a projected 5,000 chemicals initially. Based on information submitted by industry, chemicals of concern that are produced and imported in quantities exceeding 1,000 metric tons will be evaluated first. Chemicals of concern that are produced and imported in quantities above 100 metric tons will be evaluated thereafter. Chemicals that are produced and imported in quantities exceeding one metric ton will be subject to spot checks and computerized screening.

During *authorization*, chemicals identified to be of the greatest concern will have to undergo a process in which firms need to get explicit permission from authorities before selling or using such chemicals. Firms that want to continue using a chemical subject to authorization would have to demonstrate that it can be used safely, or that the chemical is necessary for a particular use and that there is no viable alternative. The Commission estimates that authorization at first would apply to approximately 1,400 substances that are known or highly suspected to be harmful to the environment and human health. It is furthermore expected that firms producing and using chemicals requiring authorization will come under market-based and political pressures to phase out such chemicals (Layton 2008).

REACH is supported by many environmental leader states, DG Environment, green members of the EP, and environmental and public health groups. In contrast, efforts to create a strong REACH regulation were resisted by the European chemicals industry, high-level politicians from countries with large chemicals industries, and many conservative and socialist members of the EP. The final version of REACH is largely a compromise between these two groups (Selin 2007). While some stakeholders would have preferred subjecting more chemicals to more stringent controls, REACH will nevertheless significantly strengthen EU chemicals policy. In general, supporters hope that REACH will address four weaknesses in earlier chemicals legislation and regulation (but it is too early to say if that will happen or not).

First, EU chemicals management was criticized by Member States and stakeholders for having become too complex and for suffering from critical implementation deficits (European Commission 2001b; Nordbeck and Faust 2003). While EU chemicals legislation was built around the three directives and one regulation discussed earlier, over 100 pieces of legislation in some part addressed chemicals. Few, if any, European public agencies or private sector actors had the necessary overview of all these pieces of legislation and their requirements to ensure full and consistent implementation across all Member States and firms (Royal Commission on Environmental Pollution 2003, 59–60). REACH will replace approximately 40 different pieces of older chemicals legislation and thereby streamline the body of EU chemicals legislation.

Second, scientific assessments indicated that protection of human health and the environment remained inadequate (Geiser and Tickner 2003; Nordbeck and Faust 2003; Schörling and Lund 2004). For most "existing" chemicals there are only scant data on toxicity, emissions, environmental dispersion, and ecosystem and health effects (European Environment Agency and United Nations Environment Programme 1999). Many of these are still in extensive use. REACH expands registration of chemicals on the European market, abolishes regulatory distinctions between "existing" and "new" chemicals, and requires more extensive risk assessment of substances used in large quantities or those having inherent properties that make them substances of great concern (i.e. those that are carcinogenic, mutagenic and reproductive toxicants).

Third, although EU treaties and texts state that precaution should guide policymaking, the influence of precautionary thinking on EU chemicals regulation remained limited (Geiser and Tickner 2003; Eckley and Selin 2004). The burden of proof was largely on regulators to prove that a chemical is not safe, rather than the producer and/or seller having to produce data indicating that it would not have adverse environmental and human health effects. Also, much time was spent producing detailed risk assessments that were not useful to policy makers, leaving little room for precautionary decision-making. REACH is designed to lead to more effective precaution-based regulation in part by requiring producers and importers of a chemical (as opposed to public authorities) to submit basic risk assessment data for registration.

Fourth, the separation between "existing" and "new" substances acted as a barrier for innovation and substitution (European Commission 2001b; Geiser and Tickner 2003; Euractiv 2005b). For every "new" chemical that a firm wanted to bring into the market, it had to produce a risk assessment not required for "existing" substances. As such, "existing" legislation in effect acted as a disincentive for the substitution of an old chemical, for which there was little or no risk assessment data, for a "new" chemical that has been developed with more recent technology and for which there is a better understanding about its inherent properties and environmental behavior. By abolishing the regulatory difference between "existing" and "new" substances, REACH is intended to stimulate innovation and accelerate the replacement of older chemicals.

EU-US Relations and REACH

Much US and EU chemicals legislation has developed in parallel. European and US governments assume similar responsibilities for environmental and human health protection from hazardous chemicals. They have borrowed basic policy ideas from each other, and made similar policy choices even though explicit regulatory design and procedures have differed (Brickman et al. 1985). While US chemicals policy in the 1970s and the early 1980s often inspired European policy making, the EU now has emerged as a leader in chemicals policy development (Brickman et al. 1985; Vogel 2003b). Political efforts to design more effective systems for chemical management continue in Europe through REACH, whereas there have been few changes to US legislation and regulation since the early 1990s.

As European chemicals legislation expands beyond that of the US, EU-US discord has grown. A US Congressional report shows that aggressive action by the Bush administration against REACH was heavily influenced by the views of US chemicals associations and companies (United States House of Representatives 2004). The US Department of State, Department of Commerce, the EPA and the US Trade Representative jointly designed their anti-REACH lobbying strategies with US chemical associations and industries, such as the American Chemistry Council, the Synthetic Organic Chemical Manufacturers Association, the American Plastics Council, DuPont and Dow Chemicals (United States House of Representatives 2004). Together, they developed common positions and arguments opposing REACH.

US government officials and industry representatives attempted to "neutralize" environmental arguments and influence politicians in European countries with large chemicals industries, including Germany, France, the United Kingdom, Italy, the Netherlands and Ireland (United States House of Representatives 2004). US ambassadors in European capitals were repeatedly instructed by the Secretary of State and representatives of other government agencies to convey American views and proposals (Powell 2003). The American Chemistry Council moreover worked closely with the American Chamber of Commerce in Brussels against REACH. In addition, US government officials attempted to build broader international opposition to REACH in Asia and South America (United States House of Representatives 2004).

In contrast, many US advocacy groups as well as individual federal and state-level policy makers supported REACH, expressing hope that it would enduce similar changes to US chemicals legislation and regulations (United States House of Representatives 2004). US-based environmental lobbies have started to invest more time and resources in Brussels with the aim of shaping European policy in ways that they hope can have spill-over effects on future US environmental policy (Pohl 2004). While European policy makers and advocacy groups earlier borrowed many ideas and practices from America on chemicals regulation (Brickman et al. 1985), US environmental and human health groups now hope that the strengthening of EU regulations and controls through REACH will stimulate similar policy changes in the US (Wiener 2004; Layton 2008).

US Criticism of REACH

Opposition to REACH from the Bush administration and the US chemical industry was argued on regulatory-, financial-, and market-related grounds. Regulatory opposition focuses on issues of data collection, precaution and basis of regulation. Financial opposition focuses on the economic costs of expanded assessment and regulation. Market-based opposition focuses on issues of standard setting and trade. Some of these criticisms are shared by European chemical companies and European policymakers, while others are mainly American and international in character.

US *regulatory*-related opposition stems from differing regulatory cultures across the Atlantic (Brickman et al. 1985; Royal Commission on Environmental Pollution 2003; Wiener 2004). Whereas EU chemicals regulation is generally outcome oriented, US policy making and regulation are more process-oriented. In the EU, policy makers and stakeholders seek agreement around broad goals and policy principles and then incrementally specify regulations and practices during implementation (in a process known as comitology). In contrast, US legislation and management rely on detailed sets of rules for implementation and controversial issues are addressed judicially. In light of these differences, American officials and industry representatives argued that REACH in its adopted form was too vague (United States 2004).

US regulatory critiques of REACH focused on issues of data collection, precaution, and the basis for regulation. REACH attempts to gather more data on "existing" and "new" chemicals than is currently available into a single data bank. The chemical industry will be largely responsible for generating and providing these data. This mandatory data collection for chemicals that have been on the market for decades is criticized by US authorities and industry as unnecessary (American Chemistry Council 2003; United States House of

Representatives 2004). REACH also requires additional testing of high-volume chemicals. In contrast, US regulators under TSCA negotiate testing requirements with producers on a case-by-case basis. REACH thereby accentuates EU-US differences over data collection and testing of high-volume chemicals.

Closely related to efforts to expand data collection and testing, REACH strives to better integrate a precautionary approach to chemicals management. To date, implementing precautionary thinking into day-to-day chemicals assessment and regulation has been difficult (Eckley and Selin 2004). Nevertheless, REACH incorporates a higher level of precaution than in current US legislation and regulation on industrial chemicals and pesticides (Powell 1999, 89). In addition, an increased application of precaution in Europe is likely to encourage the Commission and those Member States most supportive of the precautionary principle to push harder for more precautionary policymaking under international chemicals agreements, counter to American preferences (Selin and VanDeveer 2006b).

Northern Member States, traditionally the strongest supporters of the precautionary principle, led attempts to increase regulation of chemicals based on their inherent characteristics, consistent with earlier domestic policy developments in, for example, Sweden, Denmark and the Netherlands. Under REACH, chemicals can be banned if they are persistent, bioaccumulative and toxic. That is, regulatory decisions can be taken based on the inherent characteristics of a chemical. In the US, a chemical's characteristics are only part of the regulatory decision. To these are added dimensions of exposure and risk. Even if a substance is persistent, bioaccumulative and toxic, its production and use may be allowed if it is deemed that it will not be released and pose "unreasonable risk" to the environment and human health.

Furthermore, US representatives opposed REACH on *financial* grounds, arguing that the costs are too high for chemical companies and regulatory agencies given the expected benefits. Under REACH, the cost for generating new data falls on the chemicals industry, while national and federal authorities bear the responsibility and costs of evaluation and authorization. However, cost estimates for registration, evaluation and authorization differ greatly. A 2001 Commission White Paper estimates additional costs of €2.1 billion over 11 years for the chemicals industry for the registration, evaluation and authorization of "existing" chemicals (equivalent to an annual cost of €200 million) (European Commission 2001b). In contrast, industry estimated its total additional cost as €7.8 billion (Nordbeck and Faust 2003).

Much attention has been paid to the different cost estimates for REACH. Calculations commissioned by the German chemicals industry in 2002 predicted 2.35 million job losses and a 6.4 percent reduction in German Gross Domestic Product (Arthur D. Little 2002). Another study estimated that REACH would cost the French chemicals industry €29 to €54 billion over ten years and eliminate 670,000 jobs (Mercer Management Consulting 2003). These reports, however, have been attacked as being based on "false economics" when calculating direct and indirect costs (International Chemical Secretariat 2004; Schörling and Lund 2004). Another assessment estimates that the total cost for REACH over 11 years would be €5.25 to €8.05 billion, which is less than 0.1 percent of the chemical industry's sales revenues (Ackerman and Massey 2004).

Defenders of REACH believe that the cost debate is lopsided. They argue that economic costs should be measured against gains from industry innovation and reduced costs for clean-up of contaminated sites and waste management. One estimate sets European costs of PCB decontamination as high as €15 to €75 billion (Kemikalieinspektionen 2004). REACH is also part of EU efforts to reduce the costs of handling hazardous wastes (much of which contains toxic chemicals) (Selin and VanDeveer 2006b). Supporters of REACH furthermore stress that environmental and human health benefits must be included in any cost-benefit analysis, even if these are notoriously hard to quantify (EEB 2002a; World Wide Fund for Nature 2002; Schörling and Lund 2004; Sommestad and Trittin 2004).

US *market-based* concerns with REACH are tied to issues of international standard setting. These concerns are based on the fact that a strengthening of European standards for chemicals assessment and regulation affects production and product standards also outside of Europe. Similar to the "California effect" in American politics where the green and regulatory ambitious California, because of its relative size and wealth, has pushed many national environmental regulations upward (Vogel 1995), the raising of product standards in the EU is beginning to influence international production and trade by forcing other countries to meet those standards in order to maintain their economically significant exports to the EU (Selin and VanDeveer 2006b).

This development represents an important change in global standard setting. Many early international consumer and environmental protection standards were in effect set in the US because of the size of the American economy. More recently, however, the focus of much policy innovation has shifted from the US to Europe as the EU has greatly expanded its efforts on environment and human health protection over the past two decades (Vogel 2003b). With its expanding population and economic weight due to a series of enlargements to 27 Member States, the EU is increasingly replacing the US as the setter of many global product standards, from the size of Kentucky bourbon bottles to electronics consumer goods and their recycling (Pohl 2004; Buck 2007).

In the highly globalized and competitive chemicals industry, many US and other non-European owned companies export extensively to Europe. To continue, they will be required to comply with new EU standards. Companies are likely to produce chemicals based on one standard to avoid the economic costs and bureaucratic difficulties of applying multiple diverging standards. As such, the implementation of REACH is important to US chemical companies, even as their influence over standard setting is reduced as the regulatory center gravitates from Washington DC to Brussels. The American Chemistry Council views REACH as "yet another example of the Commission's attempt to establish the de facto international standard, just as it has attempted to do with respect to genetically modified organisms" (American Chemistry Council 2003).

Related to these standard setting issues, transatlantic controversy includes intense debates around implications of REACH for the export of chemicals and products containing chemicals to Europe (American Chemistry Council 2003; United States 2004; United States House of Representatives 2004; World Wide Fund for Nature 2004; Ackerman et al. 2008). The EU market consists of almost 500 million people. The NAFTA creates a US, Canada and Mexico free trade zone covering 430 million people with the Free Trade Area of the Americas attempting to establish a larger free trade area across North, Central and South America (population 870 million). As regional trade zones expand across economies on both sides of the Atlantic, market related issues may become even more important.

The WTO allows countries to take human health and environmental protection measures into account as long as they are "proportionate" to their aim and do not create unnecessary trade obstacles. The Commission claims that REACH is designed to be consistent with the global trade rules of the WTO and thereby "trade neutral" (European Commission 2001b, 2003b; World Wide Fund for Nature 2004). This assertion, however, is challenged by the US government and chemical industry and several other non-European countries (American Chemistry Council 2003; United States 2004). Together, they view many of the data collection and testing requirements under REACH as disproportionate and unnecessarily trade restrictive (American Chemistry Council 2003; United States 2004).

REACH was first discussed in the WTO Committee on Technical Barriers to Trade in March 2003 (WTO 2008). A formal comment filed by the US in June 2004 expressed concern about the workability of REACH as well as for the trade disrupting effects on a majority of the US exports to the EU market through trade in chemicals and goods containing chemicals (United States 2004). The US estimated the total value of the US export to the 15 EU members in 2003 at over $150 billion. The value of US exports to the EU furthermore increased with recent EU enlargements. As such, chemicals is another environment and human health related area of protracted EU-US debate—REACH was debated at 16 meetings of the WTO Committee on Technical Barriers to Trade between March 2003 and March 2008, with discussions continuing (WTO 2008).

Continued Transatlantic Controversy or Policy Convergence?

The EU and the US share a long history of cooperation on chemicals management. They have exchanged information and borrowed policy ideas from each other, with the US often pioneering early policy developments. More recently, however, Europe is leading on chemicals legislation and regulation through the development of the REACH regulation. REACH is one of the largest ever pieces of EU legislation on environment and public health. In contrast, there have been few changes to US federal consumer and environmental regulations and organizational capacity since the early 1990s, including for chemicals management.

Because chemicals are such valuable commodities and chemical regulations affect much international trade in chemicals and goods, the US is closely following European policy developments, including REACH. US and EU officials are also active on both sides of the Atlantic. While the Commission keenly promotes REACH in North America, US federal agencies worked with the US chemicals industry against REACH. US criticism of REACH focused on issues of regulation (data collection, precaution and regulation), financial costs (costs of expanded assessment and regulation) and markets (standard setting and trade). In contrast, supporters of REACH view it as a critical means for improving EU chemicals management and to better protect human health and the environment.

The US chemicals industry has made it clear that it does not want REACH to influence US chemicals policy even as it has to follow REACH requirements in the EU (American Chemistry Council 2003; Layton 2008). There are nevertheless signs that EU efforts on hazardous substances and wastes are shaping debates and policy making in the US (and elsewhere). As in many environmental policy areas, environmental leader states including California were among the first to borrow policy ideas from the EU as they revise state-level standards for chemicals, products and wastes (Selin and VanDeveer 2006b). As such, this could be yet another example of a "California effect" where groups of states drive up domestic environmental and human health standards (Vogel 1995; Rabe 2006a).

Some federal policy makers are also responding. In 2005, US Senators Lautenberg (D-NJ) and Jeffords (I-VT) introduced the *Child, Worker and Consumer Safe Chemicals Act* (Environment News Service 2005). Furthermore, Senator Lautenberg and US Representatives Solis (D-CA) and Waxman (D-CA) introduced the *Kid Safe Chemical Act* in May 2008 (Lautenberg 2008). This act, which was also a response to recent reports by the US General Accounting Office (2005, 2006) criticizing TSCA, was explicitly based on many of the same policy goals and ideas underpinning REACH, including: generating more risk assessment data; reversing the burden of proof by requiring industry to demonstrate the safety of both "existing" and "new" chemicals; and targeting more chemicals for controls and bans. None of these efforts, however, passed Congress to date.

Continuing EU-US controversy may have significant implications for global chemicals management. Because of the large influence of the EU and the US on international chemicals policy, transatlantic disagreements may spill-over and act as an obstacle towards successful operation and implementation of the multilateral chemicals agreements and programs that were created during the 1990s. Diverging transatlantic positions on the future of international chemicals management may also force other countries to "take sides" and could lead to growing international polarization on chemicals issues outside the transatlantic region. In addition, EU-US disagreement over REACH is yet another issue where the two oppose each other in the WTO, and thus adds to transatlantic trade related controversy.

European officials, advocacy groups and firms after the entry into force of REACH furthermore have a shared interest in exporting EU standards to other countries and in uploading such standards into international agreements. As such, a coalition of European public and private sector actors can be relied on to try to use market forces and political initiatives to upload new EU regulatory standards internationally. This is consistent with a long-standing EU strategy of active international engagement to achieve goals that cannot be obtained solely at a regional level (Selin and VanDeveer 2006b). The EU can be expected to pursue the uploading of its "new" chemical policies in a host of multilateral forums. US domestic and international responses to these efforts will be critical for future transatlantic and global politics of chemicals management.

Chapter 5
Oceans Apart?
Policy Reversals, Transatlantic Politics, and the EU Asbestos Ban

Marcus Carson

When France announced on July 3, 1996, that it would prohibit "the production, import, and sale of asbestos containing products, notably asbestos cement" (cf. AI 1996, 1) beginning January 1, 1997, the repercussions were as dramatic as the announcement itself. In the EU, France had long supported the asbestos industry's arguments that chrysotile asbestos, considered the least dangerous of six types of asbestos, could be used and managed safely. As the abrupt French turnaround shifted the center of gravity among Member States, it revived a stalled EU effort to implement a near-total ban on asbestos—one which then proceeded with remarkable speed.

Unsurprisingly, France's asbestos ban was received disapprovingly by asbestos producers and in Canada, then the world's third largest exporter of chrysotile asbestos (Morel-ál 'Huissier 1995; Perron 1999). Actions in Europe would further erode the credibility of asbestos producers and exporters, risking reductions in the exports to developing countries that had become their bread and butter. The complaint procedure that Canada initiated against France at the WTO was set into motion seeking to win back this lost ground. Canada also hoped to send a message to others that banning asbestos would be both costly and difficult. But the WTO broke with its own pattern of decisions in previous disputes to set a striking precedent, issuing one of its first rulings prioritizing public health concerns over free-trade principles and economic interests (Olson 2000; Wirth 2002). The EU ban on virtually all types and uses of asbestos went into effect on January 1, 2005.

The developments that culminated in these dramatic policy reversals, transatlantic disagreements, and an eventual EU asbestos ban unfolded incrementally, punctuated by fits and starts. The widespread perception of asbestos has been gradually transformed from that of "miracle mineral" to "deadly dust." While knowledge about asbestos hazards and their consequences has developed along broadly similar lines in Canada, the US, and the EU, regulatory responses have unfolded along divergent paths since the 1990s—a product of differences in the configuration of interests, variations in the structure of governing institutions, and struggles over the conceptual model through which asbestos hazards are understood. As of this writing, neither the US nor Canada has banned all types of asbestos, although both regulate it. Canada advocates what it calls the "safe use" of asbestos and was the chief opponent to the EU's ban. The US played a mostly supporting role to the EU.

In the US, where a history of misrepresentation by asbestos producers has taken its enormous toll in the form of human suffering and commercial bankruptcy, the tragedies of asbestos exposure and the social and political struggles they have generated played out largely in the courts. Yet, the US Geological Survey estimated the US imported some 26 million pounds of asbestos in 2001 in roofing materials, brake linings, and other products (USGS 2007). Ironically, it was the courts that overturned an attempted American asbestos ban. The US Environmental Protection Agency (EPA) finalized regulations in 1989 to phase out most uses of asbestos under the 1976 Toxic Substances Control Act. However, responding to an asbestos industry court challenge, the conservative 5th Circuit Court of Appeals in New Orleans overturned the regulations on the grounds that the EPA had overstepped its authority.

EU leadership on banning asbestos began in the 1990s following the example set by many individual Member States, after much deliberation and political struggle. That the EU would be the first to enact a ban is surprising given that the EU's overall regulatory authority is considerably weaker than the US' or Canada's. With several asbestos-producing Member States remaining opposed to a ban, the EU also lacked the unanimity necessary to implement a Directive protecting public health. In this context, the process by which key EU actors arrived at the conclusion that a ban on all types of asbestos was necessary, and then proceeded to translate that conclusion into policy, is especially interesting.

This chapter focuses on the development of the 2005 EU ban and the roles of the US and Canada. The chapter shows how the eventual banning of asbestos in the EU was first and foremost a battle over issue conceptualization and fundamental priorities—commercial trade vs. public health—and over the related issues of burden of proof and the nature of the standards by which that charge could be met. There was considerable contention over the interpretation of relevant scientific data and methodology and how that data framed issues of asbestos hazards for public policy. An important aspect of this contention was how the few remaining scientific uncertainties, especially some aspects of hazard and risk, were understood by policymakers and the general public. And while specific interests were clearly a significant basis for the decisions made by all actors, Canada was the leading lobbyist in high-level asbestos diplomacy. The targets for its combination of political hardball and skillful persuasion included national governments and supranational governance, and even international organizations concerned with science.

Leading Actors in Europe and Across the Atlantic

Many actors have shaped policy developments connected with asbestos. A network of labor, public interest, and victim's organizations advocated a complete ban. The asbestos industry and businesses using asbestos products argued for the continuation of a "safe use" policy. National governments served as proxies for some of these interests, but in addition, pursued their own particular nationalistic concerns. International organizations that deal with health and workplace health and safety standards—the World Health Organization (WHO), the International Labor Organization (ILO), and the joint WHO/ILO sponsored International Program on Chemical Safety (IPCS) played a subtle, yet significant role through making various pronouncements and judgments regarding the scientific evidence on the hazards asbestos poses.

The European Trade Union Confederation, the UK-based transnational network International Ban Asbestos Secretariat (IBAS), and other NGOs such as the ANDEVA, a French NGO working on behalf of asbestos victims, provided a mobilized and focused European constituency to which the European Commission and Parliament felt a need to respond and which helped to legitimize EU action. They helped make the available scientific evidence relevant and understandable in human terms, enlisted the support of the EP to press the Commission into renewed action, and emphasized how future health costs could undermine Europe's economic health. Victims' organizations in France and the UK mobilized to press for bans in their respective countries, then extended their activities to Brussels to lobby the Commission and Parliament.

The Canadian government acted as global coordinator for an array of sophisticated lobbying and public relations efforts aimed at protecting international asbestos markets. Canada also contributed political clout and credibility, and helps finance the Asbestos Institute (AI 2001; NRC 2001), which organizes activities directed toward "promoting the safe use of asbestos." Canada has more at stake than keeping open international markets. While only about 2,500 jobs are directly connected with asbestos exports, virtually all of its asbestos mining and production industry lies within the politically sensitive province of Quebec. Where Canada's federal government might have otherwise found it more expedient to obscure or downplay its many lobbying activities on the world stage, it had an essential interest in advertising those activities to the people of Quebec. Lobbying activities that might ordinarily be documented only as internal memos and progress reports have been publicized as speeches, press releases, and other revealing information. Canada's arguments are representative of the most important arguments made on behalf of the "safe use" of asbestos, mirroring the arguments of commercial interests—particularly asbestos producers.

The UK and France, struggling with the domestic political and health consequences of asbestos, were key to reviving the Commission's stalled efforts to institute a complete ban. A European-level ban became important enough to the British government that it was at one point considered a top priority of the British Presidency of the EU, also making the UK a primary target of Canada in its bid to deflect such efforts. Tony Blair and his Labour Party had made unequivocal promises during 1996 elections that they intended to ban further asbestos use, but backpedaling by the Health and Safety Executive (HSE) soon after the election raised doubts about the government's resolve. In an attempt to stall the British decision, Canada sent delegations of trade union members, public officials, and scientists to the UK to argue against the ban. British journalists were flown to Canada to get the full picture on Canada's "safe use" principle. Canada's efforts appear to have had a significant affect, though not in the end result. The British ban was delayed, and eventually passed as a formality under the protection of the EU ban on asbestos.

The French turnabout on asbestos was part of a broader shift in French public policy regarding the management of risk (Marris 2000), linked to a domestic policy crisis generated by the mid-1980s contamination of France's blood supply with the AIDS virus. More than 4000 people were infected, and over 1000 of them had died as of the time of the trial and conviction of four senior French health officials deemed responsible for failing to take appropriate action based on available knowledge (Hebert 1999; Henley 1999). As with the UK, Canada attempted to intervene to forestall the ban, hoping to take advantage of cultural and language ties with France, although in all probability, the Canadians underestimated the sense of vulnerability felt by many French officials. Nothing that Canada could offer—or threaten—could quite compare to the potential legal and political consequences at home of failing to act responsibly. France's reversal helped make a wider EU ban feasible, but once France had made its decision, it also needed help. Canada made clear its intention to challenge the French decision at the WTO. A Europe-wide ban would add both legitimacy and important backing.

The European Commission had multiple concerns. Commission officials worried that regulatory differences arising from "renegade" actions of Member States to regulate or ban asbestos "form a barrier to trade and have a direct impact on the establishment and operation of the common market" (Council of the European Union 1985). Adding to regulatory inconsistency, Sweden, Finland, and Austria entered the EU in 1995 with bans in place. Challenging their asbestos bans to support market integration would have exacerbated the Commission's problems with public legitimacy. While political legitimacy and harmonized market regulation were important goals the Commission was also motivated by growing public health concerns. Even before 1990, Commission officials were persuaded that the risks posed by asbestos were unacceptable (DG-IIII—DEN 1996, 3). One might also expect an enhanced sensitivity to the hazards of asbestos on the part of long-time Commission staff—the Berlaymont Building, the Commission's headquarters and a symbol of the EU, was hurriedly abandoned in December 1991 after deteriorating asbestos building materials were discovered to be contaminating the Commission's very workspace.

Commission background documents referred to new scientific data demonstrating the need for tougher restrictions (DG-IIII—DEN 1996). However, both the European Trade Union Confederation (ETUC; http://www. etuc.org) and the International Ban Asbestos Secretariat (IBAS; http://www. ibas.btinternet.co.uk) argued that the science had been well established since at least the 1970s. The Canadian government also argued that new scientific knowledge was not the deciding factor, but from a very different direction: "We believe the ban in France, and the one proposed by the EU, is not based on sound science and was taken on strictly political grounds" (AI 1997; Benjelloun 2000). This sentiment was echoed by the Canadian-based Asbestos Institute (AI 2001). This might suggest that these policy changes were the fruits of decades of hard work on the part of the public health, labor, environmental, and victims' organizations that had long sought to ban asbestos outright. But while such organizations played an essential role in raising awareness of the hazards and by demanding reforms, little evidence supports the contention that the EU ban was "political" in the sense that the Asbestos Institute portrays— that grassroots organizations intimidated governments into approving more restrictive regulation (AI 1996).

Knowledge and Asbestos Policy

Asbestos was long considered a miracle material. A naturally occurring mineral fiber, its several beneficial properties—durability, tensile strength, and heat and fire resistance—made it historically valuable. Cement, insulation and other products containing asbestos were common building materials in workplaces and public spaces. Ironically, the very properties that make it useful also make it deadly. The human body cannot break down and remove the durable microscopic fibers from the lungs once inhaled, so they remain and cause irritation. Its fire resistance resulted in its promotion to avoid one hazard, only to create another far less visible and immediate. The substance has now become synonymous with the incurable, often-deadly lung diseases it causes: asbestosis and lung cancers—including mesothelioma, a particularly lethal and otherwise rare form of cancer (Gee and Greenberg 2001).

During the early 1880s, English and French textile industries produced fireproof fabrics woven from asbestos fibers. Asbestos use quickly expanded to other areas of manufacturing, with its most important applications including brake linings, heat resistant seals, and asbestos cement board and pipe. Asbestos insulation applied through spray techniques enjoyed a brief, though extensive popularity (AI 1996; Vogel 1999).

Around the turn of the twentieth century, widespread use of asbestos generated deadly side effects, according to British and French factory inspectors. In 1906 British doctor Montague Murray reported asbestosis, an irreversible lung fibrosis that impedes the lungs' capacity to take up oxygen (Brodeur

1985). Given the relatively long latency periods of asbestos-induced illnesses (roughly 15 years for asbestosis, 20–40 years for mesothelioma), these early observations of illness and mortality followed the expanded use of asbestos in the late 1800s. The potent carcinogenic properties of asbestos were recognized by German researchers by the late 1930s, and British cancer statistics published in the late 1940s provided epidemiological data (Castleman 2001). Nevertheless, for decades the accumulating evidence of asbestos-related health problems was deemed inconclusive—in any case, not sufficient to remove a successful and profitable product from the market. In addition to research conducted by academic researchers, asbestos producers conducted research. However, it was later revealed (often in the "discovery" process in US legal proceedings) that this company research was frequently organized to protect commercial interests; results were routinely kept confidential and company lawyers and managers exercised editorial rights (Castleman 2001).

People working in environments with high concentrations of asbestos dust were most exposed to the hazards—"canaries in the coal mine." Workers with high exposure developed asbestosis, then other lung problems including mesothelioma and other cancers. They also brought home the dust on their clothes, inadvertently exposing their families. Moreover, asbestos products eventually deteriorate, releasing the fibers. Insulation and ceiling tiles are especially prone to deterioration and subsequent release of asbestos fibers. Large public buildings such as offices, hospitals, and schools have been built and insulated with asbestos containing materials. Asbestos cement and brake linings, while clearly more stable, also deteriorate with normal use. What had come to be seen as a health hazard for a relatively circumscribed class of workers suddenly became an indiscriminate threat to office or hospital workers, teachers, and children—even public officials.

The first regulatory breakthroughs came in the early 1970s. An initial wave of governmental regulations of asbestos in North America and Western Europe began systematically replacing ineffective voluntary measures. The US's Occupational Safety and Health Act was passed in 1970, including asbestos regulation as part of a larger package of measures. Denmark initiated its ban in 1972, prohibiting spray application of asbestos insulation, and banning the import of crocidolite asbestos. Sweden followed shortly after, banning crocidolite in 1975, and prohibiting asbestos cement products in 1976. The Netherlands and France established measures regulating asbestos in 1977 (DG-IIII—DEN 1996).

EU Regulation of Asbestos Hazards

The EU first regulated the use of asbestos in 1983, marking preliminary steps toward its formal re-conceptualization of asbestos from that of a commercial product with life-saving fire-retardant and other useful properties, to that of significant health hazard for workers and others. Following an international wave of legislation that sought to ameliorate the most obvious asbestos hazards, the EU's initial regulatory action sought to establish European standards for occupational health and safety pertaining to asbestos (Council of the European Union 1983a) and introduced the first Community-wide measures restricting the marketing and use of asbestos-containing products (Council of the European Union 1983b). These Directives helped legitimize and anchor a growing European consensus defining asbestos use as a serious occupational health and safety issue, while establishing a foothold for the EU as an actor with a legitimate role to play.

The first asbestos-related Directive established a legal basis for collecting comparable data on the presence of asbestos dust. Employers were required to report on asbestos use, assess the risk to workers of exposure, measure such exposure using comparable methods, and report such information to the responsible authority of the Member State. Specific exposure limits were established, with employers required to take steps to both stay within those limits and make additional efforts to minimize overall exposure, and set standards for medical surveillance and procedures for workers who had been exposed. The Directive also prohibited the most dangerous occupational contact with the fiber—application by spraying. Asbestos contamination was now evaluated from established reference points related to allowable levels. Although characterized by Beck (1995) as meaningless—simply reinforcing the status quo, such actions began to challenge and erode the paradigm that conflates more technology with "progress" by giving the level of threat needed form and tangibility.

The second asbestos-related measure built on the precedent set by the EU directive restricting the sale and use of substances considered to present a serious danger to health or environment (Council of the European Union 1976). Anchoring itself in the established legitimacy of earlier legislation, the marketing directive (Council of the European Union 1983b) phased in a ban (with exceptions) on the marketing and use of crocidolite asbestos, the most demonstrably hazardous variety, and required warning labels on all products containing asbestos. A follow-up Directive (Council of the European Union 1985) extended that ban to include six specified uses of all other types of asbestos, including toys, materials applied by spraying, and retail products in powdered form. Part of the rationale for establishing such restrictions was that Member States had enacted a variety of different regulations governing hazardous substances and that *"these differences constitute an obstacle to trade and directly affect the establishment and functioning of the common market"* (Council of the European Union 1985, emphasis added).

What cannot be determined is the extent to which EU-level asbestos regulation was driven by the desire to protect and enhance uniform regulation within the single market on the one hand, and the extent to which arguing in terms of the market served as both legal basis and legitimating cover for very real health and environmental concerns on the other. Either way, market arguments provided the only legal basis for action, given that the first EU competence in public health did not come into force until the Maastricht Treaty (1993-Article 129). Significantly, the 1987 Directive (Council of the European Union 1987) to reduce and prevent environmental pollution by asbestos was explicit that "A Member State may, in order to protect health and the environment, introduce provisions which are more stringent than those of this Directive, in compliance with the conditions laid down by the Treaty" (Council of the European Union 1987, Article 9). Nor can it be considered a coincidence that this "environmental" measure followed the first formal EU competence to address environmental issues (the Single European Act).

The lack of obvious struggle around these early EU-level restrictions suggests that such struggles had already been resolved in other arenas, both national and international. The US Occupational Safety and Health Act had already been adopted in 1970, followed by administrative tightening of asbestos regulations. Roughly similar standards for asbestos use had also been adopted by the International Labor Organization (ILO) in June 1986. Though somewhat weaker, the ILO rules echoed the EU Directives: the principle of substitution was introduced, crocidolite asbestos was prohibited, and the spraying of all forms of asbestos was prohibited (ILO 1986), and were based in part on a report developed by a committee of scientific experts working with the International Program on Chemical Safety (IPCS), which emphasized that the recognized risks of asbestos could be minimized and managed with adequate controls (Castleman and Lemen 1998).

Regulations at the national, international and EU levels further legitimized the health hazard claims about asbestos. Commercial interests were also served; given the number and variety of asbestos regulations materializing across Europe, the prospect of greater consistency and uniformity undoubtedly made such steps more appealing to many businesses. They had an interest in embracing regulation as a pre-emptive strategy for avoiding just the sort of ban that the EU eventually concluded was necessary. This approach has been characterized as "safe use" by asbestos producers (and "controlled use" by sympathetic governments), although the global pattern of asbestos exports and monitoring of use supports the contention that producers are far more interested in "use" than "safe" or "controlled." Still, conversion to substitutes in Europe had not progressed very far, so among other barriers, opposition from companies that produced or used asbestos products remained a formidable obstacle.

Subtle Changes, Large Consequences: Restructuring Policy Priorities

The paradigmatic shift in EU asbestos policy emerged as a change in emphasis subtle in its immediate effect, but with large long-term consequences. A second round of EU legislation further tightened the European restrictions on asbestos use in 1991. These Directives were adopted in the context of new 'market disturbances' as additional Member States implemented increasingly restrictive regulations. Labor unions and public interest groups made demands similar to those lodged earlier, but "more scientific evidence about the dangers of asbestos had emerged and safer substitutes had been developed to replace asbestos in many uses" (DG-IIII—DEN 1996, 2). Of these two factors, public interest groups point to the development of substitutes as most significant.[1] The discovery of substitutes helped reduce and undermine the network of opposition to a ban, eroding asbestos producers' base of business allies within the EU. This contributed to further tightening standards for occupational exposure under Directive 91/382/EEC (Council of the European Union 1991a).

Two other steps broke new ground. *All* types of asbestos were classified as Category I carcinogens (defined as substances known to cause cancer in humans) under Directive 67/548/EEC (Council of the European Union 1967, amended to include asbestos in 1991), bringing asbestos under an established body of regulation pertaining to substances known to cause cancer in humans including polychlorinated biphenyl (PCB) and dioxins). The second, equally significant step was to introduce a complete ban on the marketing and manufacture involving all types of asbestos fibers, although the ban included *numerous exceptions*. This shift was important, however, because it represented a shift of underlying assumptions, shifting the burden of proof to make remaining uses of asbestos the exception to the rule. Rather than enumerating each restriction, remaining uses of asbestos that were acceptable became the exceptions to the rule. Chrysotile, although still permitted, was prohibited (Council of the European Union 1991b) for an extended list of 14 specified categories of products including roofing felt, low density insulating or soundproofing materials, and most textiles. Products containing chrysotile asbestos not listed in the restrictions (including asbestos cement products) were not banned, so although the scope of the Chrysotile restrictions was expanded, it remained, in effect, a blanket approval with selective restrictions.

"Even in 1991, the Commission realized that more needed to be done to restrict the marketing and use of chrysotile asbestos" (DG-IIII—DEN 1996, 3). The "Detailed Explanatory Note" produced in 1996 characterized the asbestos-related policy of the European Community in the early 1990s as one of "controlled use." But it also noted that during the 1990s, more Member States considered the Community's policy to be insufficient to protect public health. The Commission announced its intention with the passage of the 1991 Directive to complete the shift from regulating to banning asbestos. It would shift the balance of its policy on asbestos, including Chrysotile, from blanket approval with exceptions to blanket prohibition with exceptions—from banning particular types of asbestos and banning Chrysotile for specified uses, to imposing a Community-wide ban on all forms of asbestos while permitting only certain specified exceptions (DG-IIII—DEN 1996).

---

1   Interview with a trade union official.

This shift in orientation effectively transferred the burden of proof from unions, public interest groups, and public health officials to asbestos producers and manufacturers. Rather than requiring those with health concerns to demonstrate with scientific certainty in each individual case that a particular type or application of asbestos was sufficiently dangerous to prohibit, the producers and manufacturers would have to prove that there were overriding reasons to permit the hazard—and that it could be acceptably managed. The announcement proved premature, however. Plans to phase out of asbestos use foundered on the lack of the necessary qualified majority vote of Member States (DG-IIII—DEN 1996, 3).

Asbestos "Politics" and Going for the Ban

Several factors converged to permit the movement to ban asbestos to proceed. One element that appears not to have significantly changed is the science underlying the ban proposals. Rather, the politics changed, although not in the sense suggested by Canada and the Asbestos Institute. Asbestos regulation had long been "political," and two decisive political elements that changed were the alignment of conventional political power and the conceptual framework from which scientific evidence was judged.

The political constellation around asbestos had changed dramatically. On one side, the network of actors with a stake in supporting the continued use of asbestos had eroded. European firms, driven by a desire to use safer materials, but also believing that a ban was coming, switched to substitutes. For example, the Belgian-based company ETERNIT, a large producer of building materials including fiber cement and board products, converted to substitutes before the ban took place to escape economic consequences of delaying a changeover until forced. Such defections substantially undermined the network of European-based political support. ETERNIT now earns revenue handling and removing asbestos, including in the new Member States. Ironically, at least some of their current work entails the removal of products that they themselves manufactured and/or installed.

Also, several Member States acted on the available science and concerns that existing regulation was insufficient to protect public health. Member States "successively imposed further national restrictions on products not covered by harmonization, *creating disunity in the Internal Market*" (DG-IIII—DEN 1996, 3, emphasis added). This was a problem that could potentially spill over into other areas of trade policy. The EU faced a trade-off between intervening in market affairs to ban a particular commodity, or letting disorder increase as Member States independently decided that public health principles were a higher priority than common market regulations. The head count also changed, leaving the asbestos industry with fewer European allies. By the time the Commission revisited a complete ban on asbestos products, Belgium had followed France's lead in adopting its own ban in 1998, with three other Member States (the

UK, Ireland, and Luxembourg) indicating support. Three remaining holdouts, Spain, Portugal, and Greece, remained opposed to any regulatory change "for scientific and technical reasons and because the economic effects it would have on their asbestos-cement industries" (DG-IIII—DEN 1996, 3).

Common Knowledge and Politicized Science

While the history of asbestos research, human tragedy, and corporate responsibility was once the subject of intense disagreement, much of the dispute was put to rest by an ever-growing body of scientific evidence and internal company documents pried out of corporate files by class-action lawsuits. However, the question of banning asbestos had by no means become uncontested; it had simply been moved to more defensible ground. This strategic retreat is most evident in the arguments of Natural Resources Canada (NRC), the Canadian governmental department responsible for regulating and overseeing Canadian asbestos production—and maintaining its export markets abroad. The NRC's "Chrysotile Asbestos Fact Sheet" (NRC 2001) contains examples of the counter arguments targeted to the general public and policymakers. It seeks to a) establish the credibility and trustworthiness of asbestos producers as experts; b) distinguish "Canadian" chrysotile asbestos from the other even more dangerous forms; and c) inoculate against the belief that banning chrysotile asbestos will effectively remove the environmental hazard. Based on these three points, it argues there is no reasonable justification for establishing trade barriers to block asbestos imports.

The document portrays the historic lack of adequate management of asbestos hazards first, as a function of an earlier lack of knowledge—an innocent mistake—and second, as a "problem of the past": "Unfortunately, public health officials were slow to see the link—in part because illnesses could take 45 years to develop." This suggests that 1) there is no longer a danger, 2) mistakes that were made were innocent, and 3) it was public health officials who erred, but understandably so—points at odds with empirical evidence. In the next step, chrysotile asbestos is rhetorically distinguished from the other forms of the deadly mineral arguing "scientists have discovered that not all asbestos is alike ... chrysotile asbestos, the most common form of asbestos used in the world and the only kind mined in Canada, can be used safely in products such as building materials, brake linings, and water and sewer pipes." The document asserts that banning asbestos would not eliminate it from our environment. Finally, it seeks to restore asbestos producers' credibility by taking an apparently aggressive stand on safety with their "safe use" approach and describing in detail the controls in place for the production and manufacturing process. The subtle redirection of attention lies in their interpretation of science and on the fact that "safe use" strategies apply only to production processes, which are somewhat controllable.

Canada actively attempted to reframe the asbestos debate. Following up on similar visits in 1997 and 1998 by journalists from Europe, "the Canadian government organized a visit to the Canadian chrysotile industry by journalists from Latin America in January 1999." Such visits were organized as part of an international strategy to "ensure a broader dissemination of the Safe Use Principle for the benefit of consumers, regulators and industries in consuming countries" (Perron 1999, 17.6). The asbestos lobby's efforts to influence the debate on asbestos have been likened to those of the tobacco industry and actively sought to manipulate the scientific discussion on health effects and helped sympathetic scientists gain positions where their views could influence public policy (Lancet 2000). The production of a 1986 expert report on asbestos produced by the Geneva-based International Program on Chemical Safety and Asbestos (IPCS) follows this pattern. Three of five scientists preparing the report had documented close ties to the Canadian government and/or the asbestos industry, while the Secretary of the Task Group responsible for the report was a Canadian government official (Castleman and Lemen 1998). The final report mirrored the position of the Canadian asbestos industry.

A subsequent effort in 1993 to prepare an update on chrysotile asbestos was marred by further political influence. Manipulation of the outcome in favor of asbestos use drew sharp criticism from officials of the US National Institute for Occupational Safety and Health (NIOSH) and the highly respected, Collegium Ramazzini, among others, resulting in their refusal to participate. An IPCS workshop on chrysotile was conducted with financial assistance from industry organizations despite controversy, with the proceedings edited by two industry consultants and the Canadian official who had earlier served as Task Group Secretary. Developing a final report stretched into 1996 (and included additional scientists), when the Task Group chair, also an employee of the Canadian government, was forced to step down when she attempted to veto the larger group's decision warning against using asbestos in building materials (Castleman 1999).

Institutionalizing New Priorities: Closing European Borders

When in 1997 the Commission revisited the question of banning asbestos— this time armed with the necessary qualified majority—important political, economic, and social considerations remained. One was the continued opposition of Greece, Spain, and Portugal. They argued that the while the EU proposal presented little or no problem to nations that had already implemented a ban as they had been able to adopt their bans on their own terms and thus minimize disruptive effects, it would be problematic for the cohesion countries which would be faced with additional unemployment and economic disturbance. Another was the certainty of a WTO challenge by Canada.

Commission officials visited mines and factories in Member States where owners and employees who would be adversely affected could express their disagreement. On one visit, workers (under the watchful eye of their employers) requested that delegation members and the EU let them make their own choices and take their own risks,[2] as unemployment would be the likely consequence of a ban. However, such requests overlooked the lack of choice offered to people exposed "downstream" and without their knowledge—a group that has often included the families of workers. Such concerns were eventually addressed by including a five-year phase-in period to permit more time for adaptation and conversion to substitutes in Greece, Portugal, and Spain.

The Commission's handling of the scientific issues connected with the ban was directed to both internal and external concerns. Strong evidence was necessary to justify imposing a ban within the EU. Given Canada's promised WTO challenge, it was clear that the scientific evidence would eventually be judged against the standards and obligations established under international agreements, including the GATT and the WTO. The Commission marshaled science behind the ban, beginning with the report by INSERM, (the French scientific agency that had been responsible for evaluating the risks of asbestos) —a report the Canadian-based Asbestos Institute, had already labeled as both "flawed" and "political." The Commission also contracted ERM to review all existing literature on the epidemiology and risks of asbestos. This report was then referred to the Commission's Scientific Committee on Toxicity, Ecotoxicity and the Environment (CSTEE) for review and an opinion regarding the conclusions of the ERM report.

Canadian scientists succeeded in reframing the context of the questions posed by the Commission away from the uncertainties inherent in public policy decisions and toward lingering areas of scientific uncertainty. The CSTEE conclusion was that "the ERM Report provides no new evidence which indicates that a change in the risk assessment for chrysotile is appropriate" (CSTEE 1998a). This conclusion was largely irrelevant to the course of action being contemplated by the Commission because it ignored the question of whether safer substitutes were available and because the hazards had long been considered sufficient to warrant a complete ban. Nevertheless, this result put the Commission in the awkward position of appearing to ignore the conclusions of its own scientific committee and thereby weakening an important basis of its arguments.

The Commission also made a second and much more specific request for an opinion from the Committee: "on the basis of the available data, do any of the following substitute fibres pose an equal or greater risk to human health than chrysotile asbestos? Cellulose fibres? PVA fibres? P-aramid fibres?" (CSTEE 1999). The conclusion of the second CSTEE opinion reads quite differently than the first: "A major concern with fibres is their carcinogenic potential. There is sufficient evidence that all forms of asbestos, including chrysotile, are

---

2   This account was provided by an observer present at the meeting.

carcinogenic to man. No evidence of fibre-caused cancer occurrence in man is available for any of the three candidate substitutes ..." (CSTEE 1998b).

Defending Public Health Priorities at the WTO

Canada's WTO challenge was formally directed at France's ban although the decision would still have clear consequences for the EU ban; the bans were similar and based on the same scientific evidence and reasoning. Since the EU represented France in the WTO case, it was in essence defending its own ban by proxy. In the end, the WTO ruled against the Canadian complaint. The final decision upholding the prioritization of public health concerns. The WTO panel faced contradictory choices regarding which rules to apply in taking up the case. In addition, the favorable decision hinged at least in part on the unanimity of experts selected to testify in the case. According to a key EU official, this was probably a miscalculation on the part of Canada. It is also significant that the precautionary principle was *not* a factor in the legal case (Christoforou 2000)—lawyers representing the EU wanted to present the French/European decision to implement an asbestos ban as entailing no scientific uncertainty, even though the precautionary principle was relevant to the EU legislation and influenced subsequent developments.[3] Working in favor of the asbestos ban, the appellate body was concerned about the consequences of the case for the trade body itself in view of sharp criticisms being directed at the WTO.

Evolution and Revolution in Asbestos Regulation

The steps from first legislative mention of asbestos in 1980 to the initial decision to pursue an asbestos ban in 1991 were part of a larger policy learning process in which the EU redefined asbestos as a threat to workers and the general public, established its legitimacy and authority to act on the problem, and then shifted its regulatory presumptions from general acceptance with exceptions to prohibition of products containing asbestos with exceptions. In the early stages, it was a largely negotiated process in which EU regulation reflected what consensus existed at the national level and within international organizations. Here, the EU echoed the policy examples set in the US, in EU Member States, or international bodies such as the ILO. Although much of the early action on asbestos was facilitated by the influence of changes taking place elsewhere, Commission officials learned important lessons from their experience. This was particularly true as negotiations broke down in 1991, giving way to conflict and stalemate.

---

3   The preparations and strategies for the WTO defense were described in an interview with a high-level member of the EU's legal team.

Grappling with the various public health, economic, and political problems posed by asbestos use, the Commission encountered obstacles to bringing European policy in line with its own redefinition of asbestos as a serious health threat, and which by its very nature could not be managed with "safe use." These factors became part of a distinctively EU story when the Commission broke fresh ground in 1991, announcing its intention to completely ban the import and use of asbestos within European borders, following a similar ban by the US Environmental Protection Agency (although as noted above, the US ban was overturned by the courts). The EU stepped into a position of global leadership on the issue as it eventually enacted the ban and subsequently took up the defense of the French ban (and indirectly, its own) at the WTO.

Paradigm Shifts and Institutional Change

The EU effort to ban asbestos not only generated new conflicts between Canada and its European trade partners, it brought the new goals pursued by European officials into direct conflict with the institutionalized core assumptions, goals and rules of the policy paradigm that has guided the broader EU chemicals policy in which it is embedded. That paradigm has emphasized market integration, economic development, and the corrective capacity of market mechanisms. The influence of market interests tended to encourage obscuring or misrepresenting asbestos hazards. Self-interested actors chose to withhold or downplay relevant knowledge, leading to severe economic and public health consequences. In order to move forward with its ban, the EU was forced, in effect, to reorder the policy priorities it applied to asbestos and reverse its own operating assumptions, then defend that reversal.

Three prominent themes emerge in the transformation of EU policy on asbestos. The first is the reconceptualization of asbestos from durable and inexpensive fire-proof building material, to a health hazard for unprotected workers, to a widespread public health hazard and confirmed killer. The second thread is the process of institutionalization of that reconceptualization via EU regulations that first seek to mitigate the unwanted side effects of asbestos, but eventually bar its use altogether. The third is the process of organized interests aligning themselves for and against the use of asbestos, the ways in which each deployed different types of knowledge about asbestos to promote their respective positions, and the eventual erosion of the interests that benefit economically from its use. One salient element of this story is the way in which these three themes are linked through social struggle over which characteristics of asbestos are emphasized and acted upon, and over the ways in which it was given expression in EU law. Powerful interests engaged to establish or undermine various claims of hazard and risk on the one hand, and to mobilize the forms of institutional power to which they had access on the other.

In a striking paradox, regulators found themselves seeking to protect the single European market for a specific category of products by eliminating one of those products. The challenge became one of how to justify a ban in the larger context of the formal rules governing EU policymaking. What are the various rule-defined conditions that must be met in order to legitimately remove a product from the market? A policy paradigm that emphasizes the commercial characteristics of market products, and the self-correcting capacity of markets to resolve problems, sets a high hurdle to clear before adopting regulatory measures.

An asbestos ban was logically inconsistent with the institutionalized, broadly market-oriented paradigmatic framework for producing regulations in the EU, but is consistent with the revised principles announced in the Chemicals White Paper. Legislating a ban on a commercial product for public health reasons proved difficult. First, it had to be framed as an exceptional case within the broader chemicals policy—and as a threat to the single market project. This enabled the ban to be pursued via the qualified majority voting procedure rather than the unanimity required for non-market measures.

Had the Commission's effort to implement a ban in the early 1990s been successful, the matter might have ended there. However, given that the EU and its Member States are party to an international-level set of agreements about policy principles embodied in the WTO, the EU was forced to address the fundamental principles under which the policy was passed and defended at the WTO in order to have a reasonable chance to define the measures not as a barrier to trade, but as necessary steps to ensure the protection of public health.

Explaining Transatlantic Differences

While transatlantic politics have been clearly important in the unfolding of asbestos regulation in the US, Canada, and the EU, the principal drivers have arguably been domestic factors seen in the differing configurations of interests, guiding policy paradigms, and institutional arrangements for governance. Significant internal obstacles have been encountered in efforts to ban asbestos use in each of the three political entities.

In Canada, the asbestos lobby has been more successful in gaining the ear of politicians than have the environmental and health movements that have agitated for a reversal of Canadian policy from "safe use" to "discontinued use." An important reason for this is the federal nature of Canadian politics, and the special case of Quebec, where virtually all of Canada's asbestos is mined. The political sensitivity of Quebec has thus far proven an insurmountable obstacle to the significant victims' movement within Canada.

The US position is more complex. As in the EU, US government agencies sought to ban further import and use of asbestos via administrative pathways but were turned back in the beginning of the 1990s. As one major US newspaper put it, "every major effort the government made to ban asbestos has been thwarted by the US and Canadian miners and producers of the versatile but deadly fibers" (Schneider 2000). Yet, the lack of political support stands in stark contrast to the EU.

The first Bush Administration declined to appeal to the Supreme Court the Circuit Court's overturning of the EPA ban. The Clinton Administration did file a brief at the WTO supporting France's ban, but no US ban emerged from the Clinton Administration. From 1994 until 2007, Congress was dominated by Republican majorities with an ideological distaste for market intervention. In practice, they demonstrated a greater interest in limiting the legal exposure of companies that have manufactured or continue to use asbestos-containing materials than in limiting future asbestos exposure by prohibiting further use of the material that is the ultimate source of those lawsuits. A significant breakthrough occurred when Senator Orin Hatch expanded his bill limiting legal remedies to include Washington Senator Patty Murray's proposed complete asbestos ban. However, the trade-offs connected with sharply limiting victims' rights to be heard in the courts were considered too great to accept the gesture.

The lack of a critical mass of support in the US appears less the result of lack of interest than a product of active lobbying. Castleman (2007) reports interference by high-level administration officials in OSHA and EPA efforts to regulate and inform about asbestos hazards, as well as what he aptly characterizes as "seeding the literature"—efforts uncovered in legal proceedings in which legal consultants literally paid millions for generating publications in the scientific literature intended to foster doubts about the linkages between asbestos exposure and disease. These activities echo the patterns of manipulation of scientific discourse witnessed previously in European and international scientific organizations noted above. Political winds in the US have once again shifted, prompting Washington Senator Patty Murray to reintroduce legislation to ban asbestos in the US on March 1, 2007. In October 2007, the Senate unanimously passed Murray's bill. A year later, a similar bill, the Bruce Vento Ban Asbestos and Prevent Mesothelioma Act of 2008—named after US Congressman Bruce Vento who died of mesothelioma, was introduced into Committee by the House of Representatives.

Although each is an interesting case in itself, the evolution and revolution of the EU's asbestos regulation is most fascinating and informative. This is in part because of the ways in which key actors engineered paradigmatic and institutional change within the EU's single market under institutional conditions that would generally not be considered favorable, and in part because of the implications the EU policy shift might signal for other areas of transatlantic regulatory conflict from chemicals to GM foods.

Conclusion

The asbestos case reflects a major reorientation of EU policy on risk regulation. The struggles over asbestos were part of a broader long-term process that included disputes over the interpretation of scientific evidence, the power of economic interests, and the Commission's interests in furthering European integration. Among the more interesting factors was the confluence of interests that put the EU in the position of protecting the integrity of its single market regulatory framework by prioritizing and protecting public health. In legal terms, framing of the issue as protecting the regulatory consistency of the internal market permitted the ban to be passed using a qualified majority rather than having to clear the higher—and in this case impossible—threshold of unanimity. These changes would have been interesting even if they had simply represented isolated changes. However, they have been part of a broader, highly significant policy reorientation of EU risk regulation impelled by developments including the BSE crisis, and by a long-term evolution of EU environmental policy and chemicals regulation (Carson 2004).

The development of EU asbestos regulation has shadowed the development of the larger body of chemicals regulation of which it is a part. As a rule, it has been subject to the same policy assumptions and standards of proof, the same policy authority as defined in the successive Treaties. Changes in both of these areas have been hotly contested. As was the case with asbestos, the vast majority of the chemical products and substances in use today were placed on the market under an assumption of acceptability: that the various benefits of a new product outweighed the risks of its use in the absence of definitive evidence to the contrary. That assumption has been supported and reinforced by the general lack of comprehensive and publicly available information regarding the hazard and risk properties of the substances. It is reinforced by the fact that many of the harmful effects are long term in nature, generated by overall bioaccumulation or by the fact that these substances can escape over time from the products containing them.

Chapter 6

Targeting Consumer Product Environmental Impacts across the Atlantic

Alastair Iles

The manufacturing, use, and end-of-life fates of cars, appliances, and electronics can cause significant environmental and health problems. This chapter compares differences in vehicle and electronic regulation in the US and the EU. Since July 2006, electronics manufacturers in the EU have been required to remove lead, mercury, cadmium and brominated flame retardants from their products. In the US, in contrast, manufacturers can voluntarily take their products back, but few companies have schemes in place. Manufacturers are not obliged to eliminate toxic substances from cars and electronics except in a few states that have implemented bans. Most electronics governance is voluntary and centers on energy conservation.

This chapter begins by summarizing the environmental and health problems that cars and electronics pose. It then outlines the challenges that regulators face when targeting products and considers the importance of the product life cycle approach used in the EU. Regulatory differences between the US and EU are explained with a comparative policy framework that combines the government system, regulatory philosophy and processes, the politics of product risk perceptions, the participation (or non-participation) of non-state actors (including industry, environmental and consumer groups, and citizens) and the pressures of harmonization within a multi-level political entity.

The EU has moved towards product regulation because of changing views of product risks, the development of greater regulatory legitimacy, the ability of a more centralized government system to press for significant changes while withstanding industry pressures, industry willingness to endorse intensifying regulation, and the need to harmonize production across the region. In the US, despite the concerns of environmental and health groups and the actions of some state governments, no federal regulations governing electronic and vehicle production or recycling have been introduced.

Industry and the Environment

Europe and the US have the largest vehicle populations in the world. In 2002, at least 214 million vehicles were on the road in the EU (European Monitoring Center on Change 2004) and at least 231 million in the US. Cell phones, computers, televisions, and home entertainment systems are also ubiquitous. The US is estimated to have at least 178 million personal computers (PCs). In the EU zone, the numbers of PCs grew 99 million in 1999, with rapid growth continuing into the 2000s (Eurostat).

European and US companies are meshed in a global network of production chains. They are among the world's largest businesses, along with Japanese car and electronics makers. In Europe, companies such as Volvo, Renault, Mercedes, Daimler, Phillips, Nokia, and Electrolux are leading corporations globally. In the US, General Motors, Ford, Dell, Westinghouse, General Electric, and Hewlett Packard are among the largest corporations worldwide. Industry and regulatory actions occurring in each region can propagate globally.

The US and Europe face similar, growing environmental problems related to the consumption of vehicles and electronic goods, although statistics for waste, recycling and disposal are limited. The disposal and scrapping of cars and electronics can lead to the dissipation of persistent, bioaccumulative and toxic (PBT) materials into the environment. Electronics can contain up to a thousand materials, many of which have not been screened for toxicity, and others which are known to cause human health problems, such as lead, mercury, hexavalent chromium, beryllium, cadmium, brominated chemicals and chlorinated plastics (Schmidt 2002). Computer monitors and television sets contain significant amounts of lead which can affect cognition and development in children and cause damage to kidneys and the circulatory system (European Commission 2000).

Cars have mercury in their light switches. The mercury can enter the ecosystem through dissipation during car crushing or by vaporization when contaminated scrap is burned in steel furnaces (Ecology Center et al. 2001). Car batteries are perhaps the single largest source of lead in consumer products. Lead weights are also used to balance car wheels. It has been estimated that 1630 metric tons may fall onto US roads yearly (Root 2000). Asbestos is used in some brakes.

Technical experts, NGOs and, increasingly, governments have argued that producers should be responsible for recycling and redesign. Because cars and electronics have complex designs, use many integrated materials and have been difficult to disassemble cheaply, their recycling poses many challenges. Recycling cars and electronics requires the creation of elaborate, costly collection and processing systems. European governments have responded to these challenges differently from the US government. They have imposed responsibility related to product content and recycling on manufacturers whereas the US has not.

Comparing Regulation in the United States and Europe

The greater willingness of European governments to impose manufacturer responsibility can be understood in terms of regulatory politics. Major differences exist between the US and European countries in environmental policymaking and regulation (some works include: Rose-Ackerman 1995; Vig and Faure 2004b). The politics and processes of chemicals regulation differ between the US, Britain, France and Germany on several dimensions. These include the: system of government (whether the government is unitary or federal and the frequency of elections); role of science expert advisors; participation of industry, trade unions and environmentalists; bureaucracy (whether political appointees prevail or a permanent public service exists); regulatory process (whether public notice and comment is required); and extent to which the courts are involved (Brickman et al. 1985).

The US and Britain diverge markedly in their national regulatory styles in controlling air pollution (Vogel 1986). Whereas the US built a direct regulatory system based on uniform national technology-forcing standards, Britain relied on factory-by-factory negotiation of industry controls between inspectors and companies. Within Europe, research reveals significant diversity among European countries in administrative arrangements, degree of legalism, policy implementation, perceptions of problems and regulatory modes in the environmental domain (Hajer 1995; Héritier et al. 1996; Knill 2001).

There are also political cultural differences (Jasanoff 1986). While the US adopts a heavily quantitative approach to risk assessment, European countries typically rely more on expert advisers to judge risks. In the past, US governments felt compelled to provide public information to watchful citizens but European governments maintained greater, magisterial secrecy. Both institutional and constructivist elements need to be combined, Jasanoff contends, to better account for the full texture of regulatory differences. In the biotechnology area, Jasanoff (2005) has demonstrated that the US, Germany and Britain diverge in their societal interpretations of the risks associated with GM foods and other biotechnology products. The ways in which citizens perceive risks in products have changed much more in Europe than in the US.

Risk perceptions in the US and Europe have diverged markedly since the early 1990s, in part because of the confluence of food safety scares (Vogel 2004). Europeans are now less inclined to trust in government authorities to make correct regulatory decisions. Because there have been no major environmental crises in the US for decades, Americans are less likely to challenge governmental action (or inaction). European environmental policymaking has become more public and open to citizen and industry participation. Many more actors are demanding a say in regulatory decisions. Conversely, US federal policymaking is becoming increasingly secretive and closed to citizen involvement, encouraging a revival of state government activity.

Recent developments in transatlantic product regulation reflect these changes. In particular, the greater legalism and state-driven intervention in Europe is reminiscent of US environmental policy in the 1970s. Yet European politics also have features that grow out of early twenty-first century environmental fears and ongoing EU institution-building. Products, therefore, provide a lens on changes in each region's regulatory politics and related transatlantic interactions.

Product Regulation in the EU

Since the mid-1990s, the EU has taken an increasingly regulatory stance toward products, rather than depending on voluntary industry measures. This shift reflects changing societal views of intrinsic product risks, European government concerns about escalating waste disposal problems and costs, a new awareness of how the precautionary principle can be implemented through product take-back and a greater industry tolerance of product-centered regulation. The EU also recognizes that it can expand markets for its products by stimulating a global demand for environmentally compatible products.

As its primary governance tool, the EU has adopted the concept of "extended producer responsibility" (EPR) on the theory that making manufacturers accountable for the environmental impacts of their products will engender behavioral changes. Forcing manufacturers to recycle obsolete products will make them cognizant of their costs and will better internalize product environmental costs. Therefore, the EU has mandated the take-back of vehicles and electronics for recycling. To a lesser extent, the EU has aimed at eliminating toxic substances and forcing design changes. In 1996, the Commission planned "EC-wide rules to limit the presence of heavy metals in products or in the production process or the ban of specific substances in order to prevent, at a later stage, the generation of hazardous waste" (European Commission 1996b). However, the Commission has moved more slowly because of industry antipathy to chemical substitutions.

The EPR concept originated in the Ordinance on Avoidance of Packaging Waste enacted in 1991 in Germany (Fishbein 1998). Unlike many other national EPR schemes, Germany imposes full financial responsibility on manufacturers and retailers to take back and recycle packaging. Starting in 1993, a consortium of over 500 corporations created the Duales System Deutschland (DSD) to collect and sort paper, glass, metal and other wastes. The following year, the EU introduced the Packaging Directive, one of the first Europe-wide laws to target product environmental impacts (European Parliament and Council of the European Union 1994; Bailey 2003). Following the German model, the directive set recycling targets ranging from 25 to 45 percent depending on the resource category.

The key features of an EPR system are: 1) the imposition of producer responsibility for the end-of-life phase; 2) the requirement that producers either physically take back their products for reuse and recycling or pay another party to do it; and 3) mandates for recycling rates, definitions of recycling and data reporting on progress (Fishbein 1998). Collection and recycling can be done differently between countries. For instance, industry takes the collection responsibility for packaging in Germany whereas local government (paid by industry) does so in Britain.

Initially most EPR activity in Europe centered on seeking voluntary commitments from industry (Thorpe and Kruszewska 1999). Early attention aimed at cars because governments recognized that landfill capacity was quickly diminishing and car waste constituted one of the greatest sources. In the early 1990s, France and Germany negotiated End-of-Life Vehicle (ELV) compacts with their domestic car manufacturing sectors (Lucas 2001). Because of the voluntary compacts, the German car industry began to design cars to be recycled. They reduced the number of plastics, labeled the plastics that were used, and found alternative ways of fastening components together (Fishbein 1998). This led to growing acceptance of redesign and recycling. It is harder for industry to assert that it cannot practice take-back if companies are already well into production changes.

During the 1990s, however, the Commission became more disillusioned with voluntary industry strategies for cars and electronics as a result of various problems. The Commission, for example, tried to negotiate a Dutch-style covenant with the European electronics sector but this process collapsed in 1995 (Thorpe and Kruszewska 1999). Powerful decision-makers in the Commission's Environment Directorate-General, such as Ludwig Krämer, director of waste management, decided to push for innovative product regulation: "We cannot rely alone on market forces, or on command-and-control legislation ... it might well be that we have to differentiate more clearly between products" (ENDS 1996). This marked a key change in European regulatory philosophy.

Several countries gave the Commission an opening to enter the field by making their own laws for cars and electronics. In 1996, Germany decided to enforce its voluntary car compact through legislation following industry lethargy. In 1998, Switzerland (though not an EU member) mandated EPR for many products, including cars and electronics. The Netherlands, Italy and Norway introduced ELV schemes (Thorpe and Kruszewska 1999; Tojo 2000). These countries urged that their regulations should be adopted at the European level. The Commission was able to argue that European interests demanded the development of a harmonized system instead of multiple national schemes. With disparities between countries, waste trade would gravitate towards those with weaker standards, manufacturers would have to produce for multiple markets instead of one corporate market and producers might face exacerbated recycling costs (European Commission 2000).

The ELV directive, which has been in force since October 2000, obliges car manufacturers to take responsibility for their products (European Parliament and Council of the European Union 2000). EU members must enact national laws to implement a directive, but they have some freedom to choose which methods and instruments to use. They must, however, ensure that collection systems are built to funnel obsolete cars from owners and vehicle repair shops to treatment facilities, at producer expense. Cars must be stripped of components and hazardous substances before treatment. Since January 1, 2006, countries have been required to attain reuse and recovery levels of 85 percent by car weight; by 2015, they must reach 90 percent. Since July 2003 vehicles sold cannot contain lead, mercury, cadmium or chromium except where substitutes cannot be found.

As the first product take-back law, the ELV directive encountered significant industry and national opposition. In the early 1990s, the Commission established a priority waste expert group (including car manufacturers) which called for manufacturers to take primary responsibility for cars and proposed targets and deadlines (ENDS). In 1997, following several national laws and compacts, the Commission proposed the new directive. Britain opposed the directive but a majority of EU nations supported it. In subsequent drafts, the Commission introduced hazardous substance bans and resisted industry efforts to weaken the responsibility clauses (ENDS 1997a). But the Commission bowed to industry pressure and dropped a ban on the use of PVC plastic. Industry was particularly antipathetic to controls on materials.

Industry's stance was mixed. The US car industry opposed the ELV directive through lobbying and argued that the rules would be a barrier to entry to the EU market. In contrast, the continental European car industry, particularly Scandinavian and German makers, was largely supportive of the directive except in relation to the fixed recycling and reuse targets. Industry support for take-back grew from a view that building recycling systems would strengthen ties with consumers. In 2000, Philip Frey of DaimlerChrysler's Center of Competence for Recycling Processes said: "We will have a chance to get in touch with lost customers" (Conference Board Europe 2003). BMW and Mercedes already had had recycling systems since the early 1990s. Ford's German subsidiary created a take-back system to compete more effectively.

The European industry was reluctant to pay for recycling, however. The European Car Manufacturers' Association, ACEA, was concerned about the costs of recycling older cars produced before the take-back scheme began. As a result of lobbying by Volkswagen and other national manufacturers, Germany suddenly withdrew its support in early 1999 and enlisted Spain and Britain to hold up the directive for months (ENDS). The Commission and Council agreed to defer the operative date until 2006. Pushed by environmentalist parties, however, the Parliament forced the operative date back to 2003 and retained free take-back of cars (ENDS 2000).

The Waste Electronic and Electrical Equipment (WEEE) directive, in force since 2003, mandates that makers and importers (not retailers unless they sell equipment under their own brands) must pay for collection, recycling, reuse and disposal of products marketed after 2005 (European Parliament and

Council of the European Union 2003b). Take-back must be free of charge to consumers, who can return products to manufacturers, or to retailers who deliver to the take-back system. Since August 2005, EU countries have banned electronic wastes from landfill facilities and created take-back systems, which can be industry-directed or government-operated. Manufacturers have also been required to collect four kilograms of equipment per person per year. They must begin designing products for disassembly and recovery, label components to aid identification of contents and accomplish recycling targets, such as 70 percent for computers.

The introduction of the electronics take-back scheme was more politicized than the ELV directive because the electronics industry had never faced stringent product or process regulation and there was a greater emphasis on phasing substances out. Following the Commission's decision to look into regulating electronics, the British consulting firm, AEA Technology, reported in July 1997 that electronics recycling would be viable but only if regulation occurred (ENDS 1997b). By 2000, the Netherlands, Belgium, Sweden, Denmark, Austria and Italy had created WEEE systems, with Germany following (Hanisch 2000; Halluite et al. 2005). That year, the Commission proposed the directive for negotiation between EU governments. The Commission subsequently rebuffed most industry's efforts on watering down key provisions (ENDS 1999, 2002a). The Parliament also demanded that the law be strengthened by bringing substance bans into force sooner and insisting on producer responsibility (ENDS 2002c). In January 2003, the Council and the Parliament agreed on a final text.

The initial industry position was predictable: both US and European electronics industries preferred a voluntary scheme but for different reasons. The US government focused on free trade and competition, complaining that the WEEE law discouraged entry into the European market (ENDS 1996). The US ambassador to the EU, Stuart Eizenstat, unsuccessfully lobbied European manufacturers to oppose the law. US manufacturers argued that a ban on flame retardants would increase fire risks. The US Electronics Industry Association attacked recycling targets as too rigid (Electronics Industry Association 2001). The dispute was really about who would get to set global standards. However, the electronics lobby was not versed in lobbying against regulation (in contrast to the US chemical industry) because US electronics makers were not subject to environmental regulatory pressure or public scrutiny.

Conversely, EU industry wanted a voluntary scheme on the grounds that regulation would be too complex. In 1999, EU industry began to convert to the Commission position as they realized they might achieve economic benefits and create new markets. Fujitsu manager, Joy Boyce, declared: "Some suppliers of electrical and electronic goods have now realised that there may be marketing advantages to be gained from producer responsibility. Companies are already talking about how they can get products back by incentivising take-back and getting consumers or businesses to buy your products again" (ENDS 2002d). Corporations understood that they could use take-back to strengthen customer relationships, boost purchases of new products and control the size of the market for reused or recycled products. It was EU industry's partial acceptance that was important to moving electronics regulation along.

The Restriction of Hazardous Substances (ROHS) directive, which came into force in 2003, has only two major clauses (European Parliament and Council of the European Union 2003a). One calls for chromium, lead, mercury, PBBs, PBDEs and cadmium to be eliminated from electrical and electronic products sold after July 1, 2006; the other provides for periodical updating (at the Commission's initiative) of the list of controlled substances. Because manufacturers have hundreds of suppliers providing thousands of materials and components, they need to start screening upstream for materials already or likely to be prohibited. They must change their supplier specifications and relationships, not just product design. ROHS marks a dramatic step in product regulation by aiming at design rather than recycling. Unlike the WEEE law, ROHS emerged from the Commission. The Commission was following technical developments closely and saw the potential for substitutes to be used.

Product Regulation in the United States

The US has taken a largely voluntary approach to governing vehicles and electronics. While product environmental impacts are recognized, the federal government prefers to allow industry to take the lead, believing that product risks do not warrant regulatory intervention and that industry knows best how to manage products. Product liability laws and litigation are thought to be adequate disciplining forces. Yet it is difficult to link exposures to toxic materials with adverse health outcomes. Workers at US semiconductor factories have failed thus far in lawsuits against their employers for negligence for not taking precautions against toxics.

In the US, vehicles are subject to two types of environmental regulations. Vehicles face emission controls for nitrogen oxides, carbon monoxide, and particulate matter. In addition, the federal government has focused on fuel economy through the Corporate Average Fuel Economy (CAFE) standards. Conversely, almost no federal or state laws exist that control or restrict the use of substances in vehicles, let alone require design for recycling and reuse. One exception is mercury, where obsolete cars are among the most significant industrial sources of contamination. In 1995, US car makers agreed to voluntarily phase out mercury from use in lighting switches (Ecology Center et al. 2001). They pledged that no new car manufactured after January 1, 2003, would use mercury in switches. But countless old cars still contain mercury switches.

In July 2002, Maine became the first US state to prohibit the landfilling of mercury switches (Goldberg 2004; Recycling Today 2004). The Act to Prevent Mercury Emissions When Recycling and Disposing of Motor Vehicles requires automakers to create facilities to collect mercury switches, pay a minimum of $1 per switch, phase out some uses of mercury and label mercury components in new cars. New Jersey, Arkansas, Texas, Washington State, Illinois, New Hampshire, California and Rhode Island have followed, to industry's annoyance. However, in August 2005, the US EPA refused to ban lead weights under the Toxic Substances Control Act despite petitions by environmentalists (Ecology Center 2005). This contrasts with an EU decision in May 2005 to reject industry requests for a delay in banning such weights under the ELV directive, using similar scientific evidence (Committee for the Adaptation to Scientific and Technical Progress of EC Legislation on Waste 2005). In 2006, EPA announced a low-profile, voluntary agreement with the steel and car industries to encourage the removal of light switches before cars are shredded.

In the absence of federal and state government action, environmental NGOs have led the call for improved car design, focusing on consumer campaigns. For example, the Cleaner Car Campaign, a joint venture of Environmental Defense, the Union of Concerned Scientists, the Ecology Center and other NGOs that has operated since 1995 focuses on harnessing consumer buying power to target fuel economy, air pollution and GHG emissions and sustainable design (http://www.cleancarcampaign.org). Its Cleaner Car Standard calls for the elimination of hazardous substances and the use of sustainable plastics, which are recyclable, biodegradable and made from biomass feedstock and renewable energy (which are recyclable, biodegradable, and made from biomass feedstock and renewable energy: Clean Production Action 2005). Its Clean Car Pledge asks consumers to agree to "buy green" and to avoid unsustainable products. However, the results appear ineffectual and there are no signs of attempting to mobilize consumers around environmental and health risks. The campaign does not demand product regulation and instead aims to ally with industry. It seeks to provide consumers with information tools to aid them to compare emissions and fuel economy performance between vehicles, yet omits toxins and materials use. This US focus on consumption differs greatly from the EU emphasis on production changes.

Virtually no federal environmental regulations apply to electronic and electrical equipment. For electronics, governance focuses on energy efficiency as a voluntary industry action. Starting in 1992, EPA has promoted the Energy Star eco-label, encouraging the electronics industry to enhance product energy efficiency (www.energystar.gov). Computers and monitors were the first product categories to be awarded the Energy Star. Manufacturers can gain access to the label if they design their computers to enter a "sleep" mode during periods of inactivity, to use less than specified amounts of electricity and to have other features. EPA claims that PCs with the label use up to 70 percent less electricity than other computers, saving consumers money and reducing GHG emissions. Focusing on energy efficiency is an US innovation that the EU has adopted. But the US approach remains voluntary while the EU approach has morphed into regulation.

Electronic wastes are emerging as a significant political issue in the US following several exposés of recycling practices and impacts of obsolete PC exports overseas by the Silicon Valley Toxics Coalition (SVTC) (SVTC, for example: Silicon Valley Toxics Coalition and Basel Action Network 2002). However, EPA has still not regulated toxics in electronics, mandated recycling, or imposed restrictions on waste disposal and export, asserting that electronics do not constitute hazardous waste. Instead, in 2002, EPA created the Resource Conservation Challenge for product manufacturers to reduce toxics use and divert materials from waste disposal. The program depends on industry agreeing to make public-private partnerships. A small number of electronics manufacturers, including Dell and Intel, have participated in the "Plug-into-E-cycling" project, which recycled 26.4 million pounds of obsolete electronics in 2003 (http://www.epa.gov/epaoswer/osw/conserve/plugin). The emphasis is on individual consumer responsibility to initiate recycling.

Several state governments, including Massachusetts, Minnesota, Arkansas and California, have prohibited the disposal of certain lead-contaminated components (notably cathode ray tubes and printed circuits) from disposal in landfills (Halluite et al. 2005). However, this ban does not provide incentives for recycling and reuse since the waste can still go overseas or to other states, or be stored.

In July 2003, California became the first state to require limited producer responsibility for recycling (Korenstein 2005). California obliges manufacturers to develop, finance and implement an e-waste recovery system for reuse and recycling of some products, such as CRT and LCD monitors and TV sets. Alternatively they can pay a fee of $6 to $10 to a third party to provide this service. Manufacturers were required to attain recycling and reuse targets of 50 percent by 2005 and 70 percent by 2007. They are expected to obtain 90 percent by 2010. Maine has already followed by enacting the Act to Protect Public Health and the Environment by Providing for a System of Shared Responsibility for the Safe Collection and Recycling of Electronic Waste (Recycling Today 2004). This law also focuses on computer monitors and TVs. Governments and NGOs tend to highlight these product categories rather than other types of electronics like cell phones or home entertainment systems.

In 2003, reflecting a successful NGO campaign aimed at legislators, California became the first US state to restrict substance use in electronics by banning two brominated flame retardants (PBDEs) as of 2008 (Sissel 2003). This followed evidence produced by scientists at the California Department of Health of exponentially growing PBDE levels in the breast milk of San Francisco Bay Area women (Hooper and She 2003). The California e-waste law was later amended to require manufacturers to eliminate the ROHS substances from CRT and LCD monitors by the start of 2007. Interestingly, this law authorizes the Department of Toxic Substances Control to study, by 2008, whether further hazardous substances should be phased out from electronics and, if so, to set deadlines. This may lead to an EU-style approach and may also result in more substances being targeted than in Europe.

As of 2008, at least 25 other states were deliberating on their own electronic waste laws, signaling an important political shift at the sub-national level (see the legislative update at http://www.computertakeback.com). Maine, Washington State, Maryland, Connecticut, Minnesota, Oregon, Texas, and North Carolina have introduced electronics recycling laws based on EPR. In 2008, New Jersey, Oklahoma, Virginia, West Virginia, and New York City followed. Many state governments are far more open to citizen and NGO participation and pressure than the federal government. The impending transition of the US television industry to a digital standard in February 2009, making tens of millions of TV sets obsolete, is intensifying this pressure.

Until around 2001, US environmentalists were not very active in developing campaigns against electronics manufacturers. Now, SVTC and many state NGOs have marshaled campaigns against individual manufacturers to try to convince them to introduce or expand their own recycling infrastructure. They targeted Dell through the Toxic Dude campaign to stop the practice of recycling PCs with prison labor (O'Rourke 2005). They persuaded eBay to help build a voluntary take-back infrastructure. SVTC directs the new Cleaner Computer Campaign, emphasizing toxics phase-outs and sponsoring research to demonstrate the risks of using or recycling electronics (www.svtc.org/cleancc). Compared to car campaigns, US groups are highlighting the risks inherent in products, and are pressing for legislative action on e-waste and hazardous materials. In Europe, by contrast, environmental groups have largely supported EU actions or demanded higher standards.

The pattern is that in Europe, product policies are increasingly driven and made at the EU level whereas in the US, most progressive policies are developed at the state government level. In Europe, governments and EU institutions are taking the lead and propelling policy reform. In the US, environmental NGOs are far more important in pushing for industry action in the face of lagging federal authorities. They gain inspiration from observing EU developments. In the US, California and other state governments provide the key arenas where political action is aimed at and where policymaking occurs. In the EU, the key arena is the Commission and the Council, not so much national legislatures or governments. What this suggests is that transatlantic product politics must be examined at multiple levels of governance.

Analyzing Transatlantic Differences

Several factors have converged to explain the EU lead over the US in product regulation. First, *regulatory philosophy* helps create the baseline against which regulatory proposals are presented and evaluated. In the 1970s, the US was the leading precautionary regulator, acting to control water and air pollution, and to eliminate PBBs and some pesticides (Jasanoff 2003). During the deregulating Reagan and Bush administrations in the 1980s, regulatory norms began to favor voluntary industry policies and cooperation with corporations (Andrews 1999). EPA resources were progressively cut. Industry became far more aggressive in lobbying, using public relations and litigating against new rules and emerging scientific knowledge about toxics.

During the 1990s, the Clinton administration supported "regulatory reinvention" to make oversight more flexible (Rosenbaum 2005). EPA also began developing the Energy Star and Design for Environment approaches rather than proposing regulatory standards. This ratified the shift to voluntarism. With the George W. Bush administration, EPA's enforcement capacity and many environmental laws were eviscerated (Rosenbaum 2003). Few new environmental laws have been advanced or enacted at the federal level since 1990. Where new regulation is introduced, it focuses on providing information, such as enhancing the Toxics Release Inventory. The Consumer Product Safety Commission (CPSC) has encouraged voluntary industry action on reducing chemical risks in products (Iles 2007). In the US, campaigns aimed at changing individual consumer behavior and market-based measures have become dominant (Maniates 2002). Yet voluntary measures are not highly legitimate either and a vacuum in product regulation currently exists.

Conversely, the EU has become far more legalistic since the 1970s. The European Community developed as an economic entity rarely imposing environmental controls, particularly on a precautionary basis. It has evolved into an environmental and social regulatory state aiming to create Europe-wide standards as part of the process of creating a shared European identity (Waterton and Wynne 1996). Environmental laws now cover numerous areas ranging from factories, biodiversity protection, to sulfur dioxide emissions (McCormick 2001). European industry has also become far more assertive in attacking potential rules, yet has not had the pervasive impact on government decision-making found in the US. This reflects, in part, the increasingly entrenched nature of two important norms: recycling and the precautionary principle. These further strengthen the legitimacy of product regulation.

Compared to the US, European countries are far more invested in recycling activities ranging from packaging to paper and glass. Since the 1980s, European countries have built an extensive recycling infrastructure, with cities, provinces and industry sectors participating (Barlesz and Loughlin 2005). This grassroots governance may be a key invisible aspect of the product regulation area: most actors have come to agree that recycling is possible and essential. With the idea of mandated recycling targets and take-back increasingly being established through societal schemes like Germany's DSD framework, it became more politically possible to contemplate the take-back of consumer products that are harder to recycle, such as vehicles. Government could demand that industry take the lead in recycling. No shift towards society-wide recycling has occurred in the US: there are still battles to move recycling into practice after 30 years of efforts. Relatively little recycling infrastructure has been built, particularly for electronics. Individuals have primary responsibility for recycling.

The precautionary principle has developed as a key, if sometimes theoretical, concept in European environmental policy (Christoforou 2003). In 1993, the EU enshrined the principle in its policymaking process. In practice, precautionary action is more likely to happen if some significant scientific evidence of environmental and human health risks already exists (Eckley and Selin 2004). In contrast, the US government often opposes the application of precaution in environmental regulation. This precautionary split can be seen not only in how the Union has regulated substances in electronics, but in its oversight over cosmetics and toys. The Council, which has the authority to announce emergency, temporary regulatory decisions, introduced a ban on phthalates in soft toys that young children can chew. The US CPSC long refused to impose a similar restriction although Congress did enact a law in mid-2008 to prohibit the use of two phthalates in toys as of February 2009. The CSPC still argues this will not affect products that are already on the market.

Product regulation is more likely to be accepted in the EU than in the US. While industry and national governments are likely to contest some parts of a proposed law, they cannot count on societal support for deregulation. European industry is increasingly accustomed to operating with regulations aimed at products. Further evidence of the European regulatory philosophy is seen in the double-sided behavior of electronics and car manufacturers. The European subsidiaries of US corporations are increasingly enthusiastic about developing product design and phasing toxics out. GM even admits that it focuses on sustainable plastics through its European subsidiaries, such as Opel and Vauxhall (Clean Production Action 2005). Conversely, the US subsidiaries of European companies appear less motivated to practice producer responsibility in the US.

Second, the *European system of government* makes product regulation much more likely than in the US. Factors such as the centralization of decision-making, the extent to which industry and environmental groups can challenge government decisions, the ability of legislative bodies to force changes in executive government decisions and the capacity of government to make regulatory decisions can affect the prospects of product regulation. Although historically the EU operated with low transparency and participation, the EU has evolved into a highly centralized government system with growing democratic input.

The Commission initiates regulatory processes by sending proposals to the Council of (Environmental) Ministers and, more recently, to the EP (Macrory et al. 2004). Although this implies that industry should find it much easier to target new policies by exploiting top-down centralization instead of battling with all 27 member nations, centralization can work in favor of stronger regulation if the agency is strong enough to resist industry pressure. The Commission can choose to overrule industry concerns. The Environment DG is more authoritative than the US EPA because it controls the shaping of proposals, has a long history of constructing the regulatory state and has permanent expert staff from all member countries who know how to out-maneuver industry and more industry-focused DGs like Enterprise.

Vitally, legal challenges to EU regulations are much less likely because of how the EU legal process is structured (Macrory et al. 2004). Governments largely mediate disputes. The Commission and the Council do not face significant constraints (except those of perceived political and technical feasibility) on their regulatory powers: they can make a wide range of decisions without having to meet a pre-existing legal standard as long as an EU constitutional base exists. The Commission must justify its proposals on environmental, technical and economic grounds but it can control its own determinations (Zito 2000). In the product area, its ability to support regulation with more quantitative data has vastly improved since the 1990s.

The Single European Act of 1986 specified that environmental issues were part of the EU constitution and that the Commission must consult with the EP before finalizing regulations (Jordan 2005). In 1997, the Amsterdam Treaty gave more legislative powers to the EP. In particular, many regulatory decisions must now be made in the co-decision mode: the Council and the Parliament must jointly agree on laws before these can enter force, though the Council retains a final say and can reject Parliament amendments (Zito 2000). Conferences of conciliation must occur between the Council and Parliament to settle discrepancies (like the Senate and Representative conferences in the US Congress).

The EP has over 500 members, some of which are environmentalists, reflecting the development of Green Parties across the region. (In contrast, environmentalist parties have largely failed to enter legislatures in the US.) The Parliament holds committee hearings, can demand changes, may criticize the Commission and is increasingly vocal through its votes (Jordan 2005). As the recent cosmetic and toy chemical debates highlight, conservative and leftist politicians increasingly concur that product risks deserve attention, while not agreeing on how to govern them (Euractiv 2005b). In the electronics debate, the Parliament played an important part, forcing the WEEE and ROHS directives to enter force much sooner, stopping attempts to remove substances from ROHS, questioning exemptions and demanding that other substances like halogenated flame retardants and PVC should be included (ENDS). Though many of these amendments were rejected, they will return to the regulatory agenda in future policymaking rounds.

Conversely, the US has a decentralized federal government with divided responsibilities for product regulation. The Food and Drug Administration, EPA, the National Highway Traffic Safety Authority and the CPSC are just some of the relevant agencies. Industry and citizen groups have multiple points at which they can impede regulatory processes, even if the agencies did favor more than voluntary industry action (Brickman et al. 1985). Regulatory traditions are critical. The primary environmental agency, EPA, has few easily accessible strong powers to control or restrict products apart from pesticides. Under the Toxics Substances Control Act, EPA can mandate the elimination of substances from consumer products and has done so in a few cases such as n-methylolacrylamide in grouts (Powell 1999). However, following a string of successful lawsuits by industry (such as the US Court of Appeal ruling in 1991 overturning an attempt to ban asbestos from use in cars and other products), EPA is reluctant to act unless it can build a strong case (Dowie 1995).

US administrative laws also mandate public notice and hearing procedures for new product regulations. These can prolong regulatory processes for years and lead to ongoing lawsuits if industry or citizen groups think their concerns have not been addressed. As well, the Office for the Management of the Budget must sign off on all regulatory proposals but tends to reject product regulation, arguing that industry voluntary action should suffice (Elliott 1994). This is a potent bottleneck point that stifles most regulatory innovation. In turn, compared to the 1970s, the US Congress no longer acts as the leading environmental law-maker (although this may change after the 2006 Congressional elections), or even as a body that subjects executive government policies on environmental or consumer safety issues to searching critiques. Relatively few challenges to government policymaking occur. The level of Congress attention to product issues or eco-design ideas is extremely low.

Third, the *politics of risk* differ. Whether products are seen as sources of risks and environmental problems—and by whom, why and how—can affect whether product regulation is proposed at all. In turn, the types of risks that products are associated with can shape the regulatory responses that materialize. European consumers are more willing to entertain the idea that products may contain unseeable risks, and that governments have failed to safeguard their health. This is a consequence, in part, of the series of food scares that cascaded across Europe during the late 1990s (Vogel 2004; Jasanoff 2005). The electronics and car cases show that European citizens and governments increasingly perceive products as containing environmental and human health risks; they also focus more on what happens to products at life's end. Consumer campaigns based on product risks (such as phthalates in toys and cosmetics) began to have more traction in the early 2000s. Although the depth of these concerns should not be exaggerated, citizen demands have given the EU (and national governments) considerable support for laws that industry otherwise opposes and have further boosted the use of the precautionary principle.

In contrast, in the US, there is a relative lack of consumer demand for products with fewer health risks. Consumers assume that because products are on the market, they are safe and regulated by the government (Iles 2007). EPA asserts that electronic wastes do not pose risks and can be sent to landfill facilities or even overseas.

Fourth, the *role of NGOs vis-à-vis governments* is critical to whether regulations are enacted, not simply proposed, and what kinds of provisions these laws incorporate. As seen in the electronics and car cases, European industry perceives significant business advantages through regulation within the region and globally. Because European companies operate in a strong recycling culture, they may see recycling systems as entrenching their markets. Because they view product regulation as inevitable, they are concerned to maintain their corporate markets. They have opposed restrictions on product contents more than recycling because they do not yet perceive corporate market benefits.

Conversely, US industry has consistently opposed any kind of regulation for both cars and electronics, insisting on its freedom to develop products and to trade worldwide. Since the Reagan neoliberal economic revolution, the US government has worked increasingly closely with industry, leading to a more corporatist model traditionally associated with European countries.

Interestingly, citizens and environmental NGOs have played a less significant role in making and publicizing product regulatory ideas than might be expected in either the EU or the US. In Europe, this is in part because the Commission has taken the lead. NGOs have worked to reinforce EU regulations. In neither Europe nor the US has there been a swelling of citizen support for consumer product regulation (as compared to changes in perceptions of product risks). There have been no mass mobilization and community protests for consumer product changes centered on toxics and materials except for food issues.

Some environmental groups in the US and EU, such as Greenpeace, Friends of the Earth UK and SVTC, have launched campaigns on product risks. Since the early 1970s, Greenpeace groups in Europe have advocated the phase-out of PVC and other chemicals used in household products. In the US, groups like the National Environmental Trust or the Breast Cancer Alliance are campaigning for the phase-out of toxics from cosmetics, toys and other consumer products, focusing on regulatory change in California and several other states.

Looking at the pattern of citizen group behavior in each region reveals much about the underlying regulatory differences. Whereas relevant European groups tend to focus on government policy mainly at the EU level and on mobilizing citizens to demand national government intervention at the Union level, their US counterparts target state governments, individual companies and consumers, ignoring the federal government for the most part. This mirrors where the groups believe their efforts will have greatest effect, even if they have had little impact thus far.

Finally, despite both being multi-level political entities, the US and EU have followed different harmonization routes. The Union now has 27 member governments that retain the power to make their own environmental laws and policies as long as the EU has not acted. In some cases, EU environmental regulation allows countries to retain standards exceeding the negotiated EU level, to entice them to support Union-wide action. The governments can also choose to implement EU requirements in terms of their national situations under the subsidiarity principle. Therefore, environmentally powerful countries such as Germany, Sweden, Britain and the Netherlands can experiment with new regulations to target what they perceive as product challenges within their domestic political context. The EU has a built-in tendency to favor broader standards but only if powerful member nations introduce their own standards or activities that are seen as significantly distorting competition within the European market (Kelemen 2004b).

The ELV and WEEE directives originated in national regulations in Germany, Switzerland and Austria. The German and Swiss governments gave the real impetus to ELV and WEEE by expanding producer responsibility to the car and electronics sectors. They recognized that landfill capacity was rapidly diminishing and were early movers on packaging waste recycling laws or voluntary initiatives. During the 1990s, Sweden and a few other nations had several isolated regulatory mandates for the phase-out of toxics from consumer products but no equivalent of the ROHS directive, underscoring its character as a Commission idea. Germany, France and the Netherlands were economically and politically powerful enough to push the EU towards regulating car take-backs.

The US has a similar logic for federalization of laws and policies (Kelemen 2004b). Indeed, the history of US environmental law is marked by numerous instances where state governments regulated issues—and Congress eventually transformed these into national laws that other countries emulated. State governments were the pioneers of the precautionary, rights-based approach to environmental protection in the 1960s. Their actions can be an important signal of future global regulation. Often US industry eventually comes to prefer a federal law rather than up to 50 varying state laws, so that manufacturing can be standardized (Dowie 1995). If Congress perceives that the US economy is impeded by interstate differences, it may intervene. However, the states have largely been missing when regulating the environmental impacts of vehicles and electronics. California has begun to target products; its market size of 36 million people guarantees that industry will be attentive.

It is important, however, not to overstate transatlantic differences. Much European industry remains unenthusiastic about EU regulatory intervention and continues to oppose many aspects of implementation, borrowing ideas and support from their US counterparts. Many cases of transgressions and lethargy can be observed in industry responses to regulation. In July 2005, the Commission took action against eight member countries for failing to instigate a WEEE recycling scheme in time (Reuters UK 2005). The Commission has created, and is expanding, some exceptions in the ROHS regime at industry's request (European Commission 2004b). US corporations such as IBM are requesting that the Commission should not remove exemptions for electronics such as server machines.

Second, US industry is increasingly forced to consider whether its electronics and vehicles need to satisfy EU standards or whether multiple designs can be adopted for different markets. The evidence is that US electronics companies are starting to comply with ROHS in their domestic design and manufacturing activities (Forsberg 2005). Making different electronics for different regions, rather than making one product for worldwide consumption, generates additional costs eating into profitability. Companies prefer to build corporate markets extending globally, which is a powerful harmonizing pressure. Since most cars manufactured in the US are not exported to Europe but are sold domestically, the ELV Directive will likely have fewer direct impacts on US car manufacturers. Nonetheless, US companies—both car and electronics—will still need to arrange to have their products sold within Europe taken back.

Conclusions

Product laws are another layer in the shifting currents of transatlantic relations. The EU is creating systems for product take-back and recycling, and requiring industry to scrutinize its design and manufacturing practices so that toxics can be eliminated and materials reused. In the US, voluntary efforts are underway to persuade industry to develop small-scale take-back experiments. The US lags in appraising products and establishing the infrastructure needed for effective take-back, which may affect its ability to make US production more sustainable.

However, the EU trend may already be in holding status. EU product governance may be more voluntary in the future as regulators begin considering product design and life cycles. In the early 2000s, EU industry succeeded in weakening the Commission's efforts to develop an Integrated Product Policy because companies remain unconvinced that product redesign has competitive advantages (ENDS 2002b). Instead of mandating product redesign, the Commission is exploring voluntary policies. Even though the EU has introduced an Energy-Using Product directive, to be implemented by August 2007, this product life cycle law depends on principles rather than design demands (European Commission). As seen in the ongoing REACH chemical policy debates, the Environment DG has lost much influence within the EU system. Without industry support or acquiescence, continuing EU regulatory innovations may stall. Nonetheless, the take-back laws are entrenched and are stimulating structural changes worldwide through supply chain pressures.

China is copying EU environmental standards to govern its domestic industry. In 2007, the Chinese government put into force a new regulatory standard based on the ROHS and WEEE regulations (ENDS 2004). The aim is to force Chinese electronics makers, who are notorious for their poor quality control and environmental problems, to compete more effectively in the world market. Chinese policymakers calculate that the European market is now bigger than the US market and that adopting EU standards will still allow Chinese manufacturers to enter the US market. Other Asian nations are likely to follow thus amplifying the impact of EU laws beyond Europe. A Hewlett Packard manager, David Lear, says: "For us, it is good to hear. We want a harmonized standard" (Forsberg 2005).

On a global scale, the stakes of competing transatlantic approaches to product regulation are immense. Whichever region can define the dominant standard can pressure other regions to follow in their design and manufacturing practices if they are to gain entry into markets or to compete globally. The US has been successful at setting global standards in other areas, such as in the information technology, internet and financial areas, in part because of its dominance of technology governance and international financial institutions. In the area of products, however, the EU leads in the politics of standard-setting.

Product regulation is one of the key areas of transatlantic political debates, not only for electronics and cars, but for a wide array of products such as beef and human growth hormone. These debates center on fundamentally different ways of making products less environmentally damaging—through regulation aimed at forcing design changes and life cycle scrutiny, or through voluntary measures targeted at changing consumer behavior and giving manufacturers incentives to switch designs. Product regulation goes to the core of contemporary environmental politics and visions about how to achieve industrial metamorphosis. Product regulation differs greatly between the US and Europe because of underlying institutional and political cultural differences. These divergences are not static, however. State governments in the US have the potential to catalyze new regulation as they learn from EU developments. The US may even eventually move beyond the EU to lead environmental standard-setting again. Understanding how these differences and dynamics shape regulation can lead to more effective policymaking for the twenty-first century.

Chapter 7

Transatlantic Food Fights in an Era of Globalization: When Menus, Rules and Choices Collide

Patricia M. Keilbach

Why do transatlantic tensions often escalate over issues related to the trade in food products? What are the broader implications of transatlantic food disputes for regional and global integration, and policy harmonization? Although cooperation and convergence are the norm and food disputes concern only one to two percent of the total value of transatlantic trade and investment, the transatlantic agricultural trade disputes have been very important because of their affect on the broader transatlantic relationship and other regions of the world. In particular, the disputes between the EU and the US over GM food products reveal that international agricultural conflicts are becoming more complex. Old trade disputes emerged as a result of traditional barriers erected by nation-states in the forms of tariffs and quotas, but the new trade disputes have resulted from regulatory differences over managing risk. Substantially different views on precaution and science have been at the center of transatlantic disputes over GMs.

GM foods are genetically engineered (gene-spliced) through the injection of foreign proteins, antibiotic-resistant genes, growth hormones or genetic constructs. The US is a strong proponent, producer, and exporter of GM foods. In contrast, numerous EU states in the 1990s banned the import of GM foods and the EU as a whole restricted the approval of new GMs and introduced a policy requiring the labeling of foods containing GM products. These policies put the EU and the US at logger heads, and eventually the US took the EU before the WTO's trade dispute resolution board over its GM policies.

In September 2006, a WTO panel ruled that the EU's de facto moratorium from 1998 to 2004 on the approval of any new GMs and the unilateral bans enforced by some Member States on all GMs, even those approved by the EU, violated the WTO Sanitary and Phytosanitary (SPS) Agreement. This ruling does not, however, mean the end to EU-US tensions over GMs as European consumers are unlikely in the near term to accept products that are genetically modified regardless of the WTO ruling. Between May 2004 and June 2007, the Commission also only approved five biotech agricultural products. Several products have been under EU review for more than six years—compared to an average six to nine months in the US, Canada and Japan (http://useu.usmission. gov/agri/GMOs.html).

The historical roots of transatlantic agricultural disputes are deep and multifaceted. In recent years, however, the conflicts have taken on new dimensions as a result of tensions between consumers' demands for choice and right to know, and agribusinesses' desire to improve their bottom line. There has, for example, been a long-running transatlantic dispute related to hormone-treated meat. In 1988, an EU ban on the import of hormone-treated meat went into effect. The US and Canada contended there was no scientific basis for the EU restrictions, but the EP voted unanimously in 1996 to maintain the ban. As a result, both the US and Canadian governments took the EU to the WTO dispute resolution board (USDA 2005). The WTO determined in 1998 that the EU ban was inconsistent with the Sanitary and Phytosanitary Agreement as it was not based on a scientific risk assessment. The WTO ruled that EU policy must be brought into conformity with WTO-rules.

Instead of opening the market to hormone-treated beef, however the EU opted to accept being sanctioned. In the interim, the EU conducted risk assessments and based on these assessments decided in 2003 to introduce a directive 2003/74/EC concerning the prohibition on the use of hormones. In an interesting twist, the EU turned around and initiated a dispute against Canada and the US in the WTO challenging the sanctions these countries had placed on EU exports because of the EU's ban on hormone beef. The EU argued the sanctions were illegal since the EU had removed the measures found WTO-inconsistent by the WTO Appellate Body (Europa Press Release 2003, 2004).

The GM-related dispute has been equally contentious and complex as it affects a large number of actors: large and small-scale farmers, consumers, large corporations, environmental groups, scientists, farm workers, and policymakers at the domestic, regional and international levels.

In contrast with the situation in Europe, a multitude of US actors including many scientists, politicians, bureaucrats and biotech corporations advocate the spread of GMs, and argue that we are dependent on GMs delivering plentiful, and more nutritious food to meet expected exponential population growth. Proponents say GMs will lower costs, increase yields, decrease the need for chemicals, and help to feed a hungry world (Pinstrup-Andersen and Schiaoler 2000; Paarlberg 2001; Runge et al. 2003). Actors in the EU have argued instead that information about the impact of GMs on human health and the environment is relatively scarce, and the promise of GMs is uncertain. Opponents are concerned about health risks and potential threat to the environment of GMs and argue that not enough studies have been done to prove their safety (Korten 2001; Lambrecht 2001; Mikkelä et al. 2001b; Shiva 2003). The US introduced GMs with very little regulation; the EU, in contrast, has greatly restricted their development, availability, and market entry.

This conflict between the US and the EU is also important in light of the impact EU and US agricultural policies have had internationally. For example, some African countries despite facing severe food shortages hesitated to accept US food aid that might contain biotech products for fear that their own food exports would later be judged unacceptable for EU consumption. Such concerns contributed to Zambia's decision to refuse food aid from the US, and Malawi's, Mozambique's and Zimbabwe's decisions to accept shipments of milled food only so as to prevent the potential planting of GM seeds. In the wake of the EU-US trade dispute, China and India slowed down the commercialization of GM food crops even though both countries have large domestic markets and are less concerned about exports. Brazil reacted to European buyers and declared the country a GM-free zone.

The EU's rapid switch from North American suppliers to those in countries that are formally GM-free contributed to a significant change in the flows of transatlantic trade. Unless European consumers become far less skeptical towards GM crops, few developing countries will choose to grow them. Few GM food and feed crops have been approved for commercial planting in the developing countries of Africa, Asia or the Middle East. This situation derived in part from fears that a highly restrictive interpretation of the precautionary principle in the EU and Japan would severely limit if not close off export sales. Although the EU's policy was developed primarily to protect European consumers and the environment from potential dangers, it has had far-reaching implications.

This chapter analyzes the origins, depths and significance of the transatlantic agricultural disputes dealing with GMs. It highlights areas of tension and compromise and explores broader impacts and implications for the enlarged transatlantic relation, and the increasingly trade liberalized world (Bernauer and Aerni, this volume). The chapter is divided into four parts. The first section explores the technological developments in agriculture that spurred the development of GMs and rules established to deal with the new technology. The second provides an analysis of the development of the EU-US trade disputes over GMs. The third focuses on the harmonization challenge at various levels and reveals problems and potentials for coordination. The final section explores prospects for cooperation within the context of an enlarged transatlantic relation and an increasingly globalized world.

The EU-US GM Trade Dispute

The event prompting the EU-US GM trade dispute was a decision made by the EU in 1998 and entering into force in June 1999 to block the commercial introduction of all new GM products and require the labeling of all foods containing one percent or more of GM ingredients. They also restricted GM field trials. Beyond this, as consumer fears about the safey of GM foods grew in the late 1990s, Austria, France, Greece, Italy and Luxembourg banned already approved GM crops. As a result, between 1998 and 2002 the number of GM crop trials in the EU dropped by nearly 90 percent. The US, Argentina and Canada, threatened to challenge the European policies within the WTO.

In an effort to fend off a WTO case, in 2003 the Commission introduced two new regulations. Regulation (EC) 1829/2003 established requirements for safety testing of GM food and feed intended for the market and strengthened and harmonized labeling requirements for GM foods. The labeling requirements apply at the point of sale to allow customers to exercise choice over the foods they buy. Regulation (EC) 1830/2003 established a requirement to trace and identify GMOs and food and feed derived from GMOs at all stages of their placing on the market. While on paper these regulations ended the moratorium on the approval of new GMs, several EU nations remained reluctant to authorize biotech crops because of public health and environmental concerns. Thus, in practice the ban on GMs continued in several states of Europe.

EU institutions, however, are not united on the issue of GMs. The Commission has supported the introduction of GM products "as appropriate" (European Commission 2005b). On 28 January 2004, for example, the Commission approved a proposal to authorize Syngenta's genetically modified Bt-11 corn for food use. The Bt-11 corn, made by Swiss agribusiness giant Syngenta was modified to produce its own insecticide and is also resistant to an herbicide. The corn would be imported as a canned food product, not for planting (ENS 2004). Another corn—US-based Monsanto's Roundup Ready—was also approved by the European Food Safety Authority (EFSA) for use as food or feed.

European consumers are overwhelmingly opposed to GM foods (Friends of the Earth Europe 2005). The EU Council of Environment Ministers responded to these consumer concerns and resisted Commission efforts to put an end to national bans (Brown 2005). On June 24, 2005, the Council voted by a large majority against the Commission proposal to require Austria, France, Germany, Greece, and Luxembourg to give up their bans on eight GM products, including Syngenta's Bt-ll corn (Euractiv 2005a). Significantly, twenty-two EU countries voted against the Commission proposal. This decision created a challenge for the Commission (European Commission 2005b).

Again on December 18, 2006 the Council voted down an effort by the Commission to require Austria to lift its ban on the import of two GMO-maize varieties. Only the UK, the Netherlands, the Czech Republic and Sweden sided with the Commission (Euractiv 2006). The disjuncture between the Commission and the Council reveals the challenge the EU has had in finding acceptable policy solutions for states committed to the seemingly competing goals of free-trade, democratic decision-making and sustainability.

The George W. Bush administration justified its stance and the WTO case against the EU as part of the fight against world hunger accusing Europe of hindering the "great cause of ending hunger in Africa with its ban on genetically modified crops" (Sanger 2003). Others contend the US position is heavily influenced by important industry lobbying groups seeking to protect their interests (Shiva 2000; Korten 2001; Davis 2003; Shiva 2003). The primary concern of many consumers, NGOs and scientific organizations is that the WTO rulings will undermine the hard-fought health, safety and environmental standards of Member States. Indeed, food policies vary across countries and tend to reflect local concerns and interests. It is necessary to understand the forces that have created this patchwork of regulations in order to understand challenges for policy harmonization.

Diverging Policies

The United States

In the 1970s, US scientists at the forefront of genetic engineering developed self-imposed guidelines relating to research in genetic modification and then solidified these at the International Conference on Recombinant DNA Molecules at Asilomar, California. These guidelines established by the small and organized scientific community, while voluntary and temporary, were adopted by the US National Institutes of Health (NIH) as the code. It was this agency's advisory board, the Recombinant Advisory Committee (RAC) that was charged with evaluating deliberate release experiments (Cantley 1995). After a 1986 court case ruling ended the NIH's dual regulatory role, assessment was divided between the US Drug Administration (USDA), the Environmental Protection Agency (EPA) and the Food and Drug Administration (FDA). Regulatory agencies and policies in the US permitted the full-fledge introduction of GMs domestically, and the regulatory style and product-oriented system, combined with the judicial victories of the organized vested interests, largely explain how the political and public challenges were kept in check against the deliberate release and the focus on end product rather than the recombinant technology used to create it (Jasanoff 1995; Dunlop 2000).

The US legal system also created incentives for private companies to take the lead. As Paarlberg notes in his study on GMs in the developing world, "When public funding for international agricultural research faltered in the 1980s, the initiative in developing most GM crops fell to private seed and biotechnology companies" (Paarlberg 2001, 11). These companies then sought to recover the high R&D investments of their biotech inventions through intellectual property rights (IPR) and patent protection. A Supreme Court ruling in 1980 (Diamond vs. Chakrabarty) provides full protection for agricultural crop inventions and any organism altered by human intervention. This patent protection gave US corporations a strong incentive to invest and develop new seeds and to patent these inventions down to the level of the individual genes and gene sequences (Paarlberg 2001, 13).

Considerable research, development and investment positioned US firms such as Monsanto and DuPont at the cutting-edge of the industry. The US now leads the way in biotech crop production and the IPR protections provided by the US legal system explain why US companies became early leaders (Paarlberg 2001, 13). US biotech corporations eager to access global markets have found significant resistance not only in Europe but also around the world. If the US biotech corporations can convince policymakers at all levels that GMs are safe and that GMs provide a scientific breakthrough essential to food security, then they stand to make enormous profits.

The revolving-door effect also draws considerable attention. For example, an April 2003 report by Innovest Strategic Value Advisors stated: "Money flowing from Alliance genetic-engineering [GE] companies to politicians, as well as the frequency with which GE company employees take jobs with US regulatory agencies (and vice versa) creates large bias potential" (as quoted in Cheng 2003). Thus, US regulatory decisions related to GMs are a result of the research and development process as well as the influence of dominant economic actors with vested interests.

The European Union

Food politics in the EU is a multi-tiered enterprise, with decisions being made in 27 Member States, the EP and a variety of executive agencies, committees and by meetings of EU ministers in the Council (Zito 2000; Jordan 2002). GMs have been controlled by the EU since the beginning of the 1990s through recommendations, regulations, and directives. Designed to protect citizens' health and the environment, legislation has addressed authorization, labeling and traceability issues relevant to GMs. Two EU directives were initially established to address GMs. Directive 90/219/EEC focuses on contained use and directive 90/220/EEC deals with deliberate release. Directive 90/220/EEC covers the procedures for the approval of new GM products and releases. But, the directive is weak and reveals the limited power of the biotech lobby organization during the late eighties.

The Commission published guidelines, Recommendation 2003/556/EC, for the development of strategies and best practices to ensure the co-existence of GM crops with conventional and organic farming. These guidelines were intended to help Member States develop workable measures for co-existence in conformity with EU legislation. The main legislation authorizing experimental releases and the marketing of GMs in the EU is Directive 2001/18/EC, as amended. It established a step-by-step approval process for a case-by-case assessment of the risks to human health and the environment prior to authorizing the placing on the market or release into the environment of any GM or product containing GMs.

Regulation 258/97/EC on Novel Foods and Food Ingredients regulates the authorization and labeling of novel foods including food products containing, consisting of, or produced from GMs. The EU recognizes the consumer's right to information and labeling as a tool to make an informed choice

(Crespi and Marette 2003). Since 1997, EU law mandates labeling to indicate the presence of GMs as such or as a component of a product. Additional legislation include Regulation (EC) No 1830/2003 concerning the traceability and labeling of GMs and the traceability of food and feed products produced from GMs, and Regulation (EC) No 1829/2003 on GM food and feed that requires the traceability of GMs throughout the food chain from "farm to fork." Commission Regulation (EC) No 65/2004 of January 14, 2004 establishes a system for the development and assignment of unique identifiers for GMs (European Commission 2003a).

While some EU biotech corporations have lobbied for increasingly permissive policies, consumers are highly concerned about the potential health and environmental risks associated with GMs. Strict regulations on biotechnology have typically been justified on the basis of public skepticism towards the technology and heightened concerns about food safety in the wake of food scares. Moreover, Europeans have less trust in their government agencies that regulate food supply than US consumers have for similar agencies (USDA, FDA, EPA). Europeans are risk averse when it comes to food and tend to apply the precautionary principle. The divergent GM regulatory policies of the EU and the US have lead to deeper and more emotional conflicts than has been typical with more traditional disputes over trade restrictions, dealing with tariffs and quotas.

The newly enlarged EU could increase the regional policy harmonization challenge because the new Member States are only in the beginning stages of establishing GM policies, and they are susceptible to influences from a variety of actors operating outside the purview of Brussels including anti-GM NGOs and multinational corporations with considerable investment potential (Schweiger and Ritsema 2003). Slovakia, for example, authorized its first GMO field tests of MON 810 maize in May 2006. On the other hand, the Polish Parliament debated a national GMO ban in the same month (GMO Compass, http://www.gmo-compass.org/eng/news/messages/200605.docu.html).

International Organizations

At the international level, layers of legal influence lie under the umbrella structure of the United Nations (UN) system where several additional bodies and treaties monitor, regulate and attempt to govern the trade in GMs. The FAO monitors food policy and addresses food security issues. Four institutional related agencies lie under the FAO: the International Plant Protection Convention (IPPC); the Office of International Epizooties; the Commission on Plant Genetic Resources (CPGR); and the Codex Alimentarius Commission (CODEX). The most important in the context of GM issues is CODEX. CODEX was created in 1963 by FAO and World Health Organization (WHO) to develop food standards, guidelines and related texts such as codes of practice under the Joint FAO/WHO Food Standards Program. The main purposes of this program are protecting health of the consumers and ensuring fair trade practices in the food trade, and promoting coordination of all food standards work undertaken by international governmental and non-governmental organizations. CODEX approves recommended standards that determine a label for food products.

While all of these FAO agencies play a role in the governance of GMs none seems as likely to impact the outcomes of harmonization challenges as two other international instruments: the Cartagena Protocol and the WTO. The Cartagena Biosafety Protocol (CBP) entered into force on September 11, 2003 and as of June 2007 had 141 parties. It is the only international treaty governing the cross-border transport of GMOs and is a supplementary agreement to the 1992 Convention on Biological Diversity (CBD). The rules set out in the protocol are intended to promote the conservation and sustainable use of biological diversity and protect the public from the potentially harmful effects of GMs. The CBP establishes criteria allowing parties to act, even when there is scientific uncertainty, to avoid potential risks. It includes guidelines for the safe transfer, handling and use of GMs (UNEP 2000, Article 1). It also established a procedure of advanced informed agreement (AIA) for any transboundary movement of GMs.

Exporting parties must give written notice and gain consent from the importing party prior to the first movement. The CBP also called for the establishment of a Biosafety Clearing House to facilitate exchange of information between parties, collate national laws and regulations and assist in implementation of the agreement. Parties are encouraged to advance public awareness of safe transfer, handling and use (Article 23), and to help developing countries to fulfill their obligations through financial aid and the transfer of technology (Articles 22 and 28). Its aim is to assist developing countries in building their capacity for managing modern biotechnology.

The centerpiece of the CBP is the AIA and the complementary Biosafety Clearing House aimed at sharing information on existing regulatory frameworks, domestic approvals of GMs, risk assessments and bilateral and regional agreements in place to govern trade in GMs (Gupta 2000, 25). The treaty allows states to ban imports of GMs, without fear of trade sanctions, if the state believes them to be a threat to local ecosystems and even in the absence of scientific evidence. At the first conference of the parties (COPs) in February 2004, governments took the first step towards establishing an operational framework for the implementation of the agreement by making important progress on documentation requirements, compliance, liability and redress and the Biosafety Clearing House. Many delegates and observers welcomed the agreement reached, while the US and other biotech exporters criticized it for failing to take into account trade implications.

The major biotech producers and exporters including Argentina, Australia, Canada, Chile, Uruguay and the US (known as the "Miami Group") do not intend to ratify the treaty, but maintain a strong interest in influencing policy developments under the CBP. The US is the world's largest producer of GM crops and it exports over $60 billion of agricultural products a year (Food Navigator 2005). The Miami Group countries argue that compulsory labeling would create serious paperwork requirements and fear that the protocol would extend to food products containing GMs as well. The US, along with other food and ingredient exporters, is motivated by concerns that the CBP could be used as a protectionist device to favor domestic GMs over foreign ones. Nevertheless, in July 2005 119 governments adopted binding rules on papers required to accompany GM commodities such as wheat, maize and soy when they are transported across borders. These rules will ensure that only approved GMs enter the territory of respective parties.

The WTO agreements on Sanitary and Phytosanitary Measures (SPS) and on Technical Barriers to Trade (TBT) and related Uruguay Round agreements also attempt to influence GM trade. These measures and standards have been agreed to by over 140 countries with the aim of promoting the free and fair trade of good quality and safe foods among all countries. The SPS and TBT agreements rely on science-based CODEX standards, guidelines, and recommendations as benchmarks for judging international food trade disputes. However, similar to CBP, the CODEX standards are also incongruent with WTO standards promoting free trade and it is these standards, guidelines, and recommendations that are deemed to be the primary obstacles to free and fair international food trade, particularly for developing countries.

The US has consistently demanded the inclusion of a provision in the CBP that would, in effect, elevate WTO rules above those of the CBP. Finally, an additional international agreement is now being used to assist in the establishment of guidelines, criteria and precedents to deal with trade in GMs. The US is pushing other countries to adopt the International Union for the Protection of New Varieties of Plants (UPOV). The UPOV was signed in 1961 with the aim of protecting new plant varieties of signatory countries. It has since been amended three times in 1972, 1978 and 1991, each time strengthening the protection of plant breeders ("Biosafety Meeting" 2004). NGOs are warning that the UPOV agreement favors plant breeders while it prevents farmers in developing countries from saving, exchanging and reusing seeds.

An examination of the institutions involved in this transatlantic trade conflict reveals a multitude of oft-competing institutions aiming to govern the trade of GMs. Nonetheless, institutional linkages exist and these include both norm and rule overlap (Rosendal 2001; Selin and VanDeveer 2003). The norms and rules of the EU are embraced in the CBP, whereas the norms and rules of the US are supported in the WTO. Policy developments in these various arenas reflect a dialectical dynamic between these regimes. Given the legitimacy of all of the aforementioned institutions, is it possible for a synthesis to emerge? Or, will actors continue to "venue shop" for a forum receptive to their rules?

The Harmonization Challenge Ahead: Implications for the Enlarged Transatlantic Relationship

The transatlantic food fights are emblematic of a new kind of global trade conflict in which health and environmental laws are at stake. Indeed, the menu of policy choices include those aimed at addressing hunger problems, those aimed at protecting the environment and human health, and those aimed at increasing flows of international trade. The collision between the push for freer trade and an array of environmental laws implemented over the last decade have spurred activists to sharply challenge the WTO's right to stand in judgment of national laws. Trade specialists had argued that legislators were passing disingenuous laws that lacked scientific rationale, with the primary goal of keeping foreign products off their shelves. In order to prevent this kind of presumed interference with free trade, the WTO's Sanitary and Phytosanitary measures encourage countries to harmonize a wide range of relevant standards at the international level. Food safety requirements are high on that list. The WTO is assisting countries to deal with the conflicts between national legislation and international regulations.

In a departure from the WTO's regulations on trade which must be based on sound scientific knowledge, the CBP incorporates the precautionary approach of the 1992 Rio Declaration allowing parties to act, even when there is scientific uncertainty, to avoid potential risks. Although the US is not a Party to the CBD and therefore cannot become a Party to the CBP, the US participated in the negotiations as a member of the Miami Group. The US is pushing for the commercialization of biotechnology, including GMs. In addition to the Miami group, commercialization is currently being pursued mainly by major corporations, which, understandably, seek to maximize profits. Other negotiation groups included the EU, the Eastern and Central European countries, a group of like-minded developing countries, and a compromise group including Japan, Korea, Mexico, New Zealand, Norway, Singapore and Switzerland.

The GM issue lies at the nexus of trade and the environment, and harmonization of policies at all levels is proving quite difficult. States in the US and EU Member States are challenging national, regional and international regulations seeking to maintain GM-free zones. Anti-GM NGOs are also assisting in this campaign and have provided important information to key actors in order to establish GM-free zones. In seven EU nations including Austria, Germany, Greece, Italy, Spain and the UK, GM-free zones have been created. Similarly, several US counties and states are seeking similar protection zones including counties in California, and the states of Vermont, Minnesota, Montana, North Dakota, South Dakota and Wisconsin.

On both sides of the Atlantic, the overwhelming majority of citizens want mandatory labeling and tracing of GMs (Pew Initiative on Food and Biotechnology 2004). The Pew Initiative on Food and Biotechnology has to date conducted three comprehensive surveys of US consumer sentiment about the application of genetic engineering to agriculture. The January 2001, August 2003 and September 2004 survey results reveal strong support for labeling (Pew Initiative on Food and Biotechnology 2004). Gene manipulation is still in its infant stages and many worry that profit concerns overshadow health and environmental concerns. Sound science is crucial for the development of sound public policy. Public policy regulating GMs are not always congruent with the public's priorities. GMs are not required to be safety-tested or labeled by the US FDA unlike new drugs and other food additives *before* they are placed on the market.

Most Europeans and Americans want their foods labeled. The European's focus has been on process and traceability as opposed to product and tracing. The big looming worry is that national labeling and certification programs will continue to be challenged as barriers to trade because they distinguish between products based upon how they were produced. WTO rules generally frown upon such distinctions. But the right to be well informed is important for consumer's confidence and labeling is an important tool to this end because it enables the consumer to know what they are buying. Consumers want information about the health and nutritional implications of their choices and they are also interested in the environmental and ethical implications of the way food and other products are produced. As noted in Foreign Affairs, "US-European differences on these and other important issues exist, but the data on public attitudes hardly seem a sign of two societies 'living in different worlds'" (Gordon 2003).

Why are there such distinct policy stances across the Atlantic? Analysis of the transatlantic divide on multilateral environmental agreements in general, and policies related to GMs in particular, reveals how a clash of scientific perspectives and diverse political-economy orientations are leading to conflicts that are ultimately motivated by differences in calculations of self-interest on both sides of the Atlantic. Indeed, from 1998 through 2003 the EU was resistant to the full-scale introduction of GMs, but as EU corporations gained market share and competitive advantage, resistance among bureaucrats and politicians waned considerably. This contributed to the EU's lifting of its defacto moratorium on the approval of new GMs. Still, numerous bans persist in Europe.

Economic incentives create impetus for harmonization, yet ideological differences rooted in assessments of "risk" hinder full cooperation. Thus, conflicts between the EU and US have much to do with a clash of scientific perspectives and economic orientations that serve to form varying notions of self-interest among actors on both sides of the Atlantic. Different regulatory approaches across the Atlantic have led to inconsistent regulations, and the largely ideologically-based concerns of citizens in the EU have often trumped economic interests that tend to promote policy convergence. The newly enlarged EU will encounter even more problems, particularly since the new members increase the overall agricultural production area of the EU and have yet to completely harmonize their policies with those of the EU.

On the surface, transatlantic trade conflicts tend to focus our attention on Brussels and Washington and the "cafeteria rules" established at that level. However, in the case of GM disputes, there are many forces at play. Much action has taken place at the international level, within the WTO and the agencies under the United Nations (UN), and at lower levels of decision making—the levels of local and individual choice. There have also been important developments at the sub-national level. Several locales, counties and states in the US federal system are challenging the federal government's permissive stance on GMs. On the other side of the Atlantic, some EU Member States are enacting policies that diverge from those established in Brussels, including states whose entry into the EU was contingent upon the harmonization of domestic policies with those of the EU.

Resistance to GMs exists also in the US. On March 2, 2004, voters in California's Mendocino County passed Measure H, a county-wide measure to ban the "propagation, cultivation, raising and growing of genetically modified organisms in Mendocino County" (Olson 2004). Mendocino was the first county in the US to pass such legislation, but not the last. On June 28, 2004 an initiative with over 11,000 signatures was submitted to the County Clerk's Office in San Luis Obispo County petitioning the county to put on the ballot a voter decision to prevent the introduction of GMs. The Organic Consumers Association in California is convinced that staying GM free will put the county in an advantageous position as consumers and world markets continue to refuse the food (Campbell 2004). Other counties considering similar initiatives include Butte, Humboldt, Marin and Sonoma. While food safety issues are paramount, economic concerns are also driving opposition to GMs. In Japan, the influential Rice Retailers' Association has threatened to seek a ban on California rice imports if GM rice is grown commercially in the US (Lee 2004).

At the state level, Vermont has set a precedent by campaigning for the establishment of legislation requiring the labeling and registration of all GM seeds sold in the state. Republican Governor James Douglas recently signed a bill making Vermont the first state to legislate regulation of GMs (Pavolka 2004). The bill sailed through Vermont's House with 125–10 vote in support of regulating and labeling. If it was going to happen anywhere, Vermont could be expected to lead the way. Indeed, over 20 percent of farmland in Vermont is devoted to organic production (a higher percentage than any other state) and the state has a strong reputation for grassroots activism. Whether the state will set a precedent, is open for debate.

The Vermont decision contradicts federal policies aimed at promoting the production and sales of GMs. The federal government has sought to keep GMs unregulated since their introduction in 1994 and the FDA considers GMs as equivalent to unmodified foods. The Bush administration considers regulations and labeling of GMs as costly, unnecessary and unfair and defends GMs worldwide. The US is unlikely to change its stance and in the same month that Vermont passed its GM regulation, the US filed papers with the WTO demanding $1.8 billion in compensation from the EU for loss of exports over the past several years (Pavolka 2004). While the EU imposed a moratorium on GMs in 1998 (implemented in June 1999), asking for time to determine the safety of GM food, the US has argued it did so without scientific evidence and in defiance of WTO policies.

Nonetheless, the dominant biotech corporations are taking note of county and state-level concerns and even the giant Monsanto recently announced it would not release its newest GM crop, Roundup Ready wheat in Minnesota, Montana, North Dakota and South Dakota. Conventional farmers in these states threatened to refuse the GM wheat crop, fearing the loss of a lucrative market, Japan. Japan has said it will refuse to import wheat from the US if growers start planting Roundup Ready wheat.

Chapter 8

Implications of the Transatlantic Biotech Dispute for Developing Countries

Thomas Bernauer and Philipp Aerni

In September 2006, the final report of the WTO's dispute settlement panel for "EC-Measures Affecting the Approval and Marketing of Biotech Products" was released. The panel ruled that the EU's de facto moratorium on the approval of GMOs from 1998 to 2003 as well as unilateral bans by some Member States of GMOs already approved by the EU were inconsistent with the WTO Sanitary and Phytosanitary (SPS) Agreement. The US government, the main plaintiff against the EU in this case, was triumphant, claiming that the verdict favored "science-based policymaking over the unjustified, anti-biotech policies adopted in the EU." Large, international NGOs, such as Friends of the Earth and Greenpeace, on the other hand cried foul, arguing that the verdict was undermining international environmental law and the precautionary principle in particular. They demanded that environmental disputes be removed from the WTO because the WTO was not equipped to deal with these cases effectively.[1]

Despite heavy political pressure by NGOs, the EU decided not to appeal. Two considerations influenced this decision. First, the principal demand as set forth in the WTO verdict is that the EU effectively implement its own GMO approval regulations, both at the EU and the Member State level. The verdict thus backs the EU Commission in implementing the EU's new GMO regulations. Since the end of the approval moratorium in 2003, 23 new GM varieties have been approved by the EU, all designed for animal feed. Yet, the backlog in the approval process continues to be large and the few that were approved met resistance by Member States, such as Austria and Poland, that continue to invoke the safety clause to justify their national ban on GMOs. Efforts by the EU to pressure these countries to comply with WTO as well as EU law have been undermined by the decision of France in February 2008 to impose a one-year moratorium on GM crops. Even though the extended time to comply with the WTO verdict expired on January 11, 2008 the plaintiffs in the WTO case are unlikely to impose any trade sanctions on Europe for the time being. This shows that the WTO verdict is unable to change much in the way

---

1   http://www.trade-environment.org/page/theme/tewto/biotechcase.html.

EU regulations for GMOs are implemented. The influence of Member State governments in supranational decisions of the EU and the general reluctance of retailers to sell GM food will continue to hamper the adoption of GMOs in Europe. In other words, why should the EU appeal a verdict that has little effect on its policy choices and may even help the Commission come to terms with a minority of GMO-adverse member countries that are reluctant to implement the EU's GMO regulations and approval decisions?

The second consideration is more ambiguous than the first and concerns the precautionary principle. The WTO panel did not question the role the precautionary principle plays in the EU's new GMO regulations, but it concluded that unilateral bans by some Member States above and beyond the EU-wide rules and justified as safeguards measures were not supported by scientific evidence. This implies that such unilateral bans must be abandoned or justified by a scientific risk assessment. Moreover, the panel argued that in the present case it did not need to take into account international environmental treaties subscribing to the precautionary principle, notably the Cartagena Protocol of the UN Biodiversity Convention, because not all parties to the GMO dispute were parties to the Protocol (the EU is a party, the US not).

The Cartagena Protocol, which has more than 130 parties, has been the most important success of the EU in its efforts to enshrine its own precautionary approach in GMO regulation at the global level. This agreement constitutes a key vehicle for "exporting" the EU policy-approach for GMOs to other countries, notably developing countries. The decision of the EU not to appeal the WTO verdict suggests that the EU prioritized the first consideration over the second. It may also signal that the EU does not believe that the WTO verdict undermines the Cartagena Protocol—quite in contrast to the assertions of anti-biotech NGOs and legal scholars (Conrad 2007).

The EU's decision not to appeal does not mean that the transatlantic dispute is resolved. While GM crops are being cultivated and their products marketed at an increasing scale in North America, the market for imported GM crops and GM crop production in the EU remains marginal. Transatlantic differences in regulatory policy and in markets are growing unabated. While the US has won the particular legal case in the WTO, the verdict has no obvious effect in terms of removing regulatory (non-tariff) barriers that impose opportunity costs in the order of hundreds of millions of US dollars per year on US crop and technology exporters. It is only a matter of time until the US and the EU clash again over GMOs in the WTO or elsewhere.

This chapter examines the implications of the transatlantic biotech dispute for developing countries, thus complementing Chapter 7 by Keilbach on the transatlantic dispute in this book. We are primarily interested in the question why stakeholders in the transatlantic GMO dispute are concerned with shaping the global environmental policy agenda, as witnessed by the debate about the Cartagena Protocol, and introducing their preferred regulatory approach in developing countries. We are also interested in whether the EU will be able to restrict the use of GMOs in food and agriculture in developing countries in the long run.

We argue that an important motivation for extending the battleground to developing countries has been competition for public trust as a source of political influence within the countries involved in the transatlantic dispute. The EU and other GMO-adverse stakeholders have been more successful in exporting their preferences and regulatory approaches to developing countries in the past decade. It appears, however, that the tide is turning because many developing countries are experiencing strong incentives to pursue a more pragmatic approach to GMOs in response to the looming world food crisis. In the long run, this development may undermine the influence of anti-biotech stakeholders in advanced industrialized countries: these stakeholders have, in recent years, increasingly positioned themselves as protectors of poor developing countries against powerful multinational biotech companies and their host countries (notably, the US). But what will happen if those poor countries that supposedly need to be protected from the technology do not wish to be protected?

Channels of EU and US Influence on Developing Countries

Developing countries have become an important issue in the transatlantic agri-biotechnology debate over the past few years. This prominence is rather surprising given that most of them are not (or at least not yet) important developers, importers or users of the technology. It reflects, at least in part, a series of efforts by advanced industrialized countries to influence poorer countries' choices in this area. Influence on developing countries' agri-biotech policies has been exercised through a variety of mechanisms.

The EU and the US have sought to influence the position of developing countries in the context of the 2000 Cartagena Protocol on Biosafety, a protocol to the 1992 UN CBD. This protocol governs transboundary movements, handling, transit, and use of living GMOs that may have adverse effects on biological diversity. It addresses primarily environmental effects of trade in GM-products, but also takes into account public health aspects. The EU supports the protocol. The US opposes it because the protocol endorses the application of the precautionary principle in international trade with GM products (Falkner 2000, 2002).

The Cartagena Protocol has established an important multilateral legal justification for EU assistance to developing countries in the biosafety area. Its capacity building component includes scientific and technical training, help in establishing institutional and regulatory mechanisms for risk assessment and risk management, access to relevant information, and financial assistance for these purposes. These activities are supported by the UN Environment Programme and the Global Environment Facility.[2] Most of these efforts have the effect of constraining rather than promoting agri-biotechnology applications.

The US, which has not joined the Protocol, has established bilateral networks of cooperation in agricultural biotechnology research and development. It has also sought to influence GMO policies of developing countries in the context of bilateral negotiations on preferential trade agreements and development assistance. European development assistance, in turn, is promoting "citizen juries" in developing countries that put GMOs "on trial." In 2005 Germany's agency for technical assistance (GTZ) launched a €2 million project to advertise its "Model Law" on biosafety to the African Union. Norway supported Zambia with a grant of $400,000 to ensure a GMO-free policy. And many other European development agencies directly or indirectly fund anti-GMO activities in Africa. In addition to foreign assistance programs, a second formal channel for exporting European-style GMO regulation to Africa operates via the United Nations Environmental Programme (UNEP). UNEP has developed a global program on biosafety regulation that is largely funded by the Global Environmental Facility (GEF). Considering that the EU and its member countries are the main sponsors of UNEP and GEF activities it is not surprising that UNEP has advised developing countries to take the European approach (Paarlberg 2008).

The US, for its part, has sought to encourage developing countries to adopt more permissive regulations that allow for agri-biotech R&D activities beyond the laboratory. For example, the USAID Biotechnology Initiatives were launched in 2001 "to use the benefits of agricultural biotechnology throughout Africa to enhance food safety and security" (Kellerhals 2001). Other initiatives include the Collaborative Agricultural Biotechnology Initiative (USAID 2003a), the Collaborative Research Support Programs (CRSPs) (USAID 2003b), the USAID-supported African Agricultural Technology Foundation (AATF), the Bean/Cowpea Collaborative Research Support Program, USDA technical assistance for the cotton growing industry in West Africa, the regional African Center of Excellence for Biotechnology,[3] and so on.

Furthermore, NGOs, business associations, and firms from advanced industrialized countries have sought to influence state and non-state actors in developing countries in several ways (Paarlberg 2001, 2003; Cohen and Paarlberg 2004; Kremer and Zwane 2005). Examples on the proponents side include corporate donations of technology to developing country research institutes, education/instruction of stakeholders from poor countries in advanced industrialized countries, and funding for biotechnology and biosafety research. In 2006, the Alliance for a Green Revolution in Africa (the Alliance, or AGRA) was established by the Rockefeller Foundation and the Bill and Melinda Gates Foundation (with funding in the order of US$ 150 million).

---

2   http://www.biodiv.org/welcome.aspx; www.unep.ch/biosafety/.

3   http://www.usda.gov/Newsroom/0398.04.html.

It indicates that US philanthropy may also become influential in promoting agricultural biotechnology in Africa (Kleckner 2006). However, AGRA has, thus far, avoided commitments to the promotion of GM crops in Africa.

On the opponents side, activities include funding for protest campaigns, capacity building activities, and organic agriculture initiatives (e.g. Paarlberg 2001; Bob 2002; Paarlberg 2003; Cohen and Paarlberg 2004). Survey research by Aerni (2001, 2002) has shown, moreover, that stakeholders from advanced industrialized countries exert substantial influence on the most vocal participants in public debates on agricultural biotechnology in developing countries. A network analysis of stakeholders involved in the public biotech debate in the Philippines showed, for example, that domestic NGOs campaigning against agri-biotechnology were largely financed by foreign stakeholders.[4]

What is Puzzling about EU and US Behavior?

A straightforward interpretation of the extension of the transatlantic battleground to poorer countries could be that of a struggle for markets and influence on international regulatory processes. The US, in this perspective, is primarily pursuing a strategy of opening markets for its agricultural and agri-biotechnology products. The EU, in turn, is trying to block such attempts and "export" its own, more restrictive, regulatory approach. Both sides are trying to coerce and/or entice poorer countries into supporting their respective policy position in the WTO, the Cartagena Protocol, and other important international fora. To the extent they are able to win more allies in these international bodies their influence on international standard setting grows.

US efforts to "export" its regulatory approach to developing countries obviously reflect to a considerable degree economic reasoning—i.e., an interest in opening new markets for US GM technology and GM farm products. To some extent, it may also reflect a conventional modernization ideology that emphasizes new technologies as key tools for overcoming poverty, hunger, disease, and underdevelopment more generally.

Why EU countries are interested in exporting their regulatory model to developing countries is not so clear. To a very limited extent, stricter standards in developing countries may create business and employment opportunities for European firms that can cope with such regulation and/or provide assistance to developing countries in implementing such regulation. Yet, by regulating GMOs similar to toxic waste (e.g., with an equally strong application of the precautionary principle and strict liability laws), Europe is also turning against its commitment to promote biotechnology research and development (Cantley

---

4 Large multinational agribusiness companies did not feature as prominently as NGOs in the financial network. Yet they were supporting local research institutes and local companies through technology transfers and public-private partnerships.

2004). After all, biotechnology is not just a source of risk but also has the potential to generate benefits for society and the environment.

One might also reason that the EU is concerned that imports of non-GM produce may become increasingly difficult (and products more expensive) as the proportion of GM crop production grows in other countries. This argument is not very convincing. It is hard to see why a market of 300 million people with rather high purchasing power should not be able to simply set its market-access rules and impose the compliance costs on the plethora of smaller, poorer countries that heavily depend on the EU. In other words, variation in straightforward economic interests offers only a very limited explanation for variation in US and EU behavior in GMO policy *vis-à-vis* developing countries.

Why do the EU and the US invest so much effort in trying to influence the agri-biotech policies of developing countries? One possible answer lies in the link between the transatlantic trade dispute and domestic politics. US decision-makers appear to have been driven in part by fear of an uncontrollable spill-over process reminiscent of the "domino effect" associated with the much-feared spread of Communism during the Cold War. They may have thought that the European regulatory model would, in the absence of countervailing US action, first be emulated by developing countries with strong trade ties to the EU and, from there, would spread to other countries as well, notably those with strong trade ties to the US.

Eventually, as much of the world was moving towards stronger legal constraints on agri-biotechnology, domestic and international pressure for stricter regulation in the US would mount. That is, if most other countries imposed strong restrictions on agri-biotechnology, voters and consumers in the US would begin questioning the legitimacy of their domestic regulations. Trust in government and regulatory authorities would suffer as a result. Patricia Keilbach in fact points to this possibility in her chapter.[5] Also this explanation raises more questions than it answers. Why have the EU and the US been fighting so acrimoniously over Africa in particular?[6] Why should we

---

5   Currently, the public attitude toward GM food in the United States is far from being decisively positive. Even though 89 percent of total soybean production, 83 percent of total cotton production and 61 percent of total corn production in the United States is genetically modified and Americans consume food products derived from these crops since 1996, they seem to still have an ambiguous attitude toward GMOs according to a recent report of the Pew Initiative on Food and Biotechnology (2006). Moreover, as Patricia Keilbach points out in her contribution, many counties on the US East and West coast have declared themselves GMO free and states such as Vermont have debated legislation to ensure consumer choice through labeling of GMOs.

6   In the words of US Senator Chuck Grassley: "The European Union's lack of science-based biotech laws is unacceptable, and it is threatening the health of millions of Africans ... some EU Member States have warned that their relations with poorer countries, including those in Africa, could be harmed if those countries accept US

expect a strong spill-over effect emanating from Africa, given that exports and imports of agricultural products from/to Africa and its potential market for the technology are small? While we do not deny that straightforward economic arguments offer some insights into EU and US behavior, we submit that the following theoretical argument, centering on competition for public trust, sheds some new light on European and US behavior.

Competition for Public Trust

Virtually all advanced democracies, and notably the EU, its member countries, and the US, are characterized by pluralist interest group politics and substantial influence of the mass media on political processes and outcomes. That is, policies tend to be strongly shaped by "intermediary" politics, in which interest groups influence political agendas and policymakers' choices not just through behind-the-scenes lobbying (rent-seeking) but, increasingly, also through the mobilization of public pressure via public attention-seeking activities (Aerni 2003; Caduff 2005). Under such conditions public trust is a valuable political asset, particularly for non-elected non-state actors, such as firms and NGOs. Their capacity to create a supra-national public through organized synchronicity of protest events and issue convergence across national publics has in fact been quite successful in creating political resonance on the part of national and supranational decision-making in Europe in a variety of policy-areas (Seifert 2006). Their subsequent popularity as defenders of the public interest enables them to obtain continued access to media coverage. The mass media with its persistent self-referentiality of mass-publics and its need to embed events into personal dramas in sequels need media-savvy protest NGOs as much as NGOs need the mass media (Luhmann 1993).

Frequent media coverage of NGOs that act in public campaigns as the defenders of the rights of the poor and the environment and against powerful corporate interests, has enabled them to gain considerable public trust. Public trust can be defined in terms of the belief among political constituencies that

---

biotech food aid. Any such threats are unacceptable ..." (March 5, 2003, United States Mission to the European Union). The EU responded in an equally provocative fashion. For example, a commentary by the EU's delegation to the US was: "Neither of us can reasonably present GMO use by Africa as a miracle option ... EU bashing on the GMO case is mainly inspired by the will of the US farm lobby to find new outlets for exports. That's where the concept of food aid kicks in ... contrary to US practice, we do not export our surpluses to the needy in the developing countries ... when farm commodity prices are high on the world market, 'made in the US' food aid shrinks drastically. When prices are low–and developing countries can afford to pay–US food aid rises spectacularly! ... the European Union is by far the biggest importer of farm products from the developing countries ... The European Union spends three times more on what is called 'official development' aid than the US in terms of GDP ..." (June 13, 2003, EU Newsweb)

a particular actor or group is acting in the public interest rather than self-interest. In agri-biotech policy it also refers to the belief that a particular actor or group is telling the truth about the benefits and risks of the technology and its applications (Eurobarometer 2003).

Public trust equips interest groups with legitimacy in the public arena, which they otherwise may find hard to obtain because they are not elected in a formal democratic process. Such legitimacy, in turn, equips them with political influence, primarily in the form of discursive power: policymakers depending on election or re-election can usually expect to attract more votes, or public support in general, if they side with those non-state actors who enjoy a high degree of trust among the electorate. (Non-elected) public officials (e.g., decision-makers in regulatory agencies) can expect more political support if they side with those who enjoy high degrees of public trust. They are usually appointed by elected politicians and receive budgets from those. Moreover, firms that are very concerned about their image with consumers can also benefit from siding with interest groups that enjoy a high degree of public trust because this signals to consumers that they care about their interests and values (e.g., with respect to healthy food from sustainable agriculture).

Public trust and legitimacy are important sources of discursive power. Discursive power refers to the ability to influence norms, values, ideas, political agendas, the framing or definition of solutions to particular societal problems, and the political discourse (or non-discourse) on specific problems more broadly. This form of power is also referred to as the "third face of power" or "soft power" in parts of the political science literature. It differs from two other forms of power, namely instrumental and structural power. The latter two derive primarily from material sources, such as economic or military capabilities, whereas discursive power hinges much more on public trust and legitimacy (Parsons 1967; Lukes 1974; Koller 1991; Fuchs 2004). Many authors have in fact argued that influence on policy-input, and notably power over norms, ideas, political agendas, and rule making, has in many areas of policymaking become more consequential than conventional sources of power over policy-output (Bourdieu 1991; Akerlof and Kranton 2000).

The above notion of public trust ties in more closely with sociological notions of moral legitimacy as a source of authority and discursive power (e.g. Fuchs 2004) than with social-capital related notions of the concept. As to the latter, most authors (e.g. Putnam 1995; Hardin 2002) view public trust quite broadly as the backbone of economically prosperous and stable democratic societies. Others see public trust as a determinant of public support for certain policies or technologies, or as an indicator for success or failure of public policies. Priest et al. (2003), for example, show that "trust gap" variables predict national levels of encouragement for several biotech applications. They argue that this points to "an opinion formation climate in which audiences are actively choosing among competing claims. Differences between European and US reactions to biotechnology appear to be a result of different trust and especially "trust gap" patterns, rather than differences in knowledge or education." The extent to which a particular stakeholder enjoys public trust affects the extent to which this stakeholder's positions on biotechnology are supported by the public.

We are primarily interested in understanding the role that the quest for public trust and discursive power plays in agri-biotech policy and the geographic expansion of the controversy. Our argument on public trust and discursive power leads to *less* optimistic views on political processes and stakeholder behavior than in some other analyses of this nature. Fuchs (2004) and many other authors assume that NGOs (or civil society) have taken the lead in moralizing many policy issues, and that business has then followed. This has, so they argue, led to many coalitions between NGOs and business, private-public partnerships, the greening of industry, corporate citizenship, and so on.

Our argument views the quest for public trust, moral legitimacy, and discursive power as more conflictual. In fact, we assume that interest groups compete for public trust and try to manage it like a private resource (Aerni and Bernauer 2006). This competitive process tends to breed political polarization and radicalism, in part because competition for public trust, if the latter is treated as a private political resource, is based on exclusion ("trust us, not them"). Moreover, public trust, once appropriated by particular interest groups, is not fungible. If an interest group that enjoys a high degree of public trust is willing to make a political bargain with another interest group that enjoys a lower degree of public trust but more political or economic power, the former group runs the risk of losing public trust entirely. The public is likely to perceive this interest group to be acting in its private rather than the public interest. In other words, it is hard to exchange trust for political power or money and political compromises become difficult to achieve.

Making Sense of the Attention-Shift to Developing Countries

Public trust in science as an arbiter in domestic and international debates over regulatory policy choices has decreased over the past two decades. This has produced a trust gap in the global agri-biotech debate. Non-governmental organizations claiming to speak for public interests against the powerful interests of science, business, and government have, ever since, tried to fill this gap. The private sector and governments have also discovered in recent years that public trust can serve as a powerful political resource because it provides legitimacy and moral authority (Bernauer and Aerni 2006).

The ensuing competition for public trust among NGOs, industry, scientists, and governments has led to a shift in the public agri-biotech debate from risks in a scientific-technical sense towards worldviews and values (Eurobarometer 2003; Gaskell 2004). At the same time the issue of food sovereignty has grown in importance. This shift has been accelerated particularly by the campaigns of very large, globally active advocacy groups, such as Friends of the Earth and

Greenpeace. The trademark of such campaigns has been simple and forceful communication of worldviews and values that reduce complex and ambiguous scientific evidence on risks and benefits of agri-biotechnology to clear-cut and globally applicable good versus bad portrayals of the technology as well as associated stakeholders and their motives.

EU and US stakeholders in the agri-biotech debate have extended their battleground to developing countries. The competition for public trust and the associated shift of the debate from risks to worldviews, motives and moral claims regarding food sovereignty help us in understanding why. Large transnational networks of pro- and anti-biotech interest groups have emerged since the mid-1990s. These networks hardly communicate privately with each other. Their representatives tend to face each other mostly in the public arena where they try to win the hearts and minds of electorates and policymakers. Accusing the other side of being morally indifferent to the fate of the poor has in the past few years become an important discursive instrument to this end.

In principle, we should expect environmental and consumer NGOs to have been the first movers in pursuing strategies of moralizing agri-biotech issues with reference to poor countries. On average, NGOs have fewer resources than companies to invest in behind the doors lobbying, provide legislative subsidies in the form of expert information, or reward political decision-makers with campaign contributions or other material benefits. Also, unlike firms, NGOs cannot benefit from threatening policymakers with relocation to other jurisdictions (West and Loomis 1999). Hence we should expect NGOs to compensate for these comparative disadvantages in relation to corporate actors by moralizing agri-biotech issues in order to increase their legitimacy, appropriate more public trust, and thus increase their discursive power (Cashore 2002).

Empirical demonstration that competition for public trust has been a major driving force in extending the transatlantic biotech controversy to developing countries is difficult, for intentions are harder to identify than actual behavior. Similar difficulties exist in regard to showing empirically whether the pro- or anti-biotech side was first to carry the "feeding the poor" issue into the controversy. What seems clear, however, is that the uncompromising nature of positions in respect to risks and opportunities of agri-biotechnology in developing countries has made political compromise very difficult if not impossible (Gaskell 2004).

We have found that NGOs have engaged much earlier than industry and government stakeholders in depicting agri-biotechnology in broader moral categories that also include the right to national food sovereignty, whereas industry and government stakeholders have long focused on more differentiated arguments about risks and benefits. For instance, the right to national food sovereignty was invoked as a justification when the Zambian government decided in 2002 to reject food aid by the World Food Programme that contained GM corn, even when facing famine. Zambia's decision to ban the import of

GM food even in an emergency situation triggered a fierce, transatlantic moral debate. Europeans accused the US of using the World Food Programme as a channel to get rid of subsidized US corn and to get Africans used to GM food. The US accused Europe of using trade and aid pressure and spreading scary stories about the risks of GMOs to drive Africans into rejecting GM food. The EU was thus depicted as contributing to starvation in Africa.

Public debate over agri-biotechnology in developing countries has emerged as a popular media topic in Europe and the US. The competition for the moral high-ground in this debate reflects efforts by both sides to appropriate public trust by demonstrating to electorates and consumers that they act in the public interest, including the interest of the poor in other parts of the world, whereas the other side seeks private benefits at the expense of societal (public) welfare or nature. Thus far, anti-biotech groups have been more successful in Europe than in the US in this regard, in part because public confidence in regulatory authorities, business, and science in Europe was lower to start with (Bernauer 2003; Eurobarometer 2003). Superior discursive power of pro-biotech interests in the US has enabled these interests to effectively prevent a wider public debate (and controversy) over the risks and benefits of the technology.

Emerging Pragmatism in Developing Countries

Much of the controversy over risks and benefits of agri-biotechnology in developing countries is based on claims by supporters and opponents of the technology from advanced industrialized countries. How do their positions map onto the positions of stakeholders and consumers from developing countries.

Greenpeace and other large, international NGOs have usually started their opposition campaigns in developing countries by organizing and mobilizing domestic NGOs. Jointly, they have staged media-savvy protest actions that allude to national symbols of sovereignty and independence. In addition, they have sent out position papers to the mass media and politicians. Such papers, often signed by numerous domestic farmer and environmental organizations, call on the respective government agencies to stop serving foreign interests and take into account the concerns of their own citizens about GMOs. Often these papers invoke popular emotional triggers such as negative experiences with the Green Revolution, US and corporate imperialism, the destruction of indigenous knowledge, the potentially unknown risks to biodiversity and human health, and disrespect for national sovereignty. Ministries of trade have been accused of allowing uncontrolled imports of GM seeds and food, ministries of economic development and science and technology have been accused of serving business rather than people's interests, and Heads of State have been asked to show courage and strength in the face of growing foreign pressure to introduce GMOs. Foreign stakeholders such as Greenpeace, Friends of the Earth, Consumer International, and numerous European government agencies have also actively sponsored civil-society opposition groups that have asked their governments to act against such "foreign interests."

Once the opposition movement against GMOs had gained public trust, many politicians started to take its concerns seriously and have asked for public hearings and strict regulation, partly because of genuine concern, but also partly in the hope of gaining greater popularity. Within developing country governments an increasing divide has developed between those ministries that mainly deal with the potential benefits of agricultural biotechnology (economy, agriculture, science and technology, trade) and those that are mainly concerned with potential risks (environment, public health, indigenous affairs). Policy network analyses of the public debate on agricultural biotechnology in the Philippines, Mexico and South Africa (Aerni and Bernauer 2006) show that the core actors in such networks are represented by non-state actors—mostly an environmental NGO on the contra-side and a business association on the pro-side.

The contra-side network is clearly distinguishable from the pro-side network—and few ties connect the two. The contra-side network mainly consists of environmental organizations, farmer/indigenous rights organizations, organic food producers and manufacturers, green parties, supermarket chains, academic institutes associated with environmental sciences and sociology, national bodies concerned with biosafety, and government agencies related to health and the environment. On the pro-side, the network includes agribusiness organizations, seed companies, some NGOs, academic institutes associated with molecular biology, plant physiology and economics, and government bodies dealing with trade, science and technology, economic development and education. Interestingly, religious, farmer and consumer organizations could not always be attributed to the same camp in each country survey. Sometimes there was even disagreement within the countries themselves between two organizations that represented the same group.

The network analyses of financial cooperation in the surveys revealed that the respective political stakeholders' stance in the public debate strongly correlated with the source of funding. In this context, European governments have become influential indirect stakeholders in public debates on GMOs in developing countries by funding local anti-biotech NGOs to a much larger extent than the US is funding pro-biotech NGOs.

The (still sparse) data on stakeholder attitudes in developing countries indicates that there are significant differences. Evidence from stakeholder surveys is important in this context for two reasons: first, it responds most directly to the question of a potential mismatch between stakeholder positions as put forth in the theoretical argument; second, contingent on the particular political system, stakeholder attitudes may, in many developing countries, have a greater impact on government policy than public perceptions as such. Surveys of stakeholder perceptions and political influence in national public debates on agri-biotechnology in the Philippines, Mexico, and South Africa show that the majority of participants hold differentiated and pragmatic views on the risks and benefits of genetic engineering in agriculture, depending on the type of crop and key problems in domestic agriculture (Aerni 2002).[7] Most surveyed stakeholders thought that Europe and the US should assist researchers in developing countries in learning to use agri-biotechnology to address urgent problems in their respective countries.

The surveys also revealed in all three countries that academia remains the most trusted political actor in the public debate on agricultural biotechnology (ahead of NGOs, the media, business and government). This may be related to the fact that academia is considered to be least dependent on foreign donor money and most competent. Although academics in developing countries do not speak with one voice there seems to be an emerging consensus that an exclusive focus on the risks of agricultural biotechnology may prevent a country from developing its own homegrown biotechnology research capacity and to tackle particular domestic challenges in agriculture, public health and environmental management as well as to enable endogenous economic development. The surveys demonstrated a gap between the rather negative views expressed by government and civil society stakeholders that represent developing countries in the Western media and in the negotiations of the Cartagena Protocol on Biosafety, and the more pragmatic views expressed by a majority of stakeholders involved in the domestic debate on agricultural biotechnology in the developing countries themselves (Aerni and Bernauer 2006).

One major reason for this difference is the power of academia in domestic debates. In many developing countries where academia was successful in assuming public leadership in the debate on GMOs, public investment in agricultural biotechnology increased significantly. Countries that were able to create their homegrown capacity in agricultural biotechnology research and development (as a consequence of these investments) also experienced a shift in public attitudes from a hostile and defensive view that emphasized issues such as "food sovereignty" to a more self-confident view of biotechnology as a driver of domestic economic growth and a source of national pride. These observations are most conspicuous in countries such as Cuba, Chile, South Africa and China. Countries like Egypt, the Philippines and Mexico, where academia has only recently become more assertive on the political stage, may well go through the same shift of attitudes.

In contrast, in most countries in Sub-Saharan Africa, where academic institutions are weak and poorly connected, and the dependence on European

---

7 In these surveys the selected stakeholders were political actors who were, via a separate survey with key persons, identified as playing a significant role in national agricultural biotechnology debates. It was assumed that such persons would be well informed and would, therefore, have considerable influence on public opinion. This approach made it possible to conduct a survey on public attitudes in spite of a low level of public awareness of the subject.

market access and foreign aid is high, governments tend to follow more closely the advice of donor countries or well-endowed pressure/interest groups within the country. Further evidence for an increasingly pragmatic approach in many developing countries comes from recent trends in agri-biotech R&D and GM crop cultivation. Several developing countries, notably Colombia, Cuba, Brazil, South Africa, China, and India have invested in agri-biotechnology for several years already. In terms of R&D spending, China, followed by India, have become the leading biotech countries in the developing world; and countries like Mexico and the Philippines who were previously highly reluctant to embrace GMOs due their initial biodiversity-related concerns have taken note of recent risk-assessment studies (Yorobe et al. 2004; Ortiz-García et al. 2005) that were not able to confirm the initial claims (Quist and Chapela 2001) and adjusted their policies accordingly. Both countries have approved the cultivation of GM crops and are investing more in domestic biotechnology research and development (Possani 2003; NNC 2005).[8]

Most of the worldwide growth of GM crop acreage in recent years has occurred in developing countries, notably China, Argentina, Brazil, and India. Some of these countries have developed a range of transgenic crop varieties and are eager to address problems in domestic agriculture with the new tools of biotechnology. In these countries, agri-biotechnology is on the verge of being perceived by policymakers and electorates no more as an imported US technology but as a homegrown technology that is associated with national scientific reputation and pride.

In stark contrast to these national research efforts in some key developing countries, public investment in international agricultural research has remained very low (e.g. Cohen and Paarlberg 2004). For example, the Consultative Group of International Agricultural Research (CGIAR), which was the driving force of the first Green Revolution in the 1970s, has experienced drastic cuts in funding from Europe and North America (Aerni 2002).

However, the continuing fall of prices for agri-biotech toolkits may soon lead to more rapid diffusion of the technology and make it affordable to many poorly equipped universities in developing countries. This would strengthen domestic agri-biotech research capacities and eventually enable developing countries to use the technology to solve their own particular local problems in agriculture. This trend could be accelerated if combined with an ambitious global open source effort in agricultural biotechnology similar to the Human Genome Project and the Institute for OneWorld Health in health biotechnology ("Open Sesame. Editorial"). If this scenario prevailed an important argument of biotech-opponents from rich and poor countries would collapse—i.e., the argument that the technology is primarily an instrument to make large multinational companies from OECD countries richer and subject poor countries to their control.

---

8   http://www.isaaa.org/Publications/briefs/briefs_26.htm.

Conclusion

The WTO verdict of September 2006 was largely in favor of the US. Yet, it is very unlikely that the WTO ruling will contribute much to solving the transatlantic dispute over approval policies and market access of GMOs. Even before the transatlantic dispute entered its hot phase in the WTO, both the EU and the US had begun to extend the regulatory battleground to developing countries. This chapter describes some mechanisms through which this extention has taken place. Economic arguments leave considerable gaps in our interpretation of why developing countries have become an important element of the transatlantic dispute. Hence, we outlined a complementary argument that focuses on public trust as a political resource. This argument offers a novel view of EU and US behavior in the GMO policy area with respect to developing countries that appears difficult to understand through economic logic alone.

Our analysis suggests that the transatlantic biotech conflict and the associated competition for public trust as a political resource will continue to breed uncompromising behavior of stakeholders from rich countries vis-à-vis developing countries, at least in the short term. Both sides will probably continue to use the mechanisms discussed above to try and pull individual developing countries to their respective side. To varying degrees, poorer countries, and smaller developing countries in particular, will continue to pursue agricultural R&D strategies and regulatory models for agri-biotechnology that are largely imposed on them by advanced industrialized nations.

Yet, a variety of important developing countries (e.g., China, India, Brazil, South Africa, Argentina, and Mexico) have in recent years pursued their own, pragmatic policies. To the extent that bottom-up demand for agri-biotechnology in developing countries grows and successful indigenous applications of the technology emerge, it will become more difficult for anti-biotech interest groups from rich countries to sustain the argument that they are protecting poor countries from the greed of powerful pro-biotech states and multinational companies. Moreover, growing South-South collaboration and its business-oriented pragmatism may also challenge the influence of advanced industrialized countries on developing countries as their bargaining chips, e.g. trade preferences and foreign aid, may lose in value. In 2005, 35 percent of FDI flows into developing countries were from other developing countries. South-South FDI is growing five times faster than conventional North-South investment (Margolis 2006).

In any event, the global regulatory landscape in the agri-biotech realm, which at the beginning of the twenty-first century was still heavily bipolar (Bernauer 2003), is likely to become more heterogeneous in the longer term. The looming food crisis in particular is making developing countries again aware that they need to invest more in the productivity of their domestic agriculture. Turning the principal question of this chapter on its head, it will be interesting to see over the coming decades how the evolution of "home-grown" agri-biotech applications in developing countries will affect the policies of advanced industrialized nations in that area.

PART III
Governing Carbon:
Renewable Energy and Climate Change

Chapter 9

Promotion of Renewable Electricity in the United States and the European Union: Policy Progress and Prospects

Ian H. Rowlands

The reliable supply of electricity is vital to industrialized societies' well-being. The summer of 2003 revealed what happens when the power is cut off.[1] Almost all of Italy's 57 million citizens were left "in the dark" after a tree fell in bad weather and prompted a series of failures on power lines from Switzerland to France. A similar number of people were left without power in Ontario, Canada and much of the northeastern United States (US) after transmission line failures in Ohio sparked the largest blackout in North America's history. These episodes illuminate the importance of having a reliable supply of electricity.

Less dramatic, though potentially no less significant, are the broader sustainability consequences of electricity production. Across a variety of scales—from the local (smog) to the global (climate change), from the short-term (water pollution) to the long-term (nuclear waste disposal)—the system of electricity supply and use has a variety of sustainability impacts (e.g. Holdren and Smith 2000). Former United Nations Secretary-General Kofi Annan identified energy as one of five top priorities (Annan 2002, 11). It is now widely agreed that greater use of renewable resources in electricity supply should be part of any plan for increasing energy sustainability.

This chapter compares and contrasts US and European approaches to the increased use of renewable resources in their respective electricity supply systems. The US began to consider greater use of renewable electricity in the 1970s, and, with the help of innovative policy approaches, had become a global leader by the early 1980s. However, in the 1990s and the 2000s (at least through the end of the George W. Bush administration), the federal government in the US did relatively little to promote additional renewable electricity production. Instead, it has been individual states that have taken on this role.

European Union (EU) countries were spurred to consider increased use of renewable resources in their energy system by the oil crises of the 1970s.

1   This chapter developed from research conducted while the author was on sabbatical at the Robert Schuman Centre for Advanced Studies (European University Institute) on a Jean Monnet Fellowship in the Transatlantic Programme. The author is grateful for this support.

The relative rigidity of many of the Member States' electricity systems, however, meant that a pan-European approach to renewable electricity did not immediately follow. Instead, individual countries—like Denmark and Germany—took the initiative to promote innovative approaches to develop renewables. The mid-1990s, however, brought the beginning of electricity industry reform, growing concerns about climate change and greater interest in the employment prospects offered by local, knowledge-based industries. Together, these served to spur interest in renewable electricity at the level of the EU. As a result, Europe now has relatively extensive policy interest in renewable electricity at both the continental and the nation-state levels.

This chapter identifies the key similarities and differences with respect to issues surrounding policy support for renewable electricity in the US and the EU. The potential implications of these differences, with a particular focus on present and future transatlantic relations, are also investigated. The development and operation of electricity systems have huge economic, social and environmental impacts upon societies around the world. Given the need for some kind of "energy transition" to a more sustainable system of electricity service provision, it is important to reflect not only upon how developments within the US and EU could unfold, but also upon how interaction between the two could serve to promote, rather than inhibit, the prospects for sustainability.

Development of Renewable Electricity Strategies

United States

Electricity in the US is primarily generated by coal (49 percent of total electricity generation in 2007), natural gas (21 percent) and nuclear power (19 percent). Renewable electricity supplies only a modest amount of the country's electricity, with hydropower contributing 6 percent and all other renewable resources supplying less than 1 percent each (e.g., wood 0.9 percent, wind 0.8 percent, waste 0.4 percent, geothermal 0.4 percent and solar 0.01 percent) (all figures from EIA 2008a). Key states include those whose renewable electricity component is large in either absolute (Washington and California) or relative (Idaho, Washington and Oregon) terms. In terms of individual resources, key contributors, in absolute terms, to the overall US total include Washington (for hydropower), California (for biomass, geothermal, hydropower, solar and wind), Oregon (for hydropower), Florida (for biomass) and Texas (for wind) (EIA 2008b).

Until the late 1950s, the US was relatively energy-independent. "In the 1960s, energy consumption began to surpass energy production and by the 1970s, the US had become a major importer of energy, and especially petroleum" (Schreurs 2004b). In response to the oil crises of that decade, renewable electricity first received legislative support in the US in 1978. Congress passed the new

Public Utility Regulatory Policies Act (PURPA) requiring utility companies to purchase electricity from so-called "qualifying facilities"—that is, small-scale producers of commercial energy that met particular ownership, size and efficiency requirements. Not only were the utilities required to purchase the electricity generated by these qualifying facilities, but the terms of the price that they were obliged to pay were favorable to the small-scale producer (particularly in states like California and New York (EIA 2005, 7)). Thus, although the use of renewable resources was not a condition for qualifying facilities, the size prerequisite meant that facilities using renewable resources were usually covered by the PURPA requirements. The Union of Concerned Scientists (2002) argues that PURPA has been the "most effective single measure in promoting renewable energy." By some estimates, the Act has helped to bring on line over 12,000 MW of non-hydro renewable generation capacity (Union of Concerned Scientists 2002).

PURPA's success, however, was largely dependent upon relatively high costs for "conventional fuels" (coal, oil, natural gas and nuclear power). The Act obliged utilities to purchase electricity from qualifying facilities that could produce it for less than what the utility could produce it for ("avoided cost"). Falling prices for conventional fuels in the mid 1980s, however, effectively removed the obligation for utilities to use qualifying facilities. As a result, PURPA's catalyzing power for renewable electricity diminished. The 1980s were generally not kind to renewable electricity policies in the US: incoming President Ronald Reagan, while unsuccessful in his attempts to close the Department of Energy, did manage to virtually eliminate "funding for renewable energy research and [terminate] federal tax credits for renewable facilities" (Heiman and Solomon 2004, 96), which had been introduced in 1978 under the Energy Tax Act. During the 1980s, 90 percent of wind power installation, worldwide, occurred in California (Heiman and Solomon 2004, 97). In the next decade, however, the US lost its position as the global leader in renewable energy.

Greater attention to the issue of climate change and concerns about rising prices for conventional energy sources during the late 1980s brought renewed federal interest in renewable electricity. The Energy Policy Act of 1992 introduced the Production Tax Credit and the Renewable Energy Production Incentive, each of which provided financial advantage, on a per unit energy generated-basis over an extended period, to those who developed renewable electricity facilities. Additionally, in 1996, the Federal Energy Regulatory Commission "implemented the intent of [the 1992 Energy Policy Act] with regulatory orders that required open and equal access to utilities' transmission lines for all electricity producers, thus facilitating customer choice among different types of power generation" (Menz 2005).

The financial incentives that were introduced were potentially extremely significant. However, because they were subject to annual renewal, they did not achieve as much as they otherwise could have. The incentives were couched in uncertainty, so high levels of confidence among would-be investors were not forthcoming. As a result, many entrepreneurs were dissuaded from constructing renewable energy facilities. This, coupled with the fact that neither the 2005 National Energy Act nor the 2007 Energy Independence and Security Act contained strong provisions to support renewable electricity (for example, neither established a renewable electricity standard), is indicative of a broader lack of action and initiative at the federal level with respect to the promotion of renewable energy, generally, and renewable electricity in particular.

Dissatisfied with the lack of federal interest in promoting renewable electricity, numerous individual US states have developed and implemented legislation to promote the use of renewable electricity. Their preferred method has been renewable portfolio standards (RPS), whereby a government dictates that a certain percentage of electricity will be provided by renewable resources. Utilities are then obliged to generate the renewable electricity themselves, or in some way support others to do the same. As of May 2008, 26 US states had some kind of RPS (for more on renewable electricity policies in the United States see Aitken 2002; and Database of State Incentives for Renewable Energy DSIRE 2005; Menz 2005; Menz and Vachon 2006; Rabe 2006b; Chen et al. 2007; Union of Concerned Scientists 2007; Rickerson et al. 2008). It is the state-level policies that have had the most impact upon the development of renewable electricity in the US.

European Union

In the EU, more electricity is generated by nuclear power (31.0 percent) than any other resource. Coal (29.5 percent) and natural gas (19.9 percent) are the second- and third-largest contributors.[2] Renewable electricity contributes about 14 percent, with the vast majority of that being hydropower (9.3 percent of total electricity generation), most of which is large-scale. Wind (2.3 percent), biomass (2.2 percent), geothermal (0.2 percent) and solar-photovoltaics (0.04 percent) make smaller contributions.[3]

The amount of electricity generated by renewable resources, in both absolute and relative terms, varies widely across Member States. Together, six of the Union's 27 members—namely, Austria, France, Germany, Italy, Spain and Sweden—generate more than 72 percent of all renewable electricity, while another six—Cyprus, Estonia, Ireland, Lithuania, Luxembourg and Malta—together generate only 0.5 percent of all renewable electricity in the EU. Additionally, the share of Member States' electricity that comes from

---

2   Figures are for 2004 and refer to the 25 Member States of the EU (European Environment Agency 2008b).

3   Figures are for 2005 and refer to the 27 Member States of the EU (European Commission 2008c).

renewable resources differs significantly, from less than 5 percent (in the case of ten countries) to more than 50 percent for Austria and Sweden.[4]

The resources that make up the renewable electricity contribution are quite different across the EU. For Austria, Latvia and Slovenia, hydropower contributes more than 95 percent of renewable electricity output. The Netherlands' renewable electricity portfolio is dominated by biomass (almost three-quarters of all generation) while in Denmark wind is dominant, providing two-thirds of all renewable electricity.[5]

Energy issues were central to the formation of the European Community. In 1951, the Paris Treaty established the European Coal and Steel Community (ECSC). Six years later, one of the two Treaties of Rome established the European Atomic Energy Community. It was only in the wake of the 1973 oil crisis, however, that renewables entered intergovernmental discussions regarding energy. At this time, the EU actively sought "to expand the role of renewables in the EU energy mix" (Jansen and Uyterlinde 2004, 93). Collier (2002, 117) reports that environmental concerns arising from energy production and use were first mentioned in the 1973 "Guidelines and Priority Actions for Community Energy Policy," while the 1986 common objectives (to be achieved by 1995) for the first time included the objective to achieve balanced solutions between energy and the environment. High-level political attention to renewable electricity, however, was not forthcoming until the mid 1990s.[6] On 20 November 1996, the Commission released a Green Paper entitled "Energy for the future: renewable sources of energy" (European Commission 1996a), and the next year, a White Paper that proposed that a renewable energy Directive be published in 1998 (European Commission 1997, 15 and 34).

After a year's delay, and much discussion and debate (Rowlands 2005b), the Directive of the EP and of the Council on the promotion of electricity from renewable energy sources in the internal electricity market entered into force on October 27, 2001, obliging Member States to aim to increase the share of renewable electricity in their overall electricity supply. "Indicative" targets were set out for each Member State—together, the 15 Member States' targets amounted to a collective objective of 22 percent by 2010 (Eckhart 2004; Jansen and Uyterlinde 2004; Rowlands 2005b). National indicative targets were included for new Member States in the Accession Treaty. Thus, they now exist for all 27 Member States of the EU, and the collective objective became 21 percent. While the addition of the accession countries served to lower

---

4   Figures are for 2005 (European Commission 2007b).

5   Figures are for 2002 (European Commission 2005c).

6   Several individual Member States took action to promote renewable electricity before the mid-1990s. In 1979, for example, Denmark introduced a program to provide capital grants for the installation of wind turbines. Two years later, it introduced feed-in tariffs (guaranteed payments) for wind energy and biomass facilities. France, Germany, Ireland, Italy and Spain also introduced policies aimed at developing renewable electricity supply and capacity (IEA 2004; Reiche and Bechberger 2004).

the overall objective, this does not mean those targets will be easy to reach in the new Member States. Challenges regarding the collection of accurate information, the development of effective policy instruments and the fostering of local industries to support renewable electricity implementation continue (Bechberger and Reiche 2003; Patlitzianas et al. 2005). Still, the EU is close to being on track to meeting its collective target.

At the European level, the most important actors in the development of renewable electricity policy are the Commission and the Council of energy ministers (Lauber 2002, 26). The Commission—led by the Directorate-General Energy and Transport, with additional input from both the Directorate-General Competition and the Directorate-General Environment—has been responsible for stimulating debate and discussion, by drafting and presenting proposals. The Commission has set the agenda to which others have responded. The Council—often the location where the most heated arguments regarding policy approaches have arisen—is usually the ultimate arbiter (Rowlands 2003).[7]

The EU continues to have great ambitions for increased use of renewable resources in their electricity supply: some, such as those at the January 2004 European Conference for Renewable Energy, "Intelligent Policy Options," called for 33 percent of supply by 2020 for the 15 countries that were members of the EU at that time (Eckhart 2004). Concerning renewable energy more broadly, in March 2007, the Council declared that it endorsed a binding target of "a 20 percent share of renewable energies in overall EU energy consumption by 2020" (Council of the European Union 2007b, 21). In 2008, Member States were negotiating as to what share of this target would be assigned to electricity (for more about the evolution of EU policy, see Harmelink et al. 2006; Held et al. 2006).

Transatlantic Comparisons

The original initiator of renewable energy programs was the US federal government. Innovative actions during the late 1970s, buoyed by some individual state-level governments, resulted in the US being the global leader in renewable electricity. But by the mid-1990s, the tide had almost fully turned, and the initiative was now with others in the world, including the Europeans. As a federal-level entity, since the early 1990s the EU has done more to promote renewable electricity than the US. Building upon this transatlantic comparison, this section compares and contrasts the respective experiences.

---

7   This, of course, is not meant to dismiss the influence of other actors as well: the European Parliament, the European Court, the renewables industries' associations, the energy sector and environmental NGOs all exercise some degree of influence (Lauber 2002, 26; Thieme and Rudolf 2002; Rowlands 2003).

Similarities

There are at least three striking similarities between the US and the EU. First, although the EU has placed more emphasis on renewables than the US, renewable resources still contribute only a modest share to overall electricity requirements in both. Their respective electricity systems continue to be centrally-structured, largely dominated by fossil fuels and/or uranium resources for nuclear power.[8] If we exclude large-scale hydropower and focus, instead, upon "new renewables," that is, wind and solar-photovoltaics, then the contributions of renewables to electricity supply are even more modest. In the EU, only 2.3 percent of the electricity generated in the year 2005 was supplied through these two new renewables (European Commission 2007a); in the US, the corresponding figure was less than 0.8 percent (EIA 2008a). The relative contributions of renewables in the aggregate electricity supply are still modest on both sides of the Atlantic.

Second, efforts at promoting the increased use of renewable electricity in each jurisdiction are proceeding within the broader context of electricity industry restructuring, that is, the unbundling, introduction of competition and/or privatization into some (or all) parts of the electricity supply system. In the US, this has been driven by the Federal Energy Regulatory Commission's implementation of the Energy Policy Act of 1992 at the federal level (EIA 2004). In the EU, the 1996 Directive on the Liberalization of the Electricity Market was the main catalyst for action (Eising 2002), and since then, Directive 2003/54/EC has introduced common rules for an EU electricity market. Arguments abound regarding the impact that electricity industry restructuring has upon the progress of renewable electricity (e.g., Wiser et al. 1998; Wijnholds 2000). Supporters argue that opening up markets and empowering consumers and businesses to make their own choices will encourage uptake of renewable electricity. Skeptics maintain that competitive forces will serve to support the lowest-price generating options, which tend to externalize many of their costs upon others in the form of air pollution and the like. Regardless, it is an important phenomenon that has consequences for all electricity policy (renewable electricity included).

Finally, the policy context in both areas is characterized by interesting intra-level responsibilities. The EU is made up of numerous sovereign states, which interact not only with each other on a bilateral basis, but also within multinational bodies at the European level. While these multinational bodies continue to set important goals for the development of renewable electricity (in particular, the aforementioned Directive from 2001), Member States are able,

---

8   Of course, electricity systems are not static. In 1990, coal-fired power stations accounted for 72 percent of the UK's electricity supply; the role of natural gas was virtually negligible (0.5 percent of total supply). By the year 2000, coal's share had fallen to 31 percent, while natural gas had become the single most important resource, with gas-fired power stations meeting 39 percent of electricity demand (DTI 2004).

at least for now, to select the particular kind of policy strategy they prefer for their individual state. As a result different Member States are pursuing different strategies: Germany, for example, has long championed strategies based on "fixed prices," while the United Kingdom has often preferred policies that involve "fixed quantities" (Hvelplund 2001).

The US, meanwhile, is a single sovereign entity. Nevertheless, the existence of individual "states," each of which has significant responsibility for electricity policy, means that there exist various "layers" of government. While the federal government in the US has influence on renewable electricity policy particularly through the broader "Energy Acts," individual states have been taking the initiative to implement strategies that they perceive to be in their own interest. Most of the policy action at the state-level has occurred in the north-east, the south-west or the upper mid-west.

Differences

"Renewable electricity" is defined somewhat differently across the Atlantic. This is important because whether particular kinds of resources are included or excluded in definitions significantly affects those same resources' access to markets and level of implementation. The EU has developed a single definition of renewable electricity. Although there remains much debate within the EU (Rowlands 2005b), the 2001 Directive on Renewable Electricity defined "renewable energy sources" as "wind, solar, geothermal, wave, tidal, hydropower, biomass, landfill gas, sewage treatment plant gas and biogases" (European Parliament 2001). This definition largely mirrors what the International Energy Agency (IEA) has classified as "renewable" (IEA 2002, 8), with one key difference, namely, the way in which biomass is defined. In the European Directive, a broad definition is used; not only is the biodegradable fraction of municipal waste included (as it is in the IEA definition), but so too is the biodegradable fraction of industrial waste. This is not the case with the IEA definition.

The US is relatively less advanced than the EU in its efforts to develop a single definition of renewable electricity, for, at present, there exists no single, universal definition of "green electricity." Because state-level governments have recently taken the lead in developing their own sets of rules and regulations, a relative patchwork of definitions has resulted (Rowlands and Patterson 2002). Nonetheless, there is some evidence that agreement surrounding "what is green" is growing in the US, for some country-wide approaches are becoming increasingly prevalent. For one, Green-e, the country's largest "green power" labeling scheme, which began with programs at the state and regional level has since 2007 implemented a Green-e National Energy Standard for Green-e Certified Renewable Energy Certificates, Utility Green Pricing Programs, and Competitive Market Electricity Products. The US Renewable Energy Production Incentive, which was created in 1992, was amended in 2005 to promote generation and utilization of electricity from renewable sources. There have also been various (unsuccessful) attempts in the US Congress to pass a RPS. In each, there exists a definition of what would qualify as renewable. Although differences exist among these, some tentative trends can be identified.

When compared to the IEA definition, "renewable" is defined more restrictedly in the US in two key ways. First, with respect to hydropower, while the IEA classifies all hydropower resources as "renewable," "Green-e" includes only "small" or "low impact" hydropower. While "small" is determined by the rated capacity of the installed turbines (30 MW or less), "low impact" is determined by the environmental impacts of the station's operation, as determined by the Low Impact Hydropower Institute (Green-e 2003b). The legislation that was passed by the US Senate, moreover, only accepted as "renewable" "incremental hydropower," that is, additional to that already being produced by existing hydropower facilities.

Second, with respect to biomass, the US is also more restrictive. Not only is "renewable municipal solid waste" usually not included (as it is in the case of the IEA), but biomass generators must also adhere to strict emission limits in New England, New York and the Mid-Atlantic region in order to receive "Green-e" certification (Green-e 2003c). This is a requirement that is missing in the IEA definition.

There are also differences in the geographic scale used to develop a single set of requirements. The EU directive contains a single definition of "renewable electricity" for all Member States. Alternatively, although there are certainly national efforts at work within the US (particularly definitions by the federal government in different pieces of proposed legislation), the Green-e ecolabeling program permits "local possibilities." More specifically, sub-national groups, such as the Power Marketer Advisory Committees and Regional Advisory Committees, can introduce some changes to the national standard, so that local priorities are met (Green-e 2003a). Uncharacteristically, therefore, the US experience with "subsidiarity" is distinct from the EU's "melting pot" approach.

Finally with respect to defining renewable electricity, in the US individual definitions are emerging that are a function of the electricity system in which the (potentially) renewable resources are located. More specifically, in programs developed by Green-e (Green-e 2003c) and Scientific Certification Systems (SCS 2003), a particular resource is only determined to be "green" after its environmental impacts have been compared to the "average" impacts of the electricity system as a whole. Hence, what is "green" in a coal-dominated jurisdiction may not be "green" in a hydro-dominated one. In the EU, meanwhile, it does not matter where particular power plants are located; the definition of green is a function of the technology, rather than where that technology is sited. Taken together, therefore, not only are the elements of "what is renewable" different, but so are the approaches to the ways in which the characteristics of the electricity system, as a whole, affect the definition of renewable electricity.

There are major differences in the strategies being taken to increase the use of renewables as well. In the EU, "feed-in tariffs" have been and continue to be widely used. Also commonly called a "fixed price scheme," this consists of an obligation for utilities to purchase, at a set price, the electricity generated by renewable energy resources. Often, the price is a function of the particular technology used to generate the electricity with lower prices existing for more "cost-competitive" resources. There is no limit as to the quantity of electricity that can qualify for such a payment (see, for example, Haas 2001; Meyer 2003). By contrast, the RPS is particularly popular in the US. Moreover, at the policy level, there has been more coordinated "large-scale" activity in the EU. In the US, action is on a state-by-state basis, and regulation is at the state level (Eckhart 2004).

Additionally, the EU renewable electricity community appears to be much more international in its outlook, as compared to the equivalent community in the US. In terms of the public sector, European governments, individually as well as collectively, have been much more active in engaging other countries in discussions regarding renewable electricity strategies. This was clear at the Johannesburg Summit in 2002, at which time EU representatives proposed that 15 percent of the global demand for energy (in all forms) should be met by renewables by 2010 (Rowlands 2005a, 86–8). The Johannesburg Renewable Energy Coalition's aim of substantially increasing the use of renewable energy sources through "ambitious and time-bound targets" (European Commission 2004c) has largely been driven by Europeans. The Secretariat for the organization has been hosted by the Commission. The Europeans proposed holding regular international conferences on renewable energies in Johannesburg and Germany, with support from other key European governments, hosted the first conference in Bonn in June 2004. Meant to galvanize international action on renewable energy, it also served to further sensitize the global community to Europe's related technical and commercial strengths. Continuing its push for renewables, the German government hosted a conference in April 2008 to pursue international discussions on the development of an International Renewable Energy Agency (IRENA) (Federal Government of Germany 2008) with the argument that much as there is an International Atomic Energy Agency, there can and should be an international agency dedicated to the promotion of renewable energies, especially in developing countries.

The US, by contrast, has exhibited much less interest in engaging others on this issue. At the Johannesburg Summit in 2002, for example, US representatives argued that a "one size fits all" approach would not be sufficiently flexible to increase the use of renewable energy ("Summary of the World Summit" 2002). Moreover, only eight official delegates attended the Bonn conference in 2004; the meeting's profile in the US was low. By contrast, the United Kingdom, alone, had an official delegation of 43; even Mexico had 18. The US did not participate in the IRENA conference. This said, there are some indications of a growing embrace of the need for a stronger US leadership role in renewable energy. It is noteworthy, for instance, that the third International Renewable Energy Conference was held in Washington DC in March 2008. Various legislative initiatives have been pursued in the House of Representatives and the Senate, encouraging greater support for renewables, such as with HR 5529, a bill introduced into committee in March 2008 instructing the President to establish an International Renewable Energy Agency.

Turning to the private sector, European companies are becoming increasingly active in international renewable energy activities, many of which have a transatlantic character. Scottish Power, for example, is not only the UK's largest wind energy developer, but it also has two US subsidiaries, one of which (PPM Energy) is the second largest wind power developer in that country (Fischer 2004). Other European businesses are complementing their expanding European operations in renewable energy through similar activities in the US, either by retaining the European name (for example, Italy's Enel formation of Enel Green Power in Europe, but also its purchase of CHI Energy Inc. and subsequent formation of Enel North America) or by purchasing and maintaining the separation of existing entities (for example, France's EDF SSIF Energies). Perhaps driven by saturating markets in Europe, it is clear that European companies are looking west to the US much more frequently than US companies are looking east to Europe. Ambrose (2003) argues that US companies will probably "move slowly to globalize their renewables strategy .... [T]hey face significant competition within their home markets on a state-to-state basis." He concludes that they may form partnerships with European companies, but it will be mainly in order to facilitate European companies' entry into the US market.

From Policy Divergence towards Policy Convergence?

Why have these differences with regard to renewable electricity, particularly in terms of policy and prospects, emerged between the EU and the US? Many maintain that the power of the "environmental agenda" in Europe is pushing the renewable electricity agenda forward. In particular, because European countries have ratified the Kyoto Protocol, and are moving forward with carbon reduction strategies, they are more willing to consider "aggressive targets for renewables deployment" (Bird et al. 2005, 1398). By contrast, the US, which declared that it would not ratify this international treaty, was not similarly motivated to take action. Similarly, the presence of a relatively high level of environmental consciousness, including concern about local air quality issues, may be serving to catalyze action on renewable electricity in Europe much more so than in the US (Moore and Ihle 1999; Schulz 2000, 128; Johansson and Turkenburg 2004, 14).

Resource availability may also be an important explanatory factor. Schulz (2000, 128) argues that in Germany during the 1970s, the primary motivator for greater interest in renewable energy was "reduced dependency on imports and the consciousness about the limitation of the stock of fossil fuels." Johanson and Turkenburg (2004, 14) maintain that the same applies to Europe, writ large. Europe's reserves of fossil fuels are, of course, limited: EU countries, together, hold less than 2 percent of the world's proved natural gas reserves (the US has 3 percent) and less than 1 percent of the world's proved oil reserves (the US has 2.7 percent). What might be more consequential for electricity, however, is a comparison of coal reserves: while the EU countries have approximately 11 percent of the world's proved coal reserves (largely in Germany), the US has fully 25.4 percent. Indeed, as the world's second largest producer of coal (after China), the US could go on producing coal at its present rate for over 250 years. Therefore, the need to move to alternatives, at least for electricity generation, is not as immediate for security reasons (BP 2004).

Resource availability relates to costs. In 2005, households in Denmark were paying more than three times as much for electricity as those in the US; electricity prices in Italy in 2005 for industrial customers were also three times as high as they were in the US. And for companies interested in expanding electricity generating facilities, coal prices in 2005 in the US were less than half what they were in Germany (figures taken from EIA 2007). In comparison, cheaper conventional fuel prices and the reality that other countries have the edge in those alternatives may have dampened interest in renewables on the part of the US.

Yet, the biggest difference between the EU and the US may have been a question of political will and leadership. In Europe, the Commission and Council, backed by several key Member States—especially Germany—pushed hard to win support across Europe for greater investment in renewables. The German government, beginning in 1990, began a steady series of policies to promote renewables. In 1990 the Germans introduced a Feed-In Tariff Law, requiring utilities to buy renewable electricity from producers, whether private individuals or companies. In 2000, a new Renewable Energy Sources Act was passed providing investors with a guarantee of 20 years of favorable feed-in tariffs (Lauber and Mez 2006). As the largest economic power within Europe, Germany's policy decisions had a major influence on European discussions. The Commission, moreover, was more than willing to promote renewable energies. In contrast, there has been a lack of federal leadership in the US on renewable electricity since the 1990s.

Differences in governmental structures may be particularly important in explaining this different level of political will. Menz (2005, 2409) argues that it is "well known that energy industry constituencies (coal, oil, and natural gas) exert an important influence on federal policy in the United States." Moore and Ihle (1999) echo this view, arguing that, in Europe, "coal and oil firms hold less political influence than they do in the US." Similar ideas regarding transatlantic differences over climate change have been advanced (Bodansky 2003).

There are signs, however, that this is changing. With oil prices steadily rising since the mid-2000s, interest in alternative energy sources, including renewable electricity is growing. There has also been a strong push from below—from US states—and from environmental advocates worried about climate change. And since the 2006 mid-term Congressional elections, there has been a push within Congress as well. Support for a new approach to energy policy is evident in the 2008 presidential elections with both Barack Obama and John McCain calling for action on climate change and greater support for renewable energies.

This is having an impact at the federal level as well. In the 110th Congress (2007–2009) over 360 bills on energy efficiency and renewables had been introduced as of mid-2008 (Sissine et al. 2008). In contrast, in the 107th Congress (2001–2002) only about 70 bills were introduced into Congress related to climate change, including renewable energy legislation. Policy proposal are beginning to echo those seen in Europe. In May 2008 the US Department of Energy issued a report, 20 percent Wind Energy by 2030: Increasing Wind Energy's Contribution to US Electricity Supply (United States Department of Energy 2008).

Europeans still have the advantage over the US in renewables. This is most evident in the area of wind power, one of the foremost renewable electricity technologies. At the turn of the century, not only was almost three-quarters of the world's windpower capacity presently in the countries of the EU (with only 15 percent in the US) (AWEA 2003), but "90 per cent of the world's wind turbine manufacturers [were] based in Europe" (Asmus 2002). According to the European Wind Energy Agency (2008), in 2004, European companies still accounted for 82 percent of the global market for wind turbines. Germany, Denmark, and more recently Spain have been the dominant players within Europe. This early lead has given European companies an economic advantage (Rowlands 2003). Hvelplund argues (2002, 66) that Denmark's remarkable progress in renewable electricity can be credited to a "green innovation" development process. This was achieved by "means of an active collaboration between some politicians who recognised that an active energy policy was necessary and possible, new small private firms, and [a] grass root[s] energy movement." He maintains that success occurred when "Denmark had a government and Minister of Energy who were willing to listen to independent lobbyists and also resist the lobbyism of the established fossil fuel companies" (2002, 74).

But here too there are signs of change. In 2007, Germany alone accounted for 23.6 percent of global installed wind energy capacity. The US, however, has rapidly expanded its wind energy sites, dominated by developments in two states: Texas and California. In fact, in 2007 the US installed more new wind energy capacity than anywhere else in the world (Global Wind Energy Council 2008). This suggests that in the future, the gap that began to emerge between US and EU policy for renewables during the 1990s and 2000s could narrow.

The extent to which it narrows will in part be a question of economics, but also in part a question of political leadership.

Transatlantic Trade and Renewable Electricity

As renewable electricity increases in importance, there is a potential that it will become a matter of growing importance in transatlantic trade relations. Renewable electricity can be found in three different kinds of goods, namely, "electrons" (or carriers of electrons), goods with "embedded electricity" or renewable electricity technologies. Unless greater attention is given to dialogue, trade disputes in any or all of these goods could emerge.

Transatlantic trade disputes could arise, for example, as a result of the trade in "electrons" that are deemed "renewable" by some, but not by others. Electricity is not usually distinctly identified because of the characteristics of the "good" itself (that is, the physical attributes of the electrons). In other words, when the electrons arrive at the place of use, they are all identical. Instead, particular kinds of electricity are demanded because of the manner in which these same electrons were "created." In this case, "renewable electricity" comes from power plants that are deemed to use "renewable resources." Potential challenges arise, however, because not everyone defines "renewable" (or, the related term, "green") in the same manner. Consider the following hypothetical example.

Imagine that a company in jurisdiction A is producing electricity from large-scale hydropower facilities. These facilities are deemed "renewable" under the terms of that jurisdiction's electricity regulations. The company uses the electricity to produce hydrogen, which it then ships across the Atlantic Ocean, to a subsidiary in jurisdiction B. There, the hydrogen is recombined with oxygen in a fuel cell to generate electricity. With no harmful emissions of any kind, the subsidiary assumes this electricity is "renewable" and it is marketed as such. Others, however, argue that because the electron's entire "life-cycle" includes large-scale hydropower—something that does not qualify, for the sake of this example, as "renewable" under jurisdiction B's electricity regulations—the electricity should not be able to compete in any special "renewable electricity" market in jurisdiction B.

If policymakers in jurisdiction B excluded those electrons from part of their domestic market, then we could envisage policymakers in jurisdiction A launching a trade challenge at the World Trade Organization (WTO). They could adopt at least two different lines of argumentation. First, they may argue that the "products"—"their" electrons generated by the oxidization of hydrogen and the "other" electrons generated by what are deemed to be "renewable processes" in jurisdiction B—are alike. Hence, the principle and associated rules of the world trading system regarding "national treatment" (that is, treating imported and locally-produced goods equally) means, they would continue to argue, that all these electrons have to be treated identically in jurisdiction B's markets.

Second, if policymakers from jurisdiction A accepted that the entire lifecycle of their electrons could be scrutinized, they might claim that there is nothing in their electrons' processes and production methods that justifies any restriction upon their importation. In other words, they are just as "environmentally-friendly" as those various resources that qualify as "renewable" under jurisdiction B's electricity regulations. They could cite the fact that some argue that hydropower facilities, because there is no combustion needed to generate electricity, have little harmful atmospheric emissions associated with them, even when considering the entire lifecycle (Gagnon 2003). They would thus maintain that policymakers in jurisdiction B could use neither GATT's Article XX(b) exception nor its Article XX(g) exception to justify the trade restriction.[9]

For their part, policymakers in jurisdiction B may counter that the lifecycle should be subject to scrutiny. They might cite recent WTO Appelate Panel rulings to give support to their claim that "a measure in which products are distinguished based on non-product-related processes and production methods ('NPR-PPMs') could satisfy the requirements of Article XX (g)" (Appleton 1999, 492).[10] Given this, they could then argue that the hydropower facilities do have implications for climate change (e.g., McCully 2002). They could then develop a case that the global climate is an "exhaustible resource" and therefore maintain that their restriction upon electricity generated by large-scale hydropower is defensible (compare with Biermann 2001).

At this point, of course, it remains entirely academic, for there is "no GATT/WTO case law on trade in electricity" (Horlick et al. 2002, 8). Instead, the point is to show how trade disputes could arise.

Perhaps more immediate will be conflict over the trade of goods claiming to contain "embedded" renewable electricity; in other words, goods that have been manufactured with what its producers argue are renewable electricity inputs. These same producers may then advertise this fact in their marketing and/or packaging—perhaps through some kind of green label.

To illustrate this possibility, consider the large-scale hydropower example. Instead, however, of using the electricity from the hydropower dam in jurisdiction A to produce hydrogen, imagine, instead, that it is used to manufacture aluminum soft drink cans. Given that large-scale hydropower is deemed to be a "renewable" resource in this jurisdiction, the soft drink producers decide to put a "green power" label on the can (or, alternatively, they may decide to use

---

9   Article XX is entitled 'General Exceptions'; paragraph (b) considers measures 'necessary to protect human, animal or plant life or health' and (g), measures 'relating to the conservation of exhaustible natural resources if such measures are made effective in conjunction with restrictions on domestic production or consumption' (Sampson and Chambers 2002; Steinberg 2002).

10   The case that has attracted the most attention is the so-called "Shrimp/Turtle case" (Appleton 1999; Sands 2000).

a more general "eco-label," or simply some kind of environmental claim). They then sell the soft drink cans in jurisdiction B.

What may well greet them in jurisdiction B is a labeling program that only allows certain kinds of electricity to be called "renewable" or "green." Large-scale hydropower, in this example, does not qualify, so policymakers in jurisdiction B may well demand that the label on the soft drink be changed before it is sold in their markets. This is, of course, not without precedent, for various countries already have a range of different requirements with regard to labeling, such as bilingual details on all retail packaging or nutritional information on food products.

Nevertheless, the soft drink producers in jurisdiction A may well dispute this. They may maintain that the GATT's Technical Barriers to Trade Agreement clearly states that labels should not be an excuse for protectionism. Hence, they may argue that these labeling requirements are ultimately driven by market-access concerns.

Broader developments suggest that this is more than simply an academic discussion. For one, ecolabeling with respect to process and production methods is an issue in international relations. Not only did Austria's attempts to restrict the import of "unsustainable timber" during the 1980s prove controversial (Sucharipa-Behrmann 1994), but more recent debates surrounding the labeling of genetically-modified food shows that this issue continues to have a high profile (Crespi and Marette 2003). Additionally, given that Australia's green power program allows for the special labeling of products "made with" particular kinds of electricity (SEDA), many of the conditions for the kind of conflict laid out above appear already to be in place. Thus, differences in the definition of "what is renewable/green" may be of consequence.[11]

Disputes could also arise because of different standards regarding renewable electricity technologies—that is, the devices used to "create" electrons. Let me elaborate with a hypothetical example. There are potentially significant environmental burdens associated with different parts of the lifecycle of a photovoltaic (PV) panel. In particular, because toxic and hazardous substances— for example, cadmium, hydrogen selenide, phosphorous oxychloride and tellurium—are sometimes used in the manufacture of PV panels or are actually found within the PV panels, their escape into the atmosphere could have health and environmental consequences (IEA 1998). Given that all jurisdictions would have regulations regarding the proper use of these substances, there could conceivably be differences in these rules. This, in turn, could lead to conflicts, for manufacturers of PV panels in jurisdiction A, who find their panels shut

---

11   The WTO website (http://www.wto.org/english/thewto_e/whatis_e/tif_e/bey2_e. htm) reports that: "One area where the Trade and Environment Committee needs further discussion is how to handle — under the rules of the WTO Technical Barriers to Trade Agreement — labelling used to describe whether for the way a product is produced (as distinct from the product itself) is environmentally-friendly" (also: WTO 2003).

out of jurisdiction B's markets for "health and safety reasons," may claim that jurisdiction B's actions are actually disguised protectionism.

Although there are efforts afoot within international society to develop accepted standards regarding renewable electricity—for the case of PV panels, PV GAP (based in Switzerland) and the International Engineering Consortium (based in the US) have played key roles—these kinds of disputes are possible. Indeed, a similar hypothetical example from wind power, which is often identified as one of the two "greenest" possible sources for electricity, could also be developed, perhaps regarding acceptable levels of noise.

These kinds of disputes are certainly foreseeable, particularly given that we often see "technology laggards" impose different kinds of standards than those being promoted by the "technology leader" in order to protect and defend their markets (for such a discussion with examples from television standards in Europe, Austin and Milner 2001). With the stakes being as high as they are (renewable energy markets could soon be worth billions of dollars (Rowlands 2003)) and with apparent differences in technological capabilities between the EU and the US, such conflicts are distinct possibilities.

There is also the prospect for transatlantic cooperation on issues related to renewable electricity to the benefit of those on both sides of the Atlantic and the global community. Indeed, differences of the kind identified above can also be reflected upon positively, because mutual learning can result. Actions can be compared and the respective costs and benefits of alternative approaches assessed. Better decisions can then be made, on the basis of these multiple experiences. Thus, potential potholes can be avoided, or successes adopted earlier than would otherwise have been the case. Comparisons that could be drawn include:

- definition of what is "renewable," including consideration of "absolute" (strict) versus "relative" (more-nuanced) approaches;
- whether the definition of "renewable" should be a function of the resource characteristics of the local electricity grid;
- targets for renewable electricity;
- support schemes for renewable electricity.

"How might the possible advantages of transatlantic cooperation be reaped, while the possible flashpoints ameliorated or avoided altogether?"

Experience at the national level suggests that it is only when multi-stakeholder groups, with broad representation, work together to tackle renewable electricity challenges that the most sustainable results ensue (Rowlands and Patterson 2002). Given the ways in which electricity impinges upon everyone's lives and the keen interests that many have in the issue, governments, businesses and civil society all require a "seat at the table." This message should be taken to the international level. Accordingly, it should not be solely government ministers, civil servants, businesspeople, or representatives of environmental groups exploring transatlantic issues in renewable electricity. Instead, a dialogue of representatives from across these sectors is needed.[12]

To date, exploration of this issue has been relatively modest. Notwithstanding some occasional mentions (for example, by the now-disbanded Transatlantic Environment Dialogue (TAED 2000)), renewable electricity has received relatively little attention in transatlantic relations.[13] For the reasons identified in this chapter (particularly given the various possibilities elaborated earlier in this chapter), this needs to change, so that both disputes and collaborations can be effectively anticipated and acted upon.

The aforementioned "International Conference for Renewable Energies," which was held in Bonn in 2004, gave support to the notion of international collaboration on renewable energy. Policy recommendations that emerged from the conference included the need to "strengthen global cooperation," which would include a "regular exchange of information regarding programming experience, target setting and evaluations between different countries [which] would support rapid progress and reduce the risk of mistakes" (ICRE 2004, 12). Additionally, it was recommended that institutional arrangements at the international level could be strengthened so that "pooling of information" could take place and so that "common standards" could be promoted. Although there continues to be debate about whether these responsibilities (and others) should be formalized in a new international organization or not, there is wider agreement that international activity in these areas is required.[14] The subsequent international conferences—not only the aforementioned one held in the US in 2008, but also the one held in China in 2005—continued to focus attention upon international aspects (at least in a comparative sense) of renewable energy.

Some businesspeople have also echoed these sentiments. The former head of British Petroleum, for example, recommended that "US policy-makers should also consider establishing a transatlantic partnership to work toward a common market-based trading system" (Browne 2004, 30). He also recommended

---

12   My comments should not be equated with a call for transatlantic harmonization on renewable electricity. There are powerful arguments for subsidiarity (Rowlands and Patterson 2002). Ttransatlantic cooperation may involve agreement to have different approaches on each side of the Atlantic.

13   There are other transatlantic linkages. The American Council on Renewable Energy (ACORE) "is continuing to build bridges to the European renewable energy community" (Eckhart 2004). ACORE encourages cross-fertilization of ideas; it hosted a "Renewable Energy Finance Forum" on Wall Street (a joint venture with Euromoney) in order to bring many of Europe's top bankers to share experience with the US. Additionally, many groups associated with the International Energy Agency bring together representatives from the US and the EU.

14   The European Association for Renewable Agencies (EUROSOLAR) and the German government have called for an International Renewable Energy Agency (IRENA), part of a broader debate surrounding international institutions to support renewable energy (e.g., Steiner et al. 2004).

increased international collaboration on research and development (Browne 2004, 30–31). Additionally, the American Institute for Contemporary German Studies has launched a new Forum for Atlantic Climate and Energy Talks (FACET),

Notwithstanding the potential utility of a new institution like IRENA, existing organizations should not be overlooked, for they might be appropriate catalysts for this task. Consider the Global Ecolabelling Network (GEN), a non-profit association of third-party, environmental performance labeling organizations. It was founded in 1994 to "provide a forum for information exchange and cooperation between organizations operating ecolabeling programs" (GEN 2003, 2). Its membership includes representatives from 26 countries and the Commission, the US is also a member. With its technical expertise in lifecycle analysis and trade issues, as well as its procedural experience in international collaboration, it may be well placed to move this agenda forward.

The stakes regarding transatlantic relations on renewable electricity are high, environmentally, economically and politically. Environmentally, the world's present system of supply and using electricity generation is unsustainable; it is estimated that approximately 37 percent of the world's industrial-related carbon dioxide (CO_2) emissions come from electricity generation (Innovest 2003). Economically, societies depend upon a reliable supply of electricity in order to function smoothly. This US $1.5 trillion a year industry is a central cog in the global economic engine. And politically, the tone of the transatlantic relationship on all issues (renewable electricity included) has important implications not only for the world's foremost trading relationship and, by extension, the entire global economic system (Sutherland 2001), but also for international diplomatic relations.

Renewable electricity has continued to occupy an only modest role in electricity production in both the US and the EU although this is rapidly beginning to change, While there is considerable similarity in goals and even performance among individual states in the US and the EU (such as California and Germany), at the federal level, policy support for renewables has been stronger in the EU than in the US. Whereas the EU is projected to be securing between 18 and 21 percent of its electricity from renewables by 2020, and California has a target of 20 percent by 2017, the US as a whole has no such target. Global climate change, the environmental effects of conventional electricity generation, energy security concerns and energy costs have combined to encourage regulatory action in the EU. While policy progress has certainly been made in some individual US states, overall, the US lags behind the EU on promotion of renewable electricity. Greater political leadership at the federal level could, however, give the US a chance to strengthen its performance.

With rising energy prices, growing awareness of climate change, heightening concerns about energy security, and greater political attention at the highest levels of government, renewable electricity production can be expected to expand rapidly on both sides of the Atlantic. Institutions should be employed to promote multi-stakeholder discussions on this pivotal transatlantic issue. Given the environmental, economic and political stakes involved, renewable electricity deserves much greater attention in the transatlantic relationship.

Chapter 10

Conflict and Cooperation in Transatlantic Climate Politics: Different Stories at Different Levels

Miranda A. Schreurs, Henrik Selin, and Stacy D. VanDeveer

On May 31, 2002, Jaume Mata Palou, Minister of the Environment of Spain and Margot Wallström, Environment Commissioner presented European instruments of ratification of the Kyoto Protocol to the United Nations (European Commission 2002b, c). EU ratification of the Kyoto treaty was greeted with celebrations in Brussels and many other European cities (Castelfranco 2005). When the Kyoto Protocol entered into force on February 16, 2005, it had been ratified by 141 countries and continued to enjoy much European support.[1] In early 2007, the Commission furthermore stated the need to go beyond the Kyoto targets, proposing to cut EU greenhouse gas (GHG) emissions by 20 percent below 1990 levels by 2020 in an effort to contain the global average temperature increase to no more than 2°C (European Commission 2007c). Leaders from all 27 Member States accepted this plan (Council of the European Union 2007a). The EU has since added a goal of achieving a 20 percent share of renewable energies in EU energy consumption and a 20 percent improvement in energy efficiency by 2020. The combined goals are now referred to as "20 20 20 by 2020" (European Commission 2008a).

In sharp contrast to the EU policy developments, the White House Press Secretary, Ari Fleischer on March 28, 2001 announced: "The President has been unequivocal. He does not support the Kyoto treaty" (The White House 2001a). Environmental Protection Agency (EPA) Administrator Christine Todd Whitman reaffirmed this, stating that the US has "no interest" in implementing the treaty (Pianin 2001). The announcement of US withdrawal from the Kyoto Protocol was greeted with dismay and anger in Europe. President Bush's first visit to Europe in June 2001, to attend an EU summit, was met by public protests as 15,000 people demonstrated in the streets of the host city, Gothenburg, Sweden (CNN 2001). Greenpeace also blocked a tanker that was trying to deliver US owned oil to Le Havre, France (Greenpeace 2001). While protestors had broad agendas beyond climate change, it is clear that the Bush

---

1   http://unfccc.int/essential_background/kyoto_protocol/status_of_ratification/items2613.php.

administration's approach to climate change contributed to a significant cooling in transatlantic relations.

Differences between Washington and Brussels on climate change policy since the 1990s have been stark, often described as symbolic of a deep "climate divide" across the Atlantic (Schreurs 2002; Busby and Ochs 2004; Jozwiak and Crowley 2004; Schreurs 2004c, 2005b, a; Cass 2006). Yet, when the transatlantic relationship is viewed at a sub-federal level, the divide across the Atlantic becomes less visible and the similarities among a host of public and private sector actors (and the connections between them) are more profound. For example, the climate change positions and actions of Germany and California are more similar to each other than are those of Germany and Greece or California and Mississippi. Just as EU Member States can be differentiated in terms of their level of commitment to climate change action, so too can states and municipalities within the US (Selin and VanDeveer 2007): Why did the EU move forward with the Kyoto Protocol and expand action after 2000, when the US did not?

Washington's position on mandatory GHG regulations is slowly changing, however. On February 28, 2007 House Majority Leader and California Democrat, Nancy Pelosi stated: "We hold our children's future in our hands ... Scientific evidence suggests that to prevent the most severe effects of global warming, we will need to cut global GHG emissions roughly in half from today's level by 2050 ... We cannot achieve the transformation we need ... without mandatory action to reduce greenhouse gas pollution." She concluded her remarks, "For twelve years, the leadership in the House of Representatives stifled all discussion and debate on global warming. That long rejection of reality is over ..." (Pelosi 2007). In the lead-up to the 2007 G8 Summit, President George W. Bush proposed a long term climate change strategy targeting the world's 15 largest GHG emitters (Stolberg 2007). What explains this burgeoning change after many years of opposition to mandatory GHG regulations?

This chapter analyzes and compares major developments in European and American climate change policy making. More specifically, it examines the roles played by individual Member States and EU organizations in the development of EU climate policy and focuses on the role played by sub-national actors in contributing to a shift in the US climate change debate and policy making. In short, it argues that in Europe, substantial federal leadership on climate change combined with that found within several (but not all) Member States was critical in developing EU climate change policy. In the US, an absence of federal leadership (and even strong federal policy opposition) hindered the development of similar national policies. Increasingly, however, the federal policy vacuum is filled by US states and municipalities taking action in the absence of meaningful federal policy, starting to narrowing the policy gap between Europe and the US (Selin and VanDeveer 2009a).

The chapter begins by outlining key components of the climate change regime. Next, it discusses the lack of leadership by federal policy makers in

Washington that has characterized much US climate change politics since the adoption of the Kyoto Protocol over a decade ago. This is followed by an analysis of the development of European climate change policy from the 1990s to the present. The chapter continues with an examination of developing sub-national climate policy in the US in response to federal inaction, which is followed by a discussion about the ways in which European and US climate change action are, in fact, becoming increasingly related as transatlantic connections are growing at a rapid pace outside the realm of the formal relations between Washington and Brussels. The chapter concludes with a short discussion about the future of transatlantic climate change relations.

The Global Climate Change Regime

Global climate change is caused largely by growing concentrations of GHGs in the atmosphere. Current atmospheric CO_2 concentrations are approximately 390 parts per million by volume (ppmv), up from 280 ppmv at the beginning of the industrial revolution. Concentrations are currently growing at a rate of 3 to 4 ppmv annually. Since the beginning of industrialization, average temperatures at the Earth's surface have increased by approximately 0.8°C. Projections of future warming by the Intergovernmental Panel on Climate Change (IPCC) suggest a global increase of 1.4°C to 5.8°C by 2100 if GHG emissions are not reduced. Abrupt and dramatic climatic changes, increases in sea levels, and changes in precipitation, including more frequent floods and droughts, are likely effects of unabated global warming (IPCC 2007a, b).

The US and Europe are responsible for the vast majority of all historic anthropogenic GHG emissions. They remain two of the three largest emitters of GHGs from fossil fuel use and cement production, behind China (Table 10.1). The US, with a population of over 300 million (or approximately 5 percent of the global population) emitted 21 percent of global GHG emissions from these two major sources in 2007. The EU-15 (those Member States originally bound by Kyoto Protocol reductions) was responsible for 12 percent. Yet, differences in emissions between industrialized and developing countries are much greater on a per capita basis. US per capita emissions were 19.4 tons in 2007, followed by 8.6 tons in the EU-15, and 5.1 tons in China. If every person lived like the average American, annual global GHG emissions from fossil fuel use and cement production would be roughly four times higher than they are already.

Global climate change policy is outlined in two major treaties: the United Nations Framework Convention on Climate Change (UNFCCC) and the associated Kyoto Protocol.[2] The UNFCCC was adopted at the United Nations Conference on Environment and Development in Rio de Janeiro in 1992, entering into force in March 1994. It has been ratified by both the EU and the

---

2   http://unfccc.int.

Table 10.1   Top three global emitters of GHG emissions from fossil fuel use and cement production in 2007

Country/region	Percent share of global emissions	Tons per capita
China	24	5.1
US	21	19.4
EU-15	12	8.6

Source: http://www.mnp.nl/en/publications/2008/GlobalCO2emissionsthrough2007.html.

US. The UNFCCC sets the long-term objective of "stabilization of greenhouse gas concentrations in the atmosphere at a level that would prevent dangerous anthropogenic interference with the climate system" (Article 2). Article 3 established the principle of "common but differentiated responsibilities" – all countries share an obligation to act, but industrialized countries have a particular responsibility to take the lead in reducing GHG emissions.

The Kyoto Protocol, adopted in 1997, builds upon the UNFCCC and sets GHG emission reduction targets for world's major industrialized countries, which are listed in Annex I. The Protocol covers six GHGs: carbon dioxide, methane, nitrous oxide, perfluorocarbons, hydrofluorocarbons, and sulfur hexafluoride. Annex I countries agreed to different reduction commitments. The 15 states that were then members of the EU agreed to a collective 8 percent reduction in their GHG emissions relative to 1990 levels by 2008–2012. The US accepted a 7 percent reduction compared to 1990 levels within the same time frame. The Kyoto Protocol outlines five options by which Annex I parties can meet their targets: develop national policies lowering emissions; calculate benefits from domestic sinks that soak up more carbon than they emit;[3] participate in emissions trading schemes with other Annex I countries; create joint implementation programs where Annex I parties get credit for projects that lower emissions in other Annex I countries; and engage the clean development mechanism where Annex I parties get credit for lowering emissions in non-Annex I countries.

The Kyoto Protocol required the ratification of 55 parties to the UNFCCC to enter into force, including Annex I parties accounting for at least 55 percent of that group's CO_2 emissions in 1990. The Kyoto Protocol entered into force on February 16, 2005 when parties representing 61.6 percent of total CO_2 emissions by Annex I countries had committed themselves to the agreement. Following Australia's ratification in 2008, the US became the only major industrialized country that refused to ratify the treaty. Significantly, GHGs emissions in the US and the EU-15 countries with a collective Kyoto target have developed

---

3   The concept of carbon sinks is based on the natural ability of trees, other plants and the soil to absorb carbon dioxide and "store" it in wood, roots, leaves, and the soil.

along largely diverging paths since the early 1990s (Figure 10.1). By 2006, EU-15 GHG emissions were about 2.2 percent below 1990, while emissions for the whole EU-27 were down by 7.7 percent (EEA, 2008). While this remains short of the EU's Kyoto commitment, it stands in stark contrast to the 14.7 percent growth in US emissions between 1990 and 2006 (EPA 2008).

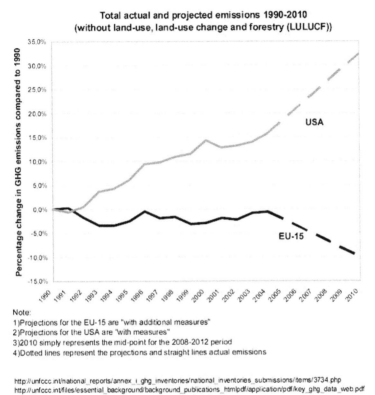

Total actual and projected emissions 1990-2010
(without land-use, land-use change and forestry (LULUCF))

Note:
1)Projections for the EU-15 are "with additional measures"
2)Projections for the USA are "with measures"
3)2010 simply represents the mid-point for the 2008-2012 period
4)Dotted lines represent the projections and straight lines actual emissions

http://unfccc.int/national_reports/annex_i_ghg_inventories/national_inventories_submissions/items/3734.php
http://unfccc.int/files/essential_background/background_publications_htmlpdf/application/pdf/key_ghg_data_web.pdf

Figure 10.1   GHG emissions trends in the United States and EU-15

Source: Bruton (2007).

The climate change regime is in contested development. At a large meeting in Bali, in December 2007, UNFCCC parties met to draw up a broad "roadmap" for the development of a follow-up agreement to the Kyoto Protocol, which expires in 2012. The Bali meeting is followed by a series of negotiation sessions. Many countries have expressed a desire to adopt a new treaty at a meeting in Copenhagen, scheduled for December 2009. However, the world's countries are deeply divided on many key issues, including the size of needed global short-term and medium-term GHG reductions; which countries should accept mandatory emission controls; the allocation of national reduction commitments; the calculation and use of carbon sinks; continued development of the joint implementation and the clean development mechanisms; and financing, technology transfer and capacity building issues.

A Lack of Leadership in Washington

Vice President Albert Gore has been credited with saving the stalled Kyoto Protocol negotiations in December 1997 (Ott 1998). After his last minute intervention in Kyoto, US negotiators shifted their negotiation stance, enabling Annex I countries to reach an agreement on emission reduction targets. US representatives entered negotiations saying that the most the US could accept was to commit to a stabilization target; they came out having agreed to a 7 percent reduction relative to 1990 levels by 2008–2012. This suggests that the Clinton administration had a genuine interest in addressing climate change. US agreement to a 7 percent reduction target at Kyoto, however, was anomalous in an otherwise fairly consistent set of signals from Washington against the kind of mandatory cuts in GHG emissions supported by the EU. Instead, US climate politics has been characterized by a lack of federal leadership.

At the United Nations Conference on Environment and Development in 1992, President George H.W. Bush opposed mandatory CO_2 emission reduction targets in the UNFCCC. In July 1997, the US Senate voted 95-0 in support of a Sense of the Senate Resolution sponsored by Republican Senator Chuck Hagel and Democratic Senator Robert Byrd stating that the US should not sign any protocol negotiated in Kyoto, or thereafter, which would: "mandate new commitments to limit or reduce greenhouse gas emissions ... unless (it) also mandates new specific scheduled commitments to limit or reduce greenhouse gas emissions for Developing Countries within the same compliance period," or "result in serious harm to the economy of the United States." (United States Senate 1997) Even though the Clinton administration signed the Kyoto agreement, there was little chance that the Senate would support ratification during the remainder of Clinton's term (Harris 2000a).

The arrival of the George W. Bush administration in the White House in January 2001 effectively ended any remaining hope that Washington might support the Kyoto Protocol after all. On March 13, 2001, soon after becoming president, Bush sent a letter to four Republican Senators stating that "I oppose the Kyoto Protocol because it exempts 80 percent of the world, including major population centers such as China and India, from compliance, and would cause serious harm to the US economy." He stated: "I do not believe ... that the government should impose on power plants mandatory emissions reductions for CO_2, which is not a 'pollutant' under the Clean Air Act" (The White House

2001b). President Bush thereby joined a majority in the US Congress opposing any kind of national mandatory GHG controls.

Opposition in Washington emanated from fossil fuel producers, automobile manufacturers, and other heavy industries (Kraft and Kamieniecki 2007). It was embodied in campaigns led by conservative think tanks, such as the Cato Institute, the Heritage Foundation, and the American Enterprise Institute, whose influence was substantially strengthened after the 1994 Republican Congressional election victories (McCright and Dunlap 2003; Jacques et al. 2008). In an illustrative summation of conservative views, a 2001 Heritage Foundation report listed five "fundamental flaws" of the Protocol: 1) it is based on "faulty science;" 2) has "unrealistic targets;" 3) has "misdirected objectives" (because it places too much emphasis on CO_2 and not enough on other heat-trapping gases); 4) "exempts developing countries;" and 5) would have "severe economic consequences" (Coon 2001).

In early 2000s, there were a few attempts in Congress to regulate GHG emissions. The 2003 Climate Stewardship Act sponsored by then-Democratic Senator Joseph Lieberman and Republican Senator John McCain, which aimed to stabilize CO_2 emissions at 2000 levels by 2010, attracted the most attention. It nevertheless failed to muster enough votes to pass (the vote was 43 to 55) (Pew Center on Global Climate Change 2003). Instead, Congressional opponents succeeded in adding amendments to appropriation bills prohibiting the use of funds "to propose or issue rules, regulations, decrees, or orders for implementation, or in preparation for implementation, of the Kyoto Protocol." This amendment effectively prohibited federal agencies from pursuing any regulatory rules related to GHGs.[4] The number of officials in the EPA working on climate change was dramatically cut, and many left their posts.[5]

In February 2002, President Bush announced the US Global Climate Change Initiative stating a goal to cut US GHG intensity, the ratio of GHG emissions to economic output, by 18 percent of 2002 levels by 2012. This was to be done through voluntary agreements with industry, development of renewable energy technologies, energy conservation, and support for nuclear energy and clean coal technology (The White House 2002). Critics quickly pointed out that, while the Global Climate Change Initiative might slow the growth in GHG emissions, it would not reverse emissions trends. Supporters of climate change action also questioned the goals and effectiveness of the Asia-Pacific Partnership on Clean Development and Climate, formulated in July 2005 on a US initiative. It brings together the US, Australia, Canada, China, India, Japan, and South Korea around voluntary approaches and technology development to address

---

4   There were numerous such amendments to appropriation bills signed into law using this language. An example is Environmental Protection Administration Office of Research and Development and Science Advisory Board Act of 1999 (HR 1742).

5   Discussion between Schreurs and an Environmental Protection Agency official, April 13, 2007.

climate change. This initiative produced few concrete initiatives or results in its first three years

Further crystallizing the mood in the White House and other parts of the executive branch against national mandatory GHG restrictions, the EPA in September 2003 rejected a petition that was filed in 1999 by 19 environmental and renewable energy groups and state level bodies, asking it to regulate "greenhouse gas emissions from new motor vehicles under §202 of the Clean Air Act." EPA declined, arguing that "the Clear Air Act does not authorize the EPA to issue mandatory regulations to address global climate change," and that even if it did "it would be unwise to do so at this time." EPA concluded that regulation of automobile emissions would conflict with President Bush's "comprehensive approach" to climate change and it could hamper the President's ability to persuade key developing countries to reduce GHG emissions.[6]

While a growing number of ambitious climate change bills including mandatory GHG reductions were introduced in Congress after 2005, they all failed to pass during the 2005–2007 and 2007–2009 sessions. President Bush also made it clear that he was likely to veto any such bill. Furthermore, the EPA continued to oppose national mandatory GHG reductions even after the Supreme Court ruled in 2007 that CO_2 could be classified as a pollutant under the Clean Air Act (Barringer and Yardley 2007). At the 2007 Bali meeting, US negotiators signed onto the Bali roadmap as a way forward toward developing the global regime, but only after protracted negotiations on language on GHG reduction and financial and technological support to developing countries. It will be up to President Barack Obama to attempt to finalize the international negotiations and decide whether a new treaty will be submitted to the Senate for possible ratification as well as to work with Congress to develop domestic GHG emission reduction policies.

Europe, Kyoto and Climate Change Policy

EU climate change policies have been shaped by a combination of domestic policy measures, bargaining among Member States, and the efforts of the Commission and the EP to stimulate policy change (Harris 2007). A shifting group of EU states have taken the lead in developing progressive national environmental policy programs and promoting a more aggressive EU climate change policy (Schreurs and Tiberghien 2007). In turn, EU climate change policies have influenced the development of national policies even in the pioneer states (Andersen and Liefferink 1997). In the early 1990s, Germany, Denmark, the Netherlands, and Sweden were among the European states that took the

---

6   Justice John Paul Stevens, Opinion of the Court, Massachusetts, et al., *Petitioners v. Environmental Protection Administration et al.*, April 2, 2007.

lead in setting GHG emission reduction goals, introducing carbon taxes, and/or promoting renewable energy generation and use.

In Germany, the collapse of the East German socialist economy led to a substantial drop in national GHG emissions. In addition, there was broad political support for climate change action. Already in 1990, Germany introduced a law requiring utilities to purchase renewable energy.[7] A coalition government between the Social Democrats and the Green Party in 1998 introduced an ecological tax reform reducing mandatory social security contributions and increasing taxes on fossil fuels (albeit with some controversial exceptions). The government also added further incentives for renewable energy expansion and promotion of public transportation (Lauber and Mez 2006). In 2007, the German government announced plans to cut domestic emissions by 40 percent below 1990 levels by 2020. By 2006, German emissions were over 18 percent below 1990 levels.

Denmark has invested much in renewable energy sources in an ongoing effort since the 1973 oil shock to break its dependence on imported fossil fuel. By the early 2000s, Denmark was one of the world's largest producers of wind turbines, holding 40 percent of the global market and obtaining 20 percent of its electricity from wind power. Denmark has also pursued fuel switching from coal and oil to natural gas and in 1992 introduced a CO_2 tax on fuels (OECD 2001a). Sweden obtains almost one third of its total energy supply from renewable sources, and has also instituted an ecotax reform and aims to reduce its CO_2 emissions by 50 percent of present levels by 2050. In addition, Sweden has introduced favorable tax policies for environmentally-friendly cars and biofuels and green certificates for promoting renewable electricity (Government Offices of Sweden 2007). Sweden reduced its GHG emissions by almost 9 percent between 1990 and 2006, while national gross domestic product grew by 44 percent (Miljödepartementet 2008).

More recently, the UK has emerged as a climate policy leader. A switch from coal to natural gas for electricity production in the 1990s helped bring down national GHG emissions. Beyond this, the UK introduced a climate change tax (raising the tax on fossil fuels and offsetting this with a reduction in National Insurance Contributions) in 2001 (DEFRA 2007a). The UK also pioneered the world's first economy-wide CO_2 emissions trading scheme in 2002 (DEFRA 2007b). In March 2007, the government of Tony Blair announced a Climate Change Bill, which calls for a reduction in CO_2 emissions of 60 percent by 2050, with an interim target of 26 to 32 percent by 2020.[8] Prime Minister Gordon Brown, succeeding Blair in the summer of 2007, continued to focus

---

7   "Act on Feeding Renewable Energies into the Grid of 7 December 1990," *Federal Law Gazette* I, 2663, amended by article 3 of the Act of 24 April 1998 (*Federal Law Gazette* I, 730, 734). http://www.wind-works.org/FeedLaws/Germany/ARTsDE.html.

8   Department for Environment, Food, and Rural Affairs, "Climate Change Bill Summary," http://www.defra.gov.uk/environment/climatechange/uk/legislation/pdf/CCBill-summary.pdf.

much attention on climate change. British GHG emissions were 15 percent below 1990 levels by 2006 (Table 10.2).

Beyond the domestic measures adopted by environmental leader states, the Commission, the EP, and the Council have each accepted the idea that climate change is a policy area where Europe can and should lead. Europe adopted a CO_2 stabilization target for 2000 relative to 1990 emission levels. In 1991 the Commission introduced a proposal for an EU strategy to limit CO_2 emissions (European Commission 1991). The Commission also played a central role in working out an EU negotiating strategy for Kyoto and in developing the burden-sharing agreement. Once the EU-15 committed to an 8 percent reduction in its combined GHG emissions relative to 1990 levels by 2008-12, a burden-sharing agreement for implementing this commitment was developed. This 1998 burden-sharing approach set individual targets for each EU Member State based on their different economic and geographic profiles (Table 10.2).

The burden-sharing agreement placed the heaviest reduction burdens on Germany (–21 percent) and the UK (–12.5 percent), the two largest fossil fuel

Table 10.2   The 1998 burden-sharing agreement and GHG emissions trends between 1990 and 2006 for the EU-15

Member State	1998 burden–sharing target (percent)	Change in GHG emissions 1990 to 2006 (percent)
Austria	–13	+15.1
Belgium	–7.5	–5.2
Denmark	–21	+2.1
Finland	0	+13.2
France	0	–3.9
Germany	–21	–18.2
Greece	+25	+27.3
Ireland	+13	+25.6
Italy	–6.5	+9.9
Luxembourg	–28	+1.0
Netherlands	–6	–2.0
Portugal	+27	+40.7
Spain	+15	+50.6
Sweden	+4	–8.7
United Kingdom	–12.5	–15.1
EU-15		–2.2

Source: European Environment Agency 2008a.

consumers in Europe. The other large emitter, Italy, assumed a smaller goal (–6.5 percent) based on its less developed economic status than the UK and Germany. Other states agreeing to large cuts are often considered among Europe's most environmentally progressive: Austria (–13 percent), Denmark (–21 percent), Luxembourg (–28 percent), and the Netherlands (–6 percent). The agreement allowed less economically well-off Member States (Greece, Ireland, Portugal, and Spain) to increase their emissions substantially, recognizing their need to grow in order to improve their economic status within the EU. It placed smaller demands on states with already high energy efficiency levels (Sweden) or low levels of CO_2 emissions (France). However, as shown in Table 10.2, Member States differ greatly in their progess toward their regional target.

Since 2004, 12 more states have become EU members: Poland, Hungary, the Czech Republic, Slovakia, Estonia, Latvia, Lithuania, Slovenia, Cyprus, Malta, Bulgaria and Romania. These countries, with the exception of Cyprus and Malta, have individual Kyoto commitments. As EU officials were preparing ratification of the Kyoto Protocol, they wanted to ensure that enlargements would not weaken the 8 percent reduction Kyoto commitment by including new Member States in the burden-sharing agreement, as most of them had experienced sharp drops in GHG emissions following the collapse of their energy intensive Soviet-style economies (see Table 10.3). Still, the accession increased incentives for Member States to invest in GHG reduction measures in these neighboring countries and to link the joint implementation mechanism to the EU's GHG trading scheme. It also increased the ability of the EU to assert global leadership on climate change (Carmin and VanDeveer 2004).

Table 10.3   Kyoto targets and emission trends in the 12 Member States joining in 2004 and 2007

Member State	Kyoto target	Change in GHG emissions 1990 to 2006 (percent)
Bulgaria	–8	–38.9
Cyprus	None	+66.0
Czech Republic	–8	–23.7
Estonia	–8	–54.6
Hungary	–6	–20.0
Latvia	–8	–56.1
Lithuania	–8	–53.0
Malta	None	+45.0
Poland	–6	–11.7
Romania	–8	–36.7
Slovakia	–8	–33.6
Slovenia	–8	+10.8

Source: European Environment Agency 2008a.

In 2000, the Commission proposed the *First European Climate Change Programme* outlining measures necessary to ratify and implement the Kyoto Protocol. It also led efforts that resulted in the adoption of a renewable energy directive in 2001 under which Member States agreed to a goal of raising their joint share of renewable energy sources in their total electricity consumption from the 1997 level of 13.9 percent in 1997 to 22.1 percent by 2010 (European Parliament 2001). In 2007, the Commission won approval from the Council for expanding renewable energy in the total energy mix from the 2007 level of 6.5 percent to 20 percent by 2020 and a binding minimum target for all Member States to have 10 percent of their transportation fuel consumption supplied by biofuels within the same time frame (Council of the European Union 2007c). The Commission also proposed limits for vehicle emissions at 130 g CO_2/km by 2012 (Euractiv 2007a). By 2008, however, EU officials were rethinking some of their biofuels goals, proposing to scale back mandates in response to concerns that rapidly increasing biofuels use was contributing to food price increases and deforestation around the globe (Kantor 2008). In December 2008, the European Union changed the target from a 10 percent biofuels target to a 10 percent renewable fuels target by 2020.

In 2005, the Commission launched the *Second European Climate Change Programme* with much activity focusing on the EU emissions trading scheme (ETS). During the Kyoto negotiations, the US was a strong supporter of emissions trading in the face of considerable European skepticism. Yet, it was the EU that launched the first international CO_2 trading scheme. The ETS went into effect in January 2005 covering over 11,500 energy-intensive installations in its first phase (2005–2007). The second trading period operates from 2008 to 2012. The ETS got off to a rough start as allowance prices spiked (initially) and then collapsed, due to the over allocation of permits. As such, the Commission has pushed states to reduce their allocation of permits and emissions caps for the second period are 13.6 percent below the first period cap for the EU-15 and 8.7 percent below these countries 2005 emissions (for the EU-27, these figures are 13.1 percent and 5.9 percent respectively) (Ellerman and Joskow 2008). The Commission has also proposed to include CO_2 emissions from aviation in the ETS and to support research on carbon capture and storage.

This account is not meant to suggest that there has been no opposition to these and other climate change related policy ideas. The Commission-led effort to create an EU-wide carbon tax in the late 1990s, for example, failed due to the inability to obtain unanimous support of all Member States (Wettestad 2005). Several Member States have strongly resisted EU efforts to lower their ETS allocations. It is also not a given that the EU can succeed in its bid to be a climate change leader (Gupta and Grubb 2000). The Commission argues that the EU-15 are on target to meet the goal of an 8 percent reduction as long as existing measures are supplemented by use of the Kyoto flexibility mechanisms.[9] Nevertheless, it is too soon to say with certainty that the EU-15 will meet their Kyoto target, as GHG emissions have increased beyond the burden-sharing targets for many Member States with Spanish emissions increasing by over 50 percent (Table 10.2).

Continued EU action on climate change is guided by a regional political consensus that global average temperature increases should not exceed +2°C (but may be higher than that in some regions). Scientists believe that this requires stabilizing atmospheric GHG concentrations below 550 ppmv, preferably close to 450 ppmv. Under this scenario, global GHG emissions would peak around 2020 and then would have to fall substantially. The Commission argues that global emissions should be cut by 50 percent from 1990 levels by 2050. To this end, the EU during the 2007 Bali meeting and subsequent negotiation sessions trying the create a Copenhagen Protocol have pushed hard to have its 20 percent reduction from 1990 levels by 2020 accepted by other countries. The Commission furthermore has stated that the EU would accept a 30 percent reduction below 1990 levels by 2020, if this goal was accepted by other industrialized countries including the US.

Growing US Sub-national Leadership on Climate Change

Transatlantic comparisons tend to focus on national level politics. Yet, in the case of climate change, much of the most interesting policy activity in the US since early 2000s has occurred at state and municipal levels and in the private sector (Selin and VanDeveer 2007, 2009a). Reactions to Washington's dismissal of the Kyoto Protocol and its opposition to mandatory GHG regulations were not only strong in Europe, but also in many environmentally progressive states and cities across the US. In the absence of federal leadership, and sometimes motivated by European examples, a substantial number of US states have taken initiatives that go beyond federal requirements and adopted numerical targets for short-term and long-term GHG reductions (Table 10.4). By 2007, more than half of all US states had formulated individual climate change action plans.

States are taking a host of energy related actions including establishing RPS requiring electricity providers to obtain a minimum percentage of their power from renewable sources. By 2008, 27 states had enacted such standards. Many states are formulating ethanol mandates and incentives, and at least ten will adopt California's CO_2 vehicle emission standards if they survive legal challenge. Californian officials are also preparing new standards designed to reduce the carbon content of fuel (Farrell and Sperling 2007). In addition, states are updating energy codes, adopting green building standards and mandating the

---

9   EU ambassador letter to Senator Boxer in 2007.

sale of more efficient appliances and electronic equipment (Selin and VanDeveer 2007; Rabe 2008). A few states, including California, Oregon, and Colorado have also established climate change and renewable energy initiatives as a result of supportive referendums (Rabe 2008).

Table 10.4   US state-wide GHG reduction targets (as of August 2008)

State	Goal	
Arizona	2000 levels by 2020	50% below 2000 levels by 2040
California	2000 levels by 2010	1990 levels by 2020; 80% below 1990 levels by 2050
Connecticut	1990 levels by 2010	10% below 1990 levels by 2020, 80% below 2001 levels by 2050
Florida	2000 levels by 2017	1990 levels by 2025, 80% below 1990 levels by 2050
Hawaii	1990 levels by 2020	
Illinois	1990 levels by 2020	60% below 1990 levels by 2050
Massachusetts	1990 levels by 2010	10% below 1990 levels by 2020
Maine	1990 levels by 2010	10% below 1990 levels by 2020
Minnesota	15% below 2005 levels by 2015	30% below 2005 levels by 2025, 80% below 2005 levels by 2050
New Hampshire	1990 levels by 2010	10% below 1990 levels by 2020
New Jersey	1990 levels by 2020	80% below 2006 levels by 2050
New Mexico	2000 levels by 2012	10% below 2000 levels by 2020; 75% below 2000 levels by 2050
New York	5% below 1990 levels by 2010	10% below 1990 levels by 2020
Oregon	Stabilize by 2010	10% below 1990 levels by 2020; 75% below 1990 levels by 2050
Rhode Island	1990 levels by 2010	10% below 1990 levels by 2020
Utah	2005 levels by 2020	
Vermont	1990 levels by 2010	10% below 1990 levels by 2020
Washington	1990 levels by 2020	25% below 1990 levels by 2035; 50% below 1990 levels by 2050

Source: Pew Center on Global Climate Change.

States also began to explore the possibility of enacting collaborative standards and policies on GHGs even before the Bush administration withdrew from the Kyoto Protocol. In July 2000, the Conference of New England Governors (Maine, New Hampshire, Vermont, Massachusetts, Rhode Island, and Connecticut) and Eastern Canadian Premiers (Nova Scotia, Newfoundland and Labrador, Prince Edward Island, New Brunswick, and Quebec) adopted a resolution recognizing climate change as a joint concern that affected their environments and economies. Out of this emerged a 2001 Climate Change Action Plan. Under this plan, states and provinces pledged to reduce their GHGs to 1990 levels by 2010 and 10 percent below 1990 levels by 2020. They also agreed to ultimately decrease emissions to levels that do not pose a threat to the climate, which according to an official estimate would require a 75–85 percent reduction from 2001 emission levels (Selin and VanDeveer 2005).

A second major multi-state initiative in the Northeast, the Regional Greenhouse Gas Initiative was proposed in 2003 (Selin and VanDeveer 2009b). Beginning in 2009, it creates a cap-and-trade scheme for CO_2 emissions from major power plants in the participating states. In 2007, Maryland became the tenth RGGI member, joining Maine, Vermont, New Hampshire, Massachusetts, Rhode Island, Connecticut, New York, New Jersey, and Delaware. RGGI is designed to stabilize CO_2 emissions from the region's power sector between 2009 and 2015. Between 2015 and 2018, each state's annual CO_2 emissions budget is expected to decline by 2.5 percent per year, achieving a total 10 percent reduction by 2019. While the goals of both the regional initiatives in the north-east are relatively modest, they are more stringent than federal policy. Significantly, many state officials explicitly framed the regional efforts in terms of influencing future federal policy and public and private sector views in the US related to climate change (Selin and VanDeveer 2007).

While the Northeast regional cooperation remains the most well developed to date, states on the West Coast – plus Arizona and New Mexico – have launched discussions about how best to facilitate further climate change cooperation as well. The discussions include possibly establishing a cap and trade system of their own. In 2007, the largest group of states to join in a climate change related policy initiative was announced: 31 states signed on as charter members of The Climate Registry. This registry is a collaborative effort to develop a common system for private and public entities to report their GHG emissions, allowing officials to measure, track, verify and publicly report such emissions in a consistent manner across states.[10] By 2008, 39 states, three native tribes and the District of Columbia were members—together with seven Canadian provinces and six Mexican states.

States' joint efforts to push Washington to act have taken other forms as well. Attorneys General from California, Connecticut, Illinois, Maine, Massachusetts, New Jersey, New Mexico, New York, Oregon, Rhode Island, Vermont and Washington, in February 2003 filed suit in federal court challenging a decision by the EPA during the Clinton administration to not classify CO_2 as a vehicle pollutant to be regulated under the US Clean Air Act. The suit was endorsed by numerous state regulatory agencies, city officials and environmental groups. In April 2007, the US Supreme Court ruled five to four that CO_2 can be classified as a pollutant under the Clean Air Act and that EPA should reconsider its

---

10   http://www.theclimateregistry.org.

decision not to regulate CO_2 emissions from vehicles. The ruling energized those in Congress and elsewhere who are pushing for the adoption of more aggressive national climate change policy as legal struggles continue (Rabe 2009; Selin and VanDeveer 2009a).

California has emerged as a leading force in climate change policy development and GHG mitigation (Farrell and Hanemann 2009). In July 2002, the California State Assembly passed the California Climate Bill (AB1493), which was signed into law by Governor Gray Davis in August. This regulation mandated the California Air Resources Board to establish a plan for achieving "maximal feasible reduction" of CO_2 emissions from vehicles, effective 2006. Car makers have been given until 2009 to meet the new standards, but US car manufacturers have challenged the standards in courts. California also established a Renewable Energy Portfolio standard in 2002, with the aim of achieving 20 percent of its energy come from renewable resources by 2017. The target date was subsequently shifted to 2010 in California's first Energy Action Plan. Governor Schwarzenegger has since announced a 33 percent renewable energy goal for electricity by 2020.

There is also considerable US action at the municipal level. By mid-2008, 830 mayors from all 50 states representing approximately 80 million Americans had signed a declaration of meeting or exceeding the reductions negotiated in Kyoto for the US. Over 260 North American municipalities are members of the International Council for Local Environmental Initiatives (ICLEI) and its Cities for Climate Protection program (CCP).[11] While many municipal climate change programs are modest, some have received impressive results (Gore and Robinson 2009). American municipalities are also increasingly developing new GHG reduction and energy efficiency programs that rely in part on innovative private financing (Foy and Healy 2007; Mishra 2007; Palmer 2007; Revkin and Healy 2007). In addition, a growing number of US firms are seeking to reduce GHG emission and investing in low carbon technology (Jones and Levy 2009).

Sub-National Influences and Transatlantic Cooperation

Two things become clear about transatlantic climate politics upon a closer look: First, policy developments and achievements in both Europe and North America are more varied than conventional wisdom suggests. The picture often emerging from journalistic and activist accounts of a "green" Europe and a "brown" North America is too oversimplified, obscuring understanding rather than enhancing it. Second, transatlantic politics of climate change are very

---

11   The CCP program functions as a tool for decentralized dissemination of policy ideas and lessons on climate change related issues across municipalities that may not otherwise interact. Betsill and Bulkeley (2004).

different at the inter-governmental level than among a host of actors at various sub-national levels and in private and civil society sectors. While relations between EU representatives and US national officials have been tense for years, many sub-national public officials, private firm representatives, and staffers of advocacy organizations are increasingly exchanging knowledge, expertise and policy ideas across the Atlantic.

As such, developing climate change policy making in many US states and European countries is increasingly based on the same combination of moral and strategic reasoning (Selin and VanDeveer 2007). Moral arguments ("it is the right thing to do") for more aggressive GHG reductions are underpinned by an acceptance of the science behind human-induced climate change and a sense of intergenerational responsibility to act for the benefit of future generations. Strategic arguments ("it is the economically sensible thing to do long-term") for expanding climate change policies are based on growing belief that local, national, and international GHG controls will only grow in scope in the future. These policy developments will create both economic limitations and possibilities for public and private sector entities in Europe, North America and elsewhere.

The growing pressures emanating from US states and municipalities, and related pressure from a growing number of civil society organizations, for the federal government to act more forcefully on climate change including setting national mandatory GHG reduction goals have intensified in recent years (Betsill and Bulkeley 2004; Rabe 2004; Selin and VanDeveer 2007, 2009a). The 2006 mid-term Congressional and state elections that brought Democrats into control of the US House and Senate, a majority of governorships, and an increased number of state legislatures across the country opened a window of opportunity for supporters of climate change policy. The US Supreme Court added to these growing national and sub-national pressures when it ruled in 2007 that the EPA has the authority to regulate CO_2 emissions under the Clean Air Act.

The expansion of state, municipal, and private sector climate change action has a growing number of public and private sector actors in the US calling for the enactment of uniform federal standards. By 2007, there were nine bills in Congress, most with multiple co-sponsors, calling for the establishment of a CO_2 cap and trade scheme (Point Carbon 2007a). In April of that year, the bi-partisan National Commission on Energy Policy revised it recommendation to government officials by increasing the GHG emissions cut in its recommendations (Point Carbon 2007b). Also in 2007, the Democratic controlled Congress and President Bush raised automobile efficiency standards – so called CAFE standards – for the first time in decades, increased a number of energy efficiency standards on other products, and played host to a few proposals for carbon taxes (Dodd 2007; Eilperin and Mufson 2007; Motavalli 2007).

Even as more aggressive climate change policies are debated in the US Congress, sub-national pressure continues. As Republican Governors Schwarzenegger (California) and Rell (Connecticut) put it in a joint op-ed in the *Washington Post*: "California, Connecticut and a host of like-minded states are proving that you can protect the environment and the economy simultaneously. It's high time the federal government becomes our partner or gets out of the way" (Schwarzenegger and Rell 2007). In addition, efforts by major European-based companies to reduce their GHG emissions and reduce energy use are increasingly being copied by North American-based companies. North American firms are also voluntarily trading CO_2 emissions credits at the Chicago Climate Exchange and the Montreal Climate Exchange, modeled in part after European trading schemes and exchanges (*The Economist* 2006).

US state action is furthermore influenced by European efforts. For example, Massachusetts officials have proposed feed-in tariffs similar to those in Germany and other European countries as a means of increasing renewable energy generation.[12] States involved in the development of RGGI are in repeated contact with European officials to mine lessons from the EU ETS on how to establish an effective GHG trading scheme.[13] The RGGI states also show that they have learned what *not* to do from the EU: in contrast to Member States handing out almost all allowances for free to participating installations, the RGGI states will auction off a substantial share of allowances. This will both generate funds to be used for public investments in cleaner technology and energy efficiency, and help the market operate more efficiently as states try to avoid European mistakes of overallocation of allowances (Selin and VanDeveer 2009b).

Like the states operating RGGI, California is drawing on European expertise in developing a cap and trade scheme to reduce GHG emissions from power plants in California as well as those beyond its borders that sell power in the state. In addition, Governor Schwarzenegger of California entered into an agreement with then British Prime Minister Blair in 2006 on cooperating on clean energy technology research and development and sharing knowledge of best practices related to Europe's emissions trading system with the goal of possibly linking developing California and European emissions trading systems in the future (Blair and Schwarzenegger 2006). Other European countries are also engaging in collaborative efforts with California, and joint ventures between European countries and US states around climate change and renewable energy are likely to grow in the future. Sub-national policy makers can furthermore cooperate with each other beyond transatlantic networks, bringing Australian or Chinese actors into cooperation arrangements, for example (Koehn 2008).

---

12   Telephone interview with David Cash, Executive Office of Energy and Environmental Affairs, Commonwealth of Massachusetts, May 18, 2007.

13   Selin and VanDeveer interviews with RGGI participants, various dates, 2005–2007; http://www.rggi.org.

By 2007, California and EU climate policies were on similar paths. Both EU and California state policies include short and long terms caps on GHG emissions, efforts to establish low and/or zero carbon fuel standards, RPS, existing or developing cap and trade schemes for utilities and some industrial emissions, large scale energy efficiency programs and goals and frequent calls for more stringent US federal policies. All of these efforts have attracted the attention of policy makers, private sector actors and NGOs across North America, Europe and parts of Asia. At the municipal level, numerous organizations such as ICLEI, The Climate Group and the Clinton Foundation were helping to build networks connecting major US cities with municipal climate-related initiatives and other exemplary programs in Europe and elsewhere.[14]

Yet the climate change mitigation impacts of US states and municipalities to date should not be exaggerated. Much state implementation remains spotty (at best) as many states are struggling to meet their relatively modest short-term GHG emission reduction goals (Rabe 2008). In addition, they cannot meet their more long-term GHG emissions reduction goals without further efforts to reduce emissions from the two major sources of GHG emissions, transportation and electricity production. Of course, state policy efforts would be greatly helped by the formulation of supportive national policies and programs. However, US sub-national and federal climate change policy will likely develop in parallel for several years to come, resulting in a situation of complex multilevel governance (Selin and VanDeveer 2007, 2009a). If so, US multilevel climate change governance will increasingly resemble that of the EU.

Concluding Remarks

The EU and the US have spent the better part of the past decade disagreeing on the appropriateness of the Kyoto Protocol as a mechanism for addressing global GHG emissions. The EU has led with the development of relatively ambitious policy goals and has limited the growth of CO_2 emissions far more effectively than the US. The EU and its Member States have initiated numerous national and supranational programs aimed at implementing the Kyoto Protocol and other EU policies out of a concern about global warming. At the same time, several Member States are struggling to bring down emissions and meet targets.

EU climate change policy leadership stems from a combination of actions by a small group of Member States and initiatives by EU level organizations. Green Member States like Germany, Denmark, Sweden, and the United Kingdom, pushed by environmentally-minded publics, pioneered the development of much European progressive energy and climate change policy in the 1990s. Over the past decade, the Commission, the Council, and the EP have exercised important leadership alongside Member States toward the implementation of the Kyoto

---

14   http://www.thelcimategroup.org and at http://www.clintonfoundation.org.

Protocol. Such leadership has been based on an acceptance of the need for precautionary action. In addition, Member States and EU organizations collaborated closely in the creation of the ETS and the formulation of more ambitious energy and GHG reduction goals beyond the Kyoto commitment period.

In the US, in contrast, there has been strong opposition to the Kyoto Protocol and mandatory GHG emissions reductions from the Bush administration and Congress since the 1990s. The federal government initiated several research and development programs into new technologies that could reduce GHG emissions in the long-term, but refused to take substantial actions that would cut emissions in the short-term. There has also been strong US industry opposition including from leading energy and automobile companies with close ties to federal policy makers opposing the Kyoto Protocol and federal regulations. There has moreover been limited public debate in the US, and the climate change issue was not until recently high on the national political agenda. In addition, much US media representation has portrayed the scientific community as much more divided than what it actually is.

Where leadership on climate change has been absent in Washington, however, local and state governments have been stepping in to fill the policy void. Many of the climate change and renewable energy policies emerging at the sub-federal level resemble the policies and programs of the EU and its pioneer states. Much as was the case in Europe years earlier when a few pioneer states pushed Europe toward action on climate change, in the US a growing number of states,and local governments are pushing from below. They are changing the dominant discourse on climate change and in the absence of federal leadership, have begun to take the policy lead and change US climate change politics from below, but these efforts have yet to result in significant reduction in GHG domestic emissions (Selin and VanDeveer 2007). Nevertheless, their efforts have been (and continue to be) significantly influenced by European debates and policy examples.

Generally, the US tends to be unwilling to act globally and accept international standards on environmental issues unless it first takes domestic action (DeSombre 2005). If the many climate change policy developments at local and regional levels and in the private sector can push the US federal government to adopt a more aggressive GHG policy, then US foreign climate change policy is likely to also be affected, as coalitions of environmentalists, public officials, and private sector actors push the US government to upload aspects of the new US domestic policy on climate change to the international level (Selin and VanDeveer 2007). If this occurs, the EU and the US may find it easier in the future to cooperate more effectively in addressing what is one of the gravest environmental and development threats ever facing the planet.

The EU continues to support a legalistic approach based on multilateral legal agreements, while the Bush administration championed the idea of voluntary partnerships, as evidenced by the Asia-Pacific Partnership on Clean Development and Climate. In the lead up to the 2007 G8 Summit, the Financial

Times reported; "Political tensions between the US and Germany over climate change have worsened sharply, with Washington threatening to no longer 'tread lightly' in negotiations on global warming … Germany wants the summit to agree on carbon emission limits but the US says climate change should be tackled with technology-based solutions rather than mandatory emissions targets …" (Williamson and Luce 2007). The 2008 G8 Summit achieved agreement on the idea that GHG emissions must decline by half by 2050 – the first time President Bush accepted a long-term numerical reduction goal. However, no baseline year was established and no shorter term goals or actions were included in the declaration.

Because of the political and economic importance of Europe and North America, and the fact that they are large emitters of GHGs, transatlantic relations are of significant importance for climate change cooperation. In fact, no effective global climate change regime can be built without transatlantic cooperation. Certainly other large emitters in the less developed world including China and India would be hard-pressed to make any meaningful GHG mitigation commitments absent transatlantic agreement and, likely, incentives funded in part by such an agreement. Moreover, it is not clear that the EU will reach all the different policy goals that it has formulated between now and 2020, or that the US will substantially reduce its GHG emissions during the same time frame. Both regions will have to undertake tough actions and enhanced transatlantic cooperation could greatly facilitate their efforts at addressing climate change.

PART IV
Governing Global Markets: Environmental Standards and Certification Approaches

Chapter 11

Export Promotion, Trade, and the Environment: Negotiating Environmental Standards for Export Credit Agencies across the Atlantic[1]

Marcus Schaper

Conflicts between trade and environmental concerns are becoming increasingly salient. GMOs (e.g. Keilbach, Bernauer and Aerni in this volume; Isaac 2002; Bernauer 2003; Toke 2004), hormones in beef (e.g. Josling et al. 1999), leg-hold traps (e.g. Princen 2002), and chemicals regulation (e.g. Selin in this volume) have all proven their potential to create transatlantic conflict. Common to these challenges are the effects of behind-the-border environmental standards on international trade. What is intended as a domestic measure of environmental or consumer protection can also represent a non-tariff barrier to trade by excluding goods, which do not meet the standards from market access. Much of the trade-environment literature focuses on the relationship of multilateral environmental agreements and the WTO rules (e.g. Eckersley 2004), the compatibility of national regulation with the WTO regime (e.g. Petersmann and Pollack 2003), or the effects of standards in one polity on those in another (cf. Vogel 1995, 1997a; Busch and Jörgens 2004; Jacob et al. 2005). Less attention is paid to the harmonization challenge on the national level in the international environmental politics literature despite experiences with regulatory harmonization in the EU context (Héritier et al. 1996; Knill and Lenschow 1998, 2000).

This chapter focuses on the US's and Germany's roles in OECD negotiations on environmental standards for export credit agencies (ECAs) and how the reconciliation of their conflicting positions paved the way for agreement during the fall 2003 negotiations of the OECD's Export Credit Group (ECG). In the case of environmental standards for ECAs such a narrowing of transatlantic relations and OECD politics is defensible because the US and Germany were the key players in these negotiations with each side representing the extremes

1   This chapter benefited from support from the German Institute for International and Security Affairs (SWP), the American Institute for Contemporary German Studies, and the Deutscher Akademischer Austauschdienst as well as from Miranda Schreurs' guidance, Peter Evans helpful discussion of the specifics of ECA politics, and Henrik Selin's comments and suggestions.

of the range of preferred outcomes. The US pressed for binding international standards and a transparent review process, while the German red-green government blocked these throughout most of the negotiations, preferring to maintain the status quo. Only when the German government's position shifted to one cautiously endorsing binding standards and transparency, could agreement be reached in December 2003. Most other OECD members' stances gravitated around either the German or the US position.

Internationally, the US has been a forerunner in the establishment of environmental standards governing export credit agencies. It had more stringent domestic standards and was the one country pushing for the international harmonization of export credit environmental standards within the OECD. Germany, on the other hand, opposed these developments, despite its strong Green Party. I argue that this distribution of roles is not a consequence of varying shades of "greenness," but can be explained with reference to the compatibility of proposed rules with domestic institutional frameworks and regulatory cultures. This provides an opportunity to assess transatlantic environmental relations beyond the often too hastily ascribed roles of environmental leader and laggard (e.g. Vogel 2003b).

It is not the idea of environmental standards per se that made agreement in the OECD so complicated, but rather the issue of institutional fit for the proposed environmental standards. Whether there is a political culture of open bureaucracy and procedural rule-making affects a nation's support for certain elements of proposed environmental standards, especially those pertaining to the transparency of the environmental assessments and the approval process. Similarly, states use their export credits to achieve non-trade policy goals, to varying degrees. As outlined below, little institutional fit existed in at least three points: 1) environmental impact statements that are not institutionalized to the same degree in Germany as they are in the US; 2) transparency of coverage decisions which is at odds with German administrative law; and 3) explicit use of export credits for non-trade related policy objectives.

Flexibility with regard to the international standard to be applied and a provision allowing the retention of environmental information for select projects prior to a coverage decision made it easier for Germany and a few other states to accept binding standards and ex-ante transparency. It seems that the loopholes created by these provisions render the Common Approaches largely ineffective in preventing bad projects, as NGO critics argue (Görlach et al. 2007).

Regulatory Harmonization

Many of these regulatory conflicts over environmental issues are of a different nature than global environmental summitry: they occur in a realm governed by the WTO and are ultimately affected by WTO rules, dispute settlement procedures, and WTO-authorized trade sanctions. In addition to the existence of enforceable international law in the trade area, the removal of such regulatory differences requires harmonization of domestic regulations, which can be more challenging than agreement to a lowest-common-denominator international treaty with no or poor enforcement mechanisms. These conflicts are about different domestic approaches to regulating a certain issue area, and they require substantial concessions by the parties that need to change their domestic regulations as a result. Regulatory harmonization is a difficult task because it questions existing domestic regulations that are not only a product of complex political bargaining processes but also reflect what is considered the proper and right way of regulating an issue area. As a result, many regulatory conflicts are allowed to simmer indefinitely (Pollack 2003, 71).

Regulatory harmonization occurs in all sectors of political activities which have an effect on trade or cross-border business activities. However, the range of these harmonization endeavors is broader than the well documented area of product and production standards. Environmental standards for export promotion instruments are a far more complicated matter than the technical devices that save dolphins and turtles from being caught in fishing nets. Implementation of environmental standards and review procedures is much more involved than setting a standard for a certain product. While product standards can be implemented more or less uniformly across varying polities and enforced by import restrictions, environmental requirements for access to government support require more coordination. Regulatory harmonization can occur on at least three levels of increasing complexity:

- *Harmonization of product standards*: prescriptions for qualitative product characteristics pose considerable barriers to trade, but in most areas harmonization is mostly a technical matter
- *Harmonization of production processes*: the question of how products are made is more challenging, but adaptation is possible provided the resulting increase in market size can compensate for the cost of adaptation
- *Harmonization of domestic regulation*: domestic rules which govern behavior in a given area are embedded in a broader institutional framework. Harmonizing these regulations can be a very challenging task when domestic institutional contexts suggest different fundamental approaches to the same regulatory task

Environmental standard-setting for export credit agencies (ECAs) is an under-researched regulatory harmonization challenge. It lies outside the focus of much of the trade literature as these environmental standards cannot be applied in a protectionist manner. It is nevertheless an intriguing topic for study as ambitious environmental standards do not help domestic firms by shielding the market from foreign importers. Instead, these standards work against domestic firms by impeding access to export support. Consequently, they provide an opportunity to assess the applicability of approaches developed to analyze standards with a protectionist potential to this non-protectionist class of environmental standards.

Government-backed export credits and export credit guarantees play an important role in providing exporters with access to emerging markets in high-risk, and especially developing and transition, countries. Transactions with firms in these countries often bear political risks that commercial insurers are not willing to cover at reasonable cost. These include large-scale infrastructure developments, like dams, which are typically financed in ways that repayment can stretch for more than a decade. Most firms cannot afford to embark on such transactions without some insurance against non-payment to gain financing from banks. Governments fill this void by acting as a bank of last resort and providing domestic exporters with export credits, or by providing insurance against political and commercial risk in the form of export credit guarantees. The category of ECAs encompasses bank-type organizations, government-backed insurers, as well as combinations of these two ideal-types (Walzenbach 1999; Evans 2005). Much of the long-term cover is granted for infrastructure projects with problematic environmental performance similar to those which triggered the development of environmental guidelines for bilateral and multilateral development assistance.

Internationalization of Domestic Environmental Standards

Peter Haas (2003, 53) suggests that the US is generally admired because of the role it played in introducing strong domestic environmental standards, which were subsequently emulated abroad. These include environmental reporting standards and environmental impact assessment requirements for public projects. Yet, he argues that the US foreign environmental policy record since 1972 "is erratic; seemingly unrelated to administration." He suggests a variety of reasons including, "bureaucratic discretion and inertia, scientific consensus, avoiding heavy economic burdens on the US economy from compliance, domestic industries' opposition to expensive pollution control regulation, and organized public concern amplified by NGOs. When consensus exists and executive branch bodies enjoy some discretion then the US is likely to be a leader. When consensus is absent, economic costs are heavy, and industry opposition is powerful than the US will be a laggard." Elizabeth DeSombre (2000, 2005) suggested that when coalitions between baptists (environmental activists) and bootleggers[2] (industry) formed, US action in international environmental politics was likely to occur. At the same time the US is likely to be a laggard

---

2   Bruce Yandle (1983) introduced the term "bootleggers and Baptists." David Vogel (1995) also applied the concept with regard to domestic US environmental policy.

internationally if it does not have corresponding domestic regulations in place prior to negotiating internationally.

Quite similar to the US role in promoting its domestic environmental regulations internationally, Germany has uploaded much of its environmental regulation to the EU level and has resisted other EU regulations which did not square with its domestic policies (Wurzel 2004). While Germany was fairly influential in shaping early EU environmental policy on its terms, other Member States have caught up in uploading their domestic regulations to the EU. The result of this regulatory competition is a mix of regulatory approaches on the European level which, at times, results in EU directives that stand in conflict with domestic regulatory traditions. Similar to US resistance to engage in international environmental agreements that are not compatible with its domestic regulations, Germany has resisted EU regulations that appeared incompatible with its existing domestic rules and traditions.

It follows that both Germany and the US have a history of internationalizing their domestic environmental regulations and resisting internationalization by others, if the proposed rules are in conflict with their own domestic rules. While this is helpful in identifying conditions under which either nation is likely to be a leader or laggard, it does not help us to understand when an internationalization attempt is likely to succeed with the rare exception of a proposal which is acceptable to all parties involved. DeSombre (2000, 247) makes the case that US market power "over the target state is an important predictor" for successful internationalization of domestic US regulation, but her argument only holds for domestic standards that can be used to limit non-conforming foreign products' access to the domestic market. Since environmental standards for export credit agencies only apply to a nation's own ECA and its clients, they cannot be used to exert pressure on target states.

In the case of environmental standards for ECAs, setting unilateral standards does not directly result in adaptation pressures on competing nations. Such unilateral standards only serve as a barrier for domestic exporters to gain support from their national ECA. Consequently, a level playing field among ECAs can only be created by scrapping domestic environmental standards resulting in a race to the bottom or by actively seeking an agreement with competing nations.

By attempting to internationalize its domestic export credit policies, the US has followed an established pattern (DeSombre 2000; Haas 2003). When both environmentalists and industry favor international environmental regulation, the US is likely to promote internationalization of its domestic policies. Similar to the import restrictions for tuna and shrimp discussed by DeSombre, binding environmental standards for ECAs implemented throughout the OECD would provide for benefits to US industry over the status quo. The difference in this case is the US' dependency on international cooperation to achieve its goal. While import barriers are problematic under WTO rules, they only require passive cooperation—i.e. no appeals to the WTO against these rules. Harmonizing environmental standards requires other states to actively cooperate by embracing the idea and working towards a common set of rules.

The US was dependent on international cooperation in its effort to harmonize environmental standards for ECAs. Given the nature of export credit rules it could not rely on its market power to promote its agenda but had to seek agreement with the other members of the OECD ECG. The other ECG member governments, however, had little incentive to concede to the American want for environmental standards harmonization with the exception of similar demands made on them domestically by activist groups and concerned constituents. Consequently, they were much concerned with the nature of the proposed rules. The US proposals consisted of procedural rules for the evaluation of environmental impacts of supported projects, binding standards to be used in the evaluation process, and public transparency of the environmental evaluation process. Especially the transparency provisions were at odds with continental European regulatory traditions.

Since there is little economic reason for a nation to agree on stricter environmental standards that would result in a competitive disadvantage for its exporters, explanations must be sought elsewhere. Among the first candidates for such alternative explanations are domestic public pressure and interest group politics. US domestic standard setting occurred in response to domestic pressure to establish such standards and most ECG members experienced some degree of domestic political pressure for the establishment of environmental standards for their ECAs. Also normative factors, for example the way an issue is framed in public discourse can influence the acceptability of various potential solutions to the harmonization challenge. Research also points to the effects of pioneer standards on the policies in other polities (Vogel 1995; Busch and Jörgens 2004).

Export Credits as a Transatlantic Regulatory Harmonization Challenge

Environmental standards for ECAs cannot be used to shield a market in a protectionist manner, but they can distort trade. OECD members first harmonized the financial aspects of export credits in the 1970s and 1980s (Moravcsik 1989). Competition among ECAs had led to a situation where ever more favorable terms offered by ECAs resulted in products being sourced in a particular country because of the ECA terms offered rather than because of product quality. Export support had turned into heavy export subsidization and strained public budgets without resulting in proportional domestic economic gains.

In response to these overly favorable export credit terms that distorted international competition, OECD Member States negotiated the Arrangement on Export Credits to provide a level playing field for national export credit agencies and exporters. Ultimately, buyers are to make their purchase decisions based on the good itself as opposed to the terms of the ECA support facilitating its export (Moravcsik 1989).

Compliance with the Arrangement has been remarkably high compared to the performance of other OECD agreements. Levit (2004, 68) argues that the Arrangement has performed so well because of being "elastic (its soft form permits experimentation and revision), pragmatic (its processes redefine compliance in a way that accommodates ECA practice with the Arrangement's rubric), measured (it embraces consensus decision-making without diluting its rules with generalities and platitudes), and ... dialogic (the camaraderie of the Participants groups and the Arrangement's unique processes assure that the Arrangement remains a vibrant and progressive discussion)." Peter Evans (2005), on the other hand, argues that Members comply with Arrangement provisions because they have an interest to do so. In his view, export credits are private goods and the Arrangement, as a cartel of suppliers, regulates access. Compliance occurs out of an interest to keep the cartel functioning and to avoid a costly race for ever more favorable financing terms. It appears that US policymakers were less optimistic about compliance with the export credit regime (United States Congress 1994): The US monitors compliance of other Members with Arrangement provisions closely and can match offers by other ECAs that do not comply with funds from the "war chest" authorized by Congress for this particular purpose (Sheppard 2002).

The Arrangement and the ECG in which it was negotiated provided a good framework for ECA cooperation. Negotiations on environmental standards for ECAs were placed in this venue with its good track record of negotiating and maintaining the Arrangements by the G7 members after their 1997 Denver summit. However, the group struggled from 1998 to 2003 to establish a common set of environmental standards. Similar to the Arrangement's goal of purging competition over financing terms, the aim of the Common Approaches on Environment and Officially Supported Export Credits was to eliminate competition over environmental requirements (OECD 2005c, 4).

The OECD ECG is comprised of 29 of 30 OECD Member States, not including Iceland. Country delegations typically include ECA representatives as well as the responsible government departments, usually finance or economics. The US delegation is headed by the Department of the Treasury and also includes Ex-Im representatives as well as a State Department official. Germany is represented by the economics ministry, Hermes employees, as well as a representative from the Foreign Office. As such the ECG is a body dominated by technical experts. Stakeholders outside of the relevant government departments have little direct influence. The ECG conducts one short consultation session with NGOs per year, but does not allow for further involvement by stakeholders (Schaper 2007).

Environmental Standards for Export Credit Agencies

The 1992 Congressional reauthorization of the US Export Import Bank (Ex-Im Bank) required this agency to devise and implement environmental standards for its operations. Implemented in 1995, Ex-Im's environmental policies were the first comprehensive set of national environmental standards for an ECA anywhere. This unilateral step resulted in disadvantages for US exporters seeking Ex-Im support, whose exports had to comply with Ex-Im's environmental policies, whereas foreign competitors could gain ECA support without those projects having to adhere to environmental standards. The Ex-Im Bank agreed to seek international harmonization of environmental guidelines to address US industry concerns about export competitiveness. Consequently, the US sought to re-level the playing field by bringing other countries' ECAs' environmental policies up to its level.

Internationally, the issue was first raised in technical OECD fora and then elevated to the heads-of-state level at the 1997 G7 summit; from 1998 on, the OECD's ECG was tasked with negotiating an agreement on environmental standards and procedures for export credit agencies. In negotiations, the US as the international leader on the issue and initiator of the harmonization endeavor, was opposed chiefly by Germany, which favored neither binding standards nor transparency of the environmental review process. In 2001, these negotiations resulted in "Draft Recommendation on Common Approaches on Environment and Officially Supported Export Credits" (Revision 6)—or commonly referred to as "Common Approaches" or "Rev. 6." These recommendations were supported by all members of the ECG, with the exception of the US and Turkey. All other members voluntarily implemented Rev. 6 from 2002 on. The US denied Rev. 6 support because it fell short of US goals for binding standards and transparency ("US Faces Uphill Battle" 2001; United Kingdom Parliament 2001, Column WA 78).

Rev. 6 of the Common Approaches called for assessment and revision in 2003. In fall 2003 the ECG set out to negotiate a text which would be acceptable to all its members. Negotiations were held from September to November 2003, the Common Approaches was passed by the OECD Council in December 2003, a review was concluded, and a new version adopted on 12 June 2007 (OECD 2007)—half a year later then scheduled (ECA Watch 2006b). At the same time, the ECG was also caught in a deadlock on anti-bribery measures when a coalition of Germany and Japan blocked an initiative spearheaded by the US (Alden et al. 2006). With the ECG devoting much of its attention to anti-bribery measures, the 2007 updates to the Common Approaches were rather limited compared to the 2003 revisions. NGOs, in fact, actually consider the 2007 Common Approaches a step back. The agreement calls for another revision by 2010.

At their core, negotiations about these environmental standards addressed two related issues: common and binding standards for project evaluation and ex-ante transparency of the environmental assessments. The publication of information on projects and their environmental impacts prior to granting cover is of central concern, because this provision enables civil society actors to monitor decisions and actions by export credit agencies and thus provides for an informal enforcement mechanism in lieu of a non-existing formal OECD compliance mechanism.

With regard to the environmental standards, controversy focused on whether entire projects or individual components needed to be evaluated and on *which* and *how* standards would be applied. Rev. 6 required international standards such as the World Bank Group's Safe Guard Policies and Pollution Abatement Handbook to be used as benchmarks, but did not provide binding standards. The 2003 Common Approaches established international standards as minimum standards and allowed for other standards to be applied if they exceed the international ones. However, instead of agreeing on one set of standards (i.e. the World Bank's), negotiators settled for a broader range of standards including regional development bank rules, some of which are considered weaker than the corresponding World Bank standards.

Before the 2003 agreement only some states allowed interested parties access to environmental assessments and consulted NGOs prior to granting cover (e.g. the US Export-Import Bank); other ECAs argued that this was not possible because it would infringe clients' commercial confidentiality. The 2003 Common Approaches required public access to environmental project information 30 days prior to making a coverage decision. Yet, they provided for deviation from this rule on a case-by-case basis (Görlach et al. 2007).

The approval of export credits for the Turkish Ilisu dam (Mossman 2007) around the time of the 2007 Common Approaches adoption by the German, Swiss, and Austrian ECAs especially put focus on the question of adherence to international standards—and thus on compliance with the Common Approaches. Approvals of export credits to the Turkish dam had been pending for a number of years due to unresolved problems with the resettlement of affected local populations and the destruction of cultural heritage in the dam's future reservoir site. The ECAs approved the dam with a list of 150 undisclosed conditions to be met by project sponsors; they refused, however, to make this list public (Ahmia 2007; Gottschlich and Kreutzfeldt 2007; Strittmatter 2007a, b). NGOs suspected that approval of this project could be possible only by derogation from international standards (ECA Watch 2006a, 2007). The range of applicable standards coupled with the flexibility with regard to publication of environmental project information facilitated agreement on the Common Approaches, but this may also render the accord less effective than originally envisioned by its promoters.

US Domestic ECA Politics

In 1992 President Bush signed the reauthorization of the Export-Import Bank. Public Law No. 102–429 contained a section establishing environmental requirements for Ex-Im.

> Directs the Bank, for any transaction involving a project for which support of $10,000,000 or more is requested and certain environmental concerns exist, to establish procedures to take into account the potential benefits and adverse environmental effects of the goods and services which it may support under its lending and guarantee programs. Authorizes the Board to withhold financing for environmental reasons or to approve financing after considering the potential environmental effects of a project. Encourages the Bank to use its programs to support the export of goods and services that have beneficial effects on the environment or mitigate potential adverse environmental effects. (United States Congress 1992a)

This provision set a precedent for other export credit agencies. Up to that point, no ECA had explicit and statutory environmental rules for its operations. Responding to a Green parliamentary inquiry, the German government asserted as early as 1985 that "the consideration of ecological and developmental aspects occurs as part of the evaluation of an individual project's supportability based on available information and development policy objectives" (Deutscher Bundestag 1985), but no formal rules and basis existed for these considerations. Ex-Im responded to its Congressional mandate by developing environmental policies which were implemented in 1995 and subsequently repeatedly refined.

The Congressional mandate came as a result of a more than 20-year-long quarrel over the applicability of the National Environmental Policy Act (NEPA) to activities outside of US borders. The *National Environmental Policy Act (NEPA)* (1969) established environmental review procedures for federal US government agencies. After NEPA's adoption, advocacy groups engaged in a drawn-out battle over its applicability to US government actions beyond US borders. NEPA itself was not clear as to whether it applied to domestic environmental effects only or also to activities by the US government beyond US borders. Most internationally active government agencies preferred a narrow interpretation of NEPA while environmental activists demanded these internationally active agencies perform the same environmental reviews of their activities as required of domestic agencies. Litigation ebbed when President Carter issued an Executive Order in 1979, which sought to clarify and limit the reach of NEPA.[3]

---

3   A lawsuit brought forth against Ex-Im by a coalition of NGOs is pending. The plaintiffs argue that through their impact on the global climate all fossil fuel projects affect US territory and therefore ought to be subject to a full environmental review conforming with NEPA.

Congressional action on environmental rules for Ex-Im was prompted by a controversial September 1976 CEQ memo to "Heads of Agencies on Applying the EIS Requirement to Environmental Impacts Abroad" (42 Fed. Reg. 61068–69). This memo sought to clarify the applicability of NEPA to international federal actions but rather than solving the uncertainty created by unclear language in NEPA and the pending lawsuits, it sparked an inter-agency dispute about the applicability of NEPA to actions beyond US borders. CEQ draft regulation dating from January 1978 further intensified controversy among federal agencies and Congress (United States Congress 1978, 86–127). The uncertainty around the external applicability of NEPA continued until President Carter's 1979 Executive Order No. 12114 ('Renewed Controversy' 1977; Schiffer 2004). However, even Carter's Executive Order did not succeed in resolving the conflicts surrounding the applicability of NEPA until implementation guidelines were issued by the CEQ under environmental champion and NRDC founder Gus Speth.[4]

Following this order, Ex-Im only had to subject its nuclear business to the stringent review under NEPA. For its remaining business, Ex-Im established "Concise Environmental Reviews" that amounted to little more than a "boiler plate form added to project documents" (Evans 2000). This review process remained largely unchallenged until the 1992 reauthorization. On the initiative of Senator Timothy Wirth (D-Colorado) section 105 "Environmental Policy" was inserted in the Senate version of the bill in order to provide "a permanent statutory authority for environmental policies and procedures currently being carried out by the Bank" (United States Congress 1992b).

Prior to the 1992 reauthorization, Ex-Im had no authority to withhold financing for environmental reasons. The 1992 standards requirement also moved the question of the applicability of NEPA to Ex-Im operations off the political agenda by establishing the basis for an own environmental review process (personal communication with Peter C. Evans, June 26, 2004). In retrospect, the Wirth amendment did much more than provide "permanent statutory authority" for what Ex-Im was already doing: it resulted in the development of the first set of comprehensive environmental review procedures and standards for project evaluation at an ECA anywhere.

Ex-Im policies implemented in 1995 require that projects are screened for their potential environmental impact. More detailed assessments are required for those projects likely to have a significant environmental impact. Summaries of the completed assessments are made available for public comment. In addition to establishing this review procedure and associated transparency provisions, Ex-Im policies set qualitative and quantitative criteria for project evaluation.

---

4 Gus Speth was NRDC plaintiff in many of these lawsuits and later founded the World Resources Institute (WRI), which plays a key role in the NGO campaign to Green international financial institutions.

Projects which fail to meet the environmental requirements can be denied Ex-Im support, if not brought into compliance.

By moving first, the US set the stage for any international agreement on the issue. Following a 20-year-old logic of requiring an environmental impact statement (EIS) when federal money is to be spent on a project, environmental standards for the Export Import Bank were an important but not ground-breaking step in the US. In fact, establishment of its own environmental review process made Ex-Im very much immune to demands to follow NEPA assessments for all its projects. For countries with different regulatory frameworks, however, requiring environmental reviews and making them publicly available constitutes a major challenge. Of particular interest here is how US political institutions and political culture have impacted the formulation of Export-Import Bank policies, and how these policies in turn have provided the core of the standards negotiated within the OECD. I argue that the OECD negotiations resulted in standards that grew out of the US regulatory context and regulatory culture. these standards require significant policy adaptations in other nations while being highly compatible with US policies. The OECD Anti-Bribery Convention displayed similar characteristics.

Opposition to exportation of US regulation is evidenced by the fact that Common Approaches required a weakening of both core US negotiation objectives binding standards and ex-ante transparency. By allowing the use of a range of standards and the exclusion of select projects from ex-ante transparency, states opposed to US initiatives retained considerable flexibility in their implementation of the Common Approaches.

Domestic German ECA Politics

Environmental concerns regarding export credits were first voiced by European NGOs working on development issues. In Switzerland, the Berne Declaration and the Swiss Coalition of Development Organizations successfully lobbied for legislation requiring application of the principles and guidelines of Swiss development policy to export credits to low-income countries (Fues 1994). This 1980 amendment marks the first incorporation of non-trade concerns into ECA policy in Europe. German development NGOs took up the issue around this time.

When the German Greens first entered Bundestag in 1983, political action followed. A 1985 inquiry (*Große Anfrage*) into the issue (Deutscher Bundestag 1985) was followed by a steady stream of inquiries (*Kleine Anfragen*) fielded by Green members of parliament. Until 1990 when the Western Greens did not win enough votes to return into Bundestag, they kept the issue of reforming Hermes, the German ECA, on the parliamentary agenda. However, with no access to this well-shielded executive agency, and with little leverage in the Bundestag, they achieved little beyond agenda-setting. After 1990, key Green faction staffers took up the issue outside of parliament. Barbara Unmüssig, former staffer to Uschi Eid and Ludger Volmer, joined the environmental think-tank WEED (World Economy, Ecology and Development), and Thomas Fues, who was behind most parliamentary inquiries on the issue from 1983 to 1990, worked with various NGOs including the Gemeinsame Konferenz Kirche und Entwicklung (GKKE), Eurodad, and WEED on Hermes reform from a developmental perspective.

Their efforts broadened the NGO community working on Hermes reform to include Urgewald, Germanwatch, and others and built up the issue's political salience over time, which led to formal political recognition in the 1998 red-green coalition agreement; this called for a reform of Hermes along "social, environmental, and sustainable principles." The rather bold policy statements on Hermes reform in the 1998 agreement are less a product of conviction than of favorable negotiation dynamics. When the Social Democratic Party (SPD) negotiators responsible for trade and economic issues worked on other sections of the agreement, language pertaining to export credits slipped into the foreign policy and development sections of the agreement because negotiators in these areas did not realize the issue's political relevance. Thus, the text on Hermes in the coalition agreement marks the first (or only?) victory of Green environmental concerns over SPD economic interests.

Advocacy work by environmental NGOs and their lobby work—especially WEED and Urgewald—resulted in higher salience of the issue, but it did not result in active policymaking by the new government in this realm. Still, it prompted the inter-ministerial coordination committee to first initiate timid environmental reform in summer 1998—even before the elections, which brought the red-green coalition to power.

Starting in 1995, exporters were required to supplement their application for Hermes export credit guarantees for projects exceeding DM 25 million with an informal memorandum outlining any environmental project externalities that they were aware of. As the format and scope of this memorandum was left mostly to the exporter's discretion, one can hardly speak of an environmental standard, as there were no established environmental criteria or environmental review process. Hermes decisions were based on these exporter-supplied memoranda and information sourced from German diplomatic representations in buyer countries (Hermes Kreditversicherungs-AG 1998). In 1995 the US Ex-Im environmental policies also went into effect.

In mid-1998 new guidelines refined the memorandum requirement by requiring applicants for Hermes coverage to answer five questions when preparing the memorandum:

1   Does this project bear significant environmental aspects?
2   Project surroundings: In what kind of surrounding or environment is the project located? Is this a specifically protected or threatened environment?

3   Are there environmental requirements by the buyer country? If so, what are they and is it guaranteed that these requirements will be followed?

4   Has there been or will there be an Environmental Impact Assessment conducted? If so, by whom and which standards are applied (local, German, or e.g. World Bank standards)?

5   Will this project substitute environmentally harmful installation, productions processes, or products?

These questions are to be answered in general terms, i.e. without technical specifications, in so far as they are relevant for the project and known to the exporter. (Hermes Kreditversicherungs-AG 1998)

The June 1998 requirement hardly qualifies as an environmental standard either, as established environmental criteria or an environmental review process were still absent; the requirements were enacted to "document to the outside, that environmental aspects are considered in the decision-making process" (Hermes Kreditversicherungs-AG 1998). The inter-ministerial coordination committee (IMA) still held wide discretionary power. Question 5 hints at the political discourse on environmental standards at that time. Industry termed the debate mostly as a debate about export of environmentally superior technology, in which Germany was seen as a world leader. Consequently, stricter standards were viewed as unnecessary, as German exports by themselves would advance environmental objectives (Drillisch et al. 1998).

In 2001, negotiations between the coalition factions and the Ministry of Economics resulted in draft environmental guidelines for Hermes credit guarantees, and prompted further parliamentary debate. All parties presented bills calling for some reform in the Hermes instrument, except for the Free Democratic Party (FDP), which favored the status-quo and called for an exclusion of explicit environmental requirements from Hermes guidelines (Deutscher Bundestag 2001b). While the Party of Democratic Socialism presented the most far-reaching—and at the same time most unrealistic— proposal which called for far-reaching inclusion of NGOs and stakeholders in the decision process, a list of project types to be excluded, and promotion of environmentally innovative technology, the Christian Democratic Union (CDU)/Christian Socialist Union presented a scenario (Deutscher Bundestag 2001a) that closely resembled the final outcome of the reform process. The SPD/ Green Party bill used rather weak language calling for the consideration of human rights in the decision process and the reporting of problematic and large-volume cases to the economics committee (in addition to the budget committee which had been the only parliamentary body with some degree of oversight of Hermes credits thus far). Furthermore, the bill called for strict adherence to the new guidelines—especially with respect to sensitive exports, such as nuclear power generation, arms, dangerous chemicals and large dam projects. Higher transparency was to be negotiated within the OECD framework (Deutscher Bundestag 2001c). The weak coalition bill is indicative of the parliamentary faction's weak standing *vis-à-vis* the Ministry of Economics (Schmid 2001). Ironically, the CDU and FDP bills mirrored positions held by the ministry and industry associations (Hagelüken 2001) thereby strengthening the ministry's position *vis-à-vis* the coalition that legitimized it.

In retrospect, the Greens now attribute the weakness of the outcome partially to the lacking power and influence of the Green parliamentarians in charge of the issue at the time. Opposition by an influential group of trade specialists within the SPD faction, and an economics ministry dominated by the non-interventionist FDP for 20-some years, the Green parliamentarians did not have the clout to bring about more decisive change.

The environmental "Guidelines for the Consideration of Ecological, Social, and Developmental Criteria for Granting Export Guarantees of the Federation" were passed by IMA on April 26, 2001. Similar to standards applied since mid-2000, they include:

- the screening of projects with a volume of €15 million and more, and a significant German share;
- review for cases in which the screening indicated the need for further investigation. More information may be requested from the buyer. Other analyses such as information from other ECAs and buyer/seller provided assessments may be considered; and
- classification of projects into three categories: A, B, and C—with category A containing projects that have considerable environmental impacts.

In contrast to US regulatory culture, German policymakers have refrained from tasking Hermes credit guarantees with tasks other than export promotion. Hermes has not been used to advance goals other than promoting German exports. Given this history, it is not surprising that the prevailing conception of export promotion worked against the incorporation of sustainability objectives into Hermes policies.

While WEED's and Urgewald's advocacy work appears to have been rather successful in terms of media coverage and political action—though more symbolic than substantial—it did not generate the broad public support for higher environmental standards needed to define the political cost calculus in ways conducive to substantial reform. All parties in Parliament took up the issue in inquiries into the government, calls for action, and policy statements on reform. However, parliament proved to be a weak negotiation partner; the government—especially the Ministry of Economics—played a highly successful two-level game (Putnam 1988). By referring to the OECD negotiations on environmental standards for export credits as the relevant international benchmark, and at the same time slowing the OECD process to bring about low international standards, the Economics Ministry successfully retained ownership of the reform process, and in the end largely dictated the standards.

Parties in parliament did not see—or ignored—opportunities to shape the OECD negotiations by clearly defining acceptable negotiation results.

German ECA politics have been mostly reactionary. The US initiated broad domestic environmental standards domestically, and set the international stage by taking its concerns to the G7 and the OECD. German politics, on the other hand, has only addressed environmental policies for Hermes when internal and external pressures required action on the topic. The domestic discourse was dominated by the importance of easy access to export credits for job creation. US calls for thorough environmental screening prior to granting cover, and a publicly transparent approval process appeared as an attempted assault on German industry's competitiveness that needed to be fended off. The Economics Ministry also provided for a great deal of inertia that worked against greening Hermes guarantees.

Negotiating the Common Approaches: Issues of Institutional Fit

Since NEPA, environmental assessments and environmental impact statements have become deeply institutionalized in the US. While much of German environmental legislation has been adapted from US domestic regulation, environmental impact assessments were only reluctantly implemented in Germany after a EU directive required them:

> Famously, the German government took five years to transpose the EU's 1985 Environmental Impact Assessment directive into national law, and another six years to introduce the administrative provisions required for its implementation— largely because, as Kraak and Pehle (2001, 6) explain, the European directive "is comparatively blind when it comes to internal German administrative structures" ... The European Commission has since sued the German government (successfully) over the inadequacies of the provisions it did manage to implement. (Dryzek et al. 2003, 116)

The US norm appears to be at odds with German regulatory culture. At their core, US Ex-Im Bank standards and the Common Approaches provide a procedure and conditions for the compilation of environmental assessments. Although a perfect fit for US domestic institutions—especially the reliance on explicit policies and procedures—this set of rules is less compatible with the German preference of retaining flexibility in corporatist arrangements with industry. Vogel (1986) identified similar institutional differences between US and British approaches in regulating industrial pollution.

The US environmental impact assessment process includes publicity of these assessments, providing stakeholders opportunities to comment on the projects' anticipated impacts and potentially challenge the adequacy of the environmental review. Like the required environmental impact statement, transparency of the process is deeply institutionalized in the US. In contrast, such information is considered commercially confidential and not to be released by public agencies in Germany. Ways are being sought to integrate the Common Approaches' 30-day transparency requirement into German regulation, but not surprisingly, German negotiators were strongly opposed to this rule for a long time.

Since the voluntary implementation of Rev. 6 in 2002, ECG Member States have experience with environmental standards that contradict many of their initial fears—be it that application of these standards helped them to better manage risk or that they found ways of following the rules without substantially adapting their review procedures and cover practice. Thus, adaptation to Rev. 6 may have facilitated agreement to the revised Common Approaches in 2003.

Implementation of Rev. 6 certainly raised the bar for what an acceptable negotiation outcome would need to look like. Negotiators knew that they had to arrive at an agreement that would be acceptable to all export credit group members—including the US—and exceed Rev. 6 in at least two aspects: role of international standards and monitoring provisions. Only a few years earlier ECAs could still dismiss environmental concerns as not relevant to their business, but the negotiations around and implementation of Rev. 6 resulted in a different discourse that was concerned with the design of suitable environmental rules rather than with more fundamental questions of their general appropriateness.

Conclusion

Many transatlantic disputes are caused by a misunderstanding or lack of knowledge about the other side's politics. In the ECA case, a number of issues got in the way of a quick agreement: environmental protection by means of an export promotion instrument ran counter to sectoral conceptions of policymaking in Germany and other states; transparency provisions were at odds with continental European regulatory cultures; and international negotiations on environmental standards for export credits were the result of US domestic politics—other states desired negotiations on the issue far less than the US did.

The time it took to agree on the Common Approaches can be attributed to the US attempt in pushing for an agreement, the terms of which were difficult to accept for many of the other parties involved. US policymakers repeatedly expressed their dismay with the German unwillingness to increase transparency in the approval process for Hermes guarantees. They perceived this as inflexibility on behalf of their colleagues aimed at bolstering their own negotiation position. For German politicians, on the other hand, provisions in administrative law against making project information available were real issues not easily solved or circumvented. Not surprisingly, agreement in 2003 was facilitated by the flexibility incorporated into the crucial provisions which at the same time weakened the agreement considerably—possibly to the point of complete ineffectiveness.

Had the Common Approaches not called for procedural standards outlining how environmental assessment was to be conducted, and had they only provided technical standards for exports and projects, agreement might have been much easier, by providing the implementing states more leeway in incorporating the new rules into their own regulatory frameworks. At the sane time, such rules may have been implemented more consistently than it is the case with the Common Approaches.

Interviews with policymakers also revealed their limited knowledge of other states' environmental standards. Most could not evaluate their own ECA's policies *vis-à-vis* other ECAs. This seems to indicate that the relative strictness of standards did not matter much—except, most likely, for US benefits from an agreement in relation to the concessions necessary by all other parties.

Both the German and the US negotiation positions throughout most of the OECD negotiations were shaped by the distribution of political power at home. The US needed an agreement that would provide for rules similar to its own to re-level the playing field and to take pressure off Ex-Im. The German delegation had little incentive to support substantial reforms that might alienate industry support needed for other reform projects. Knowing that the rules would not affect Ex-Im's largest client, Boeing, certainly helped the US initiative, while the German delegation was mindful that key production sectors of Hermes' largest client Siemens could be adversely affected by environmental requirements. For the German Greens, ECA reform was a critical topic, but the Green Environmental Ministry and Joschka Fischer's Foreign Office could only gain more influence on negotiations and thus facilitate agreement when changes in personnel in three ministries occurred almost simultaneously in 2003. The zero-sum character of the domestic German discourse on the issue, pitting the environment vs. jobs, required such a power shift for environmental concerns to play a more significant role.

The lack of institutional compatibility between the US proposed rules and the regulatory contexts of other ECG members provides a good starting point for explaining the difficulties in reaching an agreement. In comparing the German and the US regulatory context, little institutional fit existed in at least three points: 1) environmental impact statements that are not institutionalized to the same degree in Germany as they are in the US; 2) transparency of coverage decisions which is at odds with German administrative law; and 3) explicit use of export credits for non-trade related policy objectives. Not surprisingly, the combination of little political pressure to resolve the issue and proposed rules that would require substantial adaptation in Germany resulted in the German delegation's hesitant approach throughout most of the negotiations.

Regulatory harmonization is a thorny issue where negotiators need to appreciate not only political and economic costs of compromises, but also substantial regulatory costs resulting from regulatory reforms in response to harmonized rules. Still, regulatory harmonization challenges such as the one presented in this chapter have the potential for increasingly contributing to transatlantic strife. They will require more attention by academics studying harmonization processes and by negotiators educating themselves about other states' regulatory systems and cultures.

Chapter 12

The Emergence of Non-State Environmental Governance in European and North American Forest Sectors

Benjamin Cashore, Graeme Auld, Deanna Newsom, and Elizabeth Egan

Any effort to understand and explain the development of environmental forest policies across the Atlantic in the last decade and a half must pay careful attention to the emergence of two distinct paths that emerged out of the ashes of the 1992 Rio Earth Summit's failed efforts to achieve a binding global forest convention (Humphreys 2007). Though very different, both paths were strongly influenced by a widely accepted explanation for the demise of intergovernmental negotiations: concerns that *national sovereignty*—i.e. the right of each country to decide what to do within its own boundaries—was being threatened.

The first path sidestepped the sovereignty issues by focusing on developing *processes* to *define*, rather than implement "sustainable forestry," often through meetings of experts on what constituted appropriate "criteria and indicators" and the promotion of "national forest programs" through which countries, it was hoped, would see fit to address globally important concerns (Humphreys 2006). Travelers on this path included domestic forestry agencies, intergovernmental negotiators, and international agencies such as the United Nations' Food and Agricultural Organization (FAO) and, ultimately, the United Nations' Forum on Forests (UNFF) (Humphreys 2004).

The second pathway sidestepped the sovereignty issue by rejecting state-centered intergovernmental negotiations altogether, turning instead to the marketplace to address global forest deterioration by developing and demanding *global* standards with *prescriptive* requirements. Followers on this path include most of the world's leading environmental groups, their social allies, a handful of forest companies and retailers, governmental environmental and aid agencies, the World Bank, and, eventually, philanthropic foundations. They created and/or developed the Forest Stewardship Council (FSC) program which was designed in 1993 to monitor companies for the environmental stewardship and social practices, and certify those that practiced forestry in accordance with pre-established performance criteria.

The first pathway has essentially evolved as would have been expected—much progress has been made in defining the economic, environmental and social benefits of the forest, National Forest Programs have been initiated to varying degrees within Europe, and functional equivalents in Canada and the US (Humphreys 2006). Just what they do, and the practices they place attention on, varies considerably from one country to the other (Howlett and Rayner 2006). Unlike many other environmental arenas, the EU has shied away from any directives over forest management and has focused instead on giving financial incentives to poorer countries to develop national forest programs (Humphreys 2006).[1] Similarly, the EU, under the leadership of Finland, has been a strong promoter of the "Helsinki" criteria and indicators forest sustainable forest management processes, which is loosely linked with other criteria and indicators processes operating globally. At the international level, the United Nations Forum on Forests continues to bring together many of the world's countries to deliberate over the declining state of the world's forest ecosystems, but continues to produce no international agreements about what to do (Dimitrov 2005; Dimitrov 2006).

This chapter focuses on the striking, and arguably less predictable, trajectory of the second pathway. As the FSC model gained incremental support throughout the 1990s, especially in North American and Europe, industry and/or forest owners began to reverse their initial opposition to certification and instead created "FSC competitor" programs that they hoped the market place would accept as a legitimate alternative to the FSC. These competitor programs, like path one, were designed to respect national sovereignty and to give much discretion to individual forest companies and domestic forest agencies in implementing and choosing what types of forest operations ought to be certified.

This chapter analyzes these power struggles in British Columbia, Canada, the US, Germany, Sweden, the United Kingdom and Finland. Our historical review reveals a puzzle: in some countries, forest companies responded to market pressures by expressing interest in, or achieving certification according to the environmental group supported international Forest Stewardship Council (FSC) standards. However, in other countries forest companies gave the FSC little attention, supporting either no forest certification program at all, or industry and/or forest owner initiated certification programs.

Why did some countries' forest sectors support FSC forest certification while others preferred forest industry and/or forest owner initiated alternatives? We argue that attention to this conflict, and the international trends and domestic variations in support that have occurred, requires careful attention to the role of three mediating factors: the place of the country/region in the global

---

1   The development of national forest programs has been developed by the "European Cooperation in the Field of Scientific and Technical Research" (COST), which though funded by the EU is not, technically, an EU initiative (Humphreys 2006). Our thanks to David Humphreys for this clarification.

economy; the structure of the domestic forest sector; and the history of forestry on the pubic policy agenda. Taken together, these factors influence strategic choices available to FSC supporters in their efforts to use economic carrots and sticks to "convert" forest owners to support the FSC, and whether FSC strategists face an uphill battle, either failing to gain support, or being forced to "conform" by altering and changing its own program. Addressing such a question enhances our understanding of the emergence of forest certification and highlights the need for policy scholars to conduct careful sectoral-level analyses of cross-Atlantic trends in public and *private* policy innovation and adoption (Cashore et al. 2004).

Following this introduction a second section identifies the two different conceptions of forest certification vying for support in North America and Europe. A third section locates forest certification as an advanced form of a non-state market driven (NSMD) governance systems. NSMD government systems are proliferating and emerging in a range of globally important sectors including fisheries, coffee production, agriculture and eco-tourism. A fourth section reviews the historical development of forest certification in our cases and the ultimate patterns of support for the FSC and its competitor programs. A sixth section assesses seven hypotheses' explanatory power for understanding the emergence of forest certification in these countries.

Two Conceptions of Forest Certification

By 1992, ongoing frustration with domestic and international public policy approaches led many transnational environmental groups to promote eco-labeling certification institutions. In the case of forestry, the World Wide Fund for Nature (WWF) spearheaded a coalition of environmental and socially concerned environmental groups, who joined with select retailers, governmental officials, and a handful of forest company officials to create the international FSC. Officially formed in 1993, the FSC turned to the market for rule-making authority by offering forest landowners and forest companies who practiced "sustainable forestry" (in accordance with FSC policies) an environmental stamp of approval through its certification process, thus expanding the traditional "stick" approach of a boycott campaign by offering "carrots" as well.

The FSC created nine "principles" (later expanded to 10) and more detailed "criteria" that are performance-based, broad in scope and that address tenure and resource use rights, community relations, workers' rights, environmental impact, management plans, monitoring and conservation of old growth forests, and plantation management (Moffat 1998, 44; Forest Stewardship Council 1999). The FSC program also mandated the creation of national or regional working groups to develop specific standards for their regions based on the broad principles and criteria.

Table 12.1   Conceptions of Forest Certification

	Conception 1	Conception 2
National sovereignty	Belief that domestic states should be constrained through development of global requirements/standards	Respects rights of countries to determine forest policies appropriate for operations within their own borders
Who participates in rule making	Environmental and social interests participate with business interests	Business-led
Rules—substantive	Non-discretionary	Discretionary-flexible
Rules—procedural	To facilitate implementation of substantive rules	End in itself (belief that procedural rules by themselves will result in decreased environmental impact)
Policy scope	Broad (includes rules on labor and indigenous rights and wide ranging environmental impacts)	Narrower (forestry management rules and continual improvement)

Source: Cashore (2002).

As important as the rules themselves is the FSC "tripartite" conception of governance in which a three-chamber format of environmental, social, and economic actors, each with equal voting rights, has emerged. Each chamber is itself divided equally between North and South representation (Domask 2003).

The lumping together in one chamber of those economic interests (i.e., companies and non-industrial forest owners) who must actually implement SFM rules with firms along the supply chain who might demand FSC products, as well as with consulting companies created by environmental advocates, has been the source of much controversy and criticism. Along with concerns about the FSC being overly prescriptive, these governance procedures have negatively affected forest owners evaluations of the FSC (Rametsteiner 1999; Vlosky 2000; Sasser 2002) and encouraged the development of "FSC alternative" certification programs offered in all countries in North America and Europe where the FSC has emerged. In the US, the American Forest and Paper Association created the Sustainable Forestry Initiative (SFI) certification program. In Canada, the Canadian Standards Association (CSA) program was initiated by the Canadian Sustainable Forestry Certification Coalition, a group of 23 industry associations from across Canada (Lapointe 1998). And in Europe, following the Swedish and Finnish experiences with FSC-style forest certification, an "umbrella" Pan European Forest Certification (PEFC) system (renamed the Program for the Endorsement of Forest Certification in 2003) was created in 1999 by European landowner associations that felt especially excluded from the FSC processes.

These FSC competitor programs originally emphasized organizational procedures and discretionary, flexible performance guidelines and requirements (Hansen and Juslin 1999, 19). However, each of these has changed, and continues to change in response to market pressure. For instance, the SFI originally focused on performance requirements, such as following existing voluntary "best management practices" (BMPs), legal obligations, and regeneration requirements. The SFI later developed a comprehensive approach through which companies could chose to be audited by outside parties for compliance to the SFI standard, and developed a "Sustainable Forestry Board" independent of the AF&PA with which to develop ongoing standards.

Key Features of Non-State Market Driven Environmental Governance

Five key features distinguish NSMD governance from other forms of public and private authority (see Table 12.3). The most important feature of NSMD governance is that there is *no use of state sovereignty to enforce compliance*.

A second feature of NSMD governance is that its institutions constitute governing arenas in which adaptation, inclusion, and learning occur over time and across a wide range of stakeholders. The founders of NSMD approaches, including forest certification, justify these on the grounds that they are more democratic, open, and transparent than the clientelist public policy networks they seek to replace. A third key feature is that these systems govern the "social domain" (Ruggie 2004)—requiring profit-maximizing firms to undertake costly reforms that they otherwise would not pursue. This distinguishes NSMD systems from other arenas of private authority, such as business coordination over technological developments (the original reason for the creation of the International Organization for Standardization) that can be explained by profit seeking behavior and through which reduction of business costs is the ultimate objective.

Fourth, authority is granted through the market's supply chain. Much of the FSC's and its domestic competitors' efforts to promote sustainable forest management (SFM) are focused on convincing consumers and producers along the supply chain to support, and demand that its supplies come from certified forests. While landowners may be appealed to directly with the lure of a price premium or increased market access, environmental organizations may act through boycotts and other direct action initiatives to convince large retailers, such as Home Depot, to adopt purchasing policies favoring the FSC, thus placing more direct economic pressure on forest managers and landowners. The fifth key feature of NSMD governance is the existence of verification procedures designed to ensure that the regulated entity actually meets the stated standards. This distinguishes NSMD systems from many forms of corporate social responsibility initiatives that require limited or no outside monitoring (Gunningham et al. 1998, ch. 4).

Table 12.2   Comparison of FSC and FSC competitor programs

	FSC	PEFC	SFI	CSA
Origin	Environmental groups, socially concerned retailers	Landowner (and some industry)	Industry	Industry
Types of standards: performance or systems-based	Performance emphasis	Combination	Combination	Combination
Territorial focus	International	Europe origin, now international	National/ bi-national	National
Third party verification of individual ownerships	Required	Required	Optional	Required
Tracking provisions?	Yes	Yes	Yes	Yes

Sources: Hansen et al. (2006), Cashore et al. (2004), adapted from Moffat (1998, 152), Rickenbach et al. (2000), and www.pefc.org.

Notes: *Performance-based* refers to programs that focus primarily on the creation of mandatory on the ground rules governing forest management, while *systems-based* refers to the development of more flexible and often non-mandatory procedures to address environmental concerns. *Third party* means an outside organization verifies performance; *second party* means that a trade association or other industry group verifies performance; *first party* means that the company verifies its own record of compliance. *Chain of custody* refers to the tracking of wood from certified forests along the supply chain to the individual consumer.

Table 12.3   Key features of NSMD governance

Role of the state	State does not use its sovereign authority to directly require adherence to rules
Institutionalized governance mechanism	Procedures in place designed to created adaptation, inclusion, and learning over time across wide range of stakeholders
The social domain	Rules govern environmental and social problems
Role of the market	Support emanates from producers and consumers along the supply chain who evaluate the costs and benefits of joining
Enforcement	Compliance must be verified

Sources: Adapted from Cashore (2002), Cashore et al. (2004), and Bernstein and Cashore (2007).

Table 12.4   Support for FSC certification across countries

	BC (Canada)	US	UK	Germany	Sweden	Finland
Initially	Scant	Scant	Scant	Scant	Scant	Scant
After efforts to gain support	Widespread pragmatic	Scant	Significant pragmatic	Weak	Pragmatic industry; landowner opposition	Scant

The Emergence of Forest Certification in North America and Europe[2]

Our comparative research in North America and Europe revealed a puzzling divergence regarding forest owner support for forest certification (Cashore et al. 2004). Initially most forest owners balked at the idea of FSC certification, with only scant support occurring across all cases. However, after active efforts on the part of environmental groups to influence the supply chain dynamics, largely focused on boycotting and shaming large purchasers of forest products, such as lumber retailers such as B&Q in the UK and Home Depot in North America, as well as German publishing houses and others, support for certification divergence within North America and Europe. In British Columbia, Canada, the FSC made significant inroads through active legitimacy achievement strategies, with the result that initial forest company rejections of the FSC gave way to a situation in which seven of the ten largest companies in the province indicated some support for this program (Cashore et al. 2004). However, in the US, most large forest companies continue to reject the FSC and have instead strongly supported the AF&PA's Sustainable Forestry Initiative. In the UK, state forest owners reluctantly supported the FSC, while small, private landowners now support the PEFC. The PEFC has gained the support of most state and private forest landowners in Germany while the FSC is supported by a minority of state forest landowners, whose German political masters support an environmental agenda (Auld 2001; Newsom 2001). In Sweden, large industrial forest companies support the FSC, while small landowners reject it (Cashore et al. 2004). In Finland, where small forest owners dominate, the PEFC has obtained widespread support, while the FSC has failed to make significant inroads.

---

2   Interviews were conducted with key members of the forest policy communities detailed in Cashore et al. (2004) and Cashore et al. (2007). For brevity, we only refer to specific interviews when identifying key factual points.

The Analytical Framework

Converting and Conforming

We assessed this puzzle deductively and inductively, with careful attention placed on forest certification as a highly dynamic process in which active "agency" efforts by environmental groups to alter initial evaluations of forest owners against the FSC[3] were facilitated and/or debilitated by enduring features common to each country's forest sector. Our classification framework drew heavily from Suchman (1995), and focused on the distinguishing efforts of certification programs and their supporters' first attempt to influence outside audiences by "converting" forest owners to support their system. When converting fails to generate support, strategies then turn to second-best "conforming" efforts that see the certification program change its rules and procedures to address forest owner concerns, in hopes of increasing support from the audiences from whom they seek approval. We documented, and then theorized about the factors that facilitated FSC supporters' "converting" efforts (explaining that when these factors do not exist FSC supporters will have to conform, and/or fail to gain widespread interest from forest owners). We argue that successful "converting" strategies fit with what Vogel has termed elsewhere as "trading up" (Vogel 1995), where increased trade and market transactions lead to increased environmental protection, while "conforming" strategies for the FSC, they assert represent what Vogel (1995) refers to as "trading down."

The Argument

Factors Influencing FSC Converting Efforts

We argue that three structural features—place in the global economy, structure of the forest sector, and the history of forestry on the public policy agenda—work to facilitate or debilitate efforts to have forest companies and non-industrial forest owners support the FSC. These factors help us understand why the FSC has gained pragmatic support from forest companies and forest landowners in some countries/regions, but little or no pragmatic support from forest companies and landowners in other countries/regions. We developed these hypotheses by drawing on a broad set of theoretical literatures from political science, sociology, policy studies, and economics, as well as extensive inductive research. This effort was also innovative in that we undertook research for the Finnish case following analysis and write up of the other five, which permitted us to do

---

3 Cashore et al. (2004) draw on "pragmatic," "moral" and cognitive legitimacy distinctions developed by Suchman. This chapter emphasizes efforts to gain "pragmatic" support, since it was this category that informed the bulk of their attention.

what Geddes (1990, 2003) and King et al. (1994) have criticized comparative historical analyses for sometimes failing to go outside the original case studies from which a theory was developed to explore whether the causal relationships apply elsewhere. This effort, as we review below, permitted us to rigorously assess our original inductively and deductively derived framework, and resulted in qualifications to the hypotheses that we discuss below. In the following section we identify our specific hypotheses, the rationale behind them, and then discuss their applicability/relevance in the context of our six case studies.

A. Place in the Global Economy

Hypothesis 1: (Forest Sector Export Dependent)   Forest companies and non-industrial forest owners in a country/region that sells a high proportion of its forest products to foreign markets are more likely to be convinced to support the FSC than those who sell primarily in a domestic-centered market.

Hypothesis 2: (Forest Sector Import Dependent)   Forest companies and non-industrial forest owners selling wood to a domestic market in a country/region that imports a large proportion of all the forest products it consumes are more likely to be convinced to support the FSC than those in a country/region that imports a small proportion of all the forest products it consumes.

Rationale   The rationale behind Hypothesis 1, broadly supported by existing research (Keck and Sikkink 1998), is that it is often easier for environmental NGOs to wage internationally-focused boycott campaigns in countries that consume the products than in the countries where those products are manufactured (Barker and Soyez 1994; Bernstein and Cashore 2000). And at least part of the reason for this is that campaigns waged domestically are open to domestic criticism that they are hurting the domestic economy and supporting rulings from "outside" the political system, since international market campaigns in general, and NSMD certification systems in particular, originate outside any one country's domestic processes. International market pressure is largely immune to such concerns, since they suffer no sanctions or "backlash" that domestic retailers can, and do, undergo.

Hypothesis 2 identifies those cases in which domestic interest in foreign management practices is so strong that a "boomerang" effect occurs in which their own practices, which otherwise would never have made it on to a policy agenda, are also subject to scrutiny. This phenomenon has been largely underdeveloped in existing literature, and hence our justification is largely inductive. Our specific rationale for this hypothesis is that international market boycotts can reverberate to internal practices when a countries imports significant quantities of the product under scrutiny (forest products in our case). In other words, importing large amounts of forest products can influence the susceptibility of forest companies and landowners to FSC converting strategies.

There are two ways in which this susceptibility is created. First, forest companies and producers in a region that imports a large proportion of its forest products will be especially susceptible to competition from FSC-certified producers outside its borders if their own domestic market is demanding FSC-certified products. Fear of losing market share to foreign imports makes these domestic producers more susceptible to FSC converting strategies. Second, forest companies and landowners in a region that imports a large proportion of its forest products will be more susceptible to moral suasion to practice the same sustainability requirements than their foreign producers are being required to do. Otherwise they risk facing accusations of promoting a double standard.

Applicability to cases   Our initial analysis of the first five cases (British Columbia, US, United Kingdom, Germany and Sweden) revealed strong support for Hypotheses 1. Those countries that had a high level of dependence on foreign markets for their exports (British Columbia and Sweden) revealed some of the strongest forest company interest in FSC-style certification. As detailed in Cashore et al. (2004), both British Columbia and Swedish forest companies showed some of the strongest interest in, and attention to, environmental groups "converting" strategies aimed at international markets (especially manufacturers and retailers). In the BC case, domestic and transnational environmental groups pressured demand-side companies in Europe and the US to terminate their contracts with companies operating in the region that did not conform to FSC criteria (Stanbury and Vertinsky 1997; Stanbury 2000) while Sweden faced similar campaigns aimed at their critically important UK and German markets. However, our Finnish case (Cashore et al. 2007) revealed an anomaly in that while being the most export dependent of any of the original five cases, forest firms and forest owners in Finland never seriously considered the FSC, and instead vigorously worked in developing a made-in-Finland solution and quickly sought recognition from the Program for the Endorsement of Forest Certification (PEFC). The Finnish case does not actually disprove Hypothesis 1, since it did come under intense pressure, but the relative influence and push of other factors, detailed below, "tipped the scales" toward developing a made-in-Finland solution. Recognition of this highlights the importance in understanding the intersection of domestic and international pressures, which work quite differently than we would expect in the public policy and market interaction cases that Vogel and others have captured in their research.

Our cases also illustrate the validity of Hypothesis 2. We found when a region is a net importer of raw materials, domestic FSC converting strategies are enhanced, when retailers demand that foreign and domestic supply are subject to the same scrutiny. That is, the FSC and its supporters are more able to pursue converting strategies, when institutional consumers make purchasing commitments that apply to both domestic and foreign products. For instance, the UK case revealed that, when the supply-side in a region is small and cannot produce the volume of forest products required to meet local demand, it becomes susceptible to competition from FSC imports (Auld 2001). Hence, when the British home improvement retailer B&Q issued an ultimatum to it suppliers that, by the end of 1999, it intended to purchase only FSC certified wood, local processors were cast under the same net, even though they were not the source of original concerns (DIY 1998; National Home Center News 1998; Stanbury 2000). In fact, competition from FSC-certified suppliers in Sweden and the fear that countries in the Baltic States would follow suit (Hansen and Juslin 1999; Tickell 2000), made UK local producers recognize the need to protect their UK market share by conforming to FSC sustainability requirements. However, as we discuss below, this hypothesis does not exist independently and must be assessed in conjunction with the others.

The US case, which was neither import nor export dependent, is also consistent with Hypotheses 1 and 2, in that these features strongly limited the FSC efforts to implement successful converting strategies. With the exception of Finland, we witnessed the lowest degree of forest company and forest owner support for the FSC, despite widespread market-based converting efforts on the part of the FSC and its supporters. The US case revealed that it was easier to secure the commitment of a Home Depot to prefer FSC wood by focusing on "endangered" forests in BC and the tropics rather than on problems with domestic forestry practices.

Taken together, our exploration of these hypotheses is revealing about the influence of economic globalization in assisting efforts by environmental groups to force upward environmental standards. The cases support Bernstein and Cashore's argument, drawing on Vogel, that the "downward" "race to the bottom" effects of economic globalization can be reversed by efforts to link access to these markets with environmental performance requirements (2000). As a result, our analysis challenges those environmental critics who contend that economic globalization always has negative consequences. While much more research needs to be done to understand how the upward and downward effects intersect, environmental groups can, in certain cases, use the power of the global marketplace to force companies and forest owners to make choices they otherwise would not have made.

B. Structure of Domestic Forest Sector

Hypothesis 3: (Large Companies)   Large and concentrated industrial forest companies are more likely to be convinced to support the FSC than relatively small and less concentrated industrial forest companies.

Hypothesis 4: (Large Forest Ownerships)   Unfragmented non-industrial forest ownerships are more likely to be convinced to support the FSC than fragmented non-industrial forest ownerships.

Hypothesis 5: (Weak associational systems)   Forest companies and non-industrial forest owners in a country/region with diffuse or non-existent associational systems are more likely to be convinced to support the FSC than those in a country/region with relatively well-coordinated, unified associational systems.

Rationale   The rational for these three hypotheses is as follows. First, concentrated companies—companies with extensive forestland holdings and operations at all points of the supply chain, from the stump to the retail shelf—are more susceptible to converting strategies by FSC supporters. Being easily identifiable, they are more easily "targeted" by environmental campaigns than smaller, less recognizable companies (Sasser 2002). In addition, their size makes it easier to adopt FSC-style certification owing to reduced transaction costs both in terms of ease of accessing certified fiber supply and ease of tracking certified products along the market's supply chain. Second, fragmented land ownership creates obstacles for FSC style certification. Many small landholdings face diseconomies of scale in implementing adopting certification and, perhaps more importantly, small, non-industrial, and private forest owners tend to be philosophically opposed to an environmental-group initiated program creating rules for their forest lands and also opposed to a program in which non-industrial private forest owners do not have a lead role in decision-making processes (Newsom et al. 2002). All these factors mean that the more a region is characterized by fragmented small non-industrial private forest ownerships, the less susceptible its forest sector will be to FSC converting strategies.

Third, the existence of a well-developed associational structure is influential because, as existing literature has found, it has a strong affect on businesses ability to influence policymaking processes (Schmitter and Streeck 1981; Coleman 1988). Hence, in forest certification, Cashore, Auld and Newsom assert that we would expect that the more integrated an associational system, the better able it is to "fend off" pressures from the FSC by undertaking well-coordinated and strategic responses (Oliver 1991). Further, such an association is better poised to limit the ability of individual members to defect or break ranks, such as in the case of a company or landowner who wishes to take advantage of relatively high demand for FSC certified products. Well-represented and unified industries appear not only to be less fertile ground for FSC market campaigns but also able to create a cultural environment in which forest companies are not receptive to certification market pressures.

Applicability to cases   All six cases revealed the importance of the structure of each country or region's forest sector in understanding FSC efforts to increase support. Developments in countries or regions with and without large, concentrated industrial forest companies, fragmented non-industrial forest ownerships,[1] and well-integrated associational systems all provided support for the hypothesized direct effects of these factors in mediating strategic efforts by the FSC and its supporters.

In BC and Sweden, large concentrated forest companies were relatively easy and identifiable targets for campaigners focusing largely on UK and German retail markets. This increased the tendency of both BC and Swedish companies to positively evaluate the FSC; in the case of BC, it pushed them away from solely supporting the competitor program (the Canadian Standards Association), and in the case of Sweden, it caused them to withdraw support from the stalled Nordic Forest Certification competitor program. However, the Finnish and US case reveal the importance in qualifying the independent effects of Hypotheses 3, with the intersecting effects of other factors (especially Hypothesis 4). For example, Finland's forest companies are indeed globally influential, the largest being Stora-Enso, UPM-Kymmene, Metsäliitto, and Ahlstrom (in 2002, these companies were among the 10 leading industries in Europe and in the world). However, their operations in Finland do not follow the same level of vertical or horizontal integration found in British Columbia or Sweden. While they are horizontally integrated at the level of product manufacturing, they only own 9 percent of Finnish forests and are highly dependent on private non-industrial landowners for their raw material.[4] What this meant is that whereas BC and Swedish companies could make executive level decisions about whether to directly respond to pressure to become FSC certified, Finnish companies could only make such decisions from a "bottom up" approach, which would have required significant and widespread learning and changes in original positions from thousands of small forest owners—a task much more challenging than the pressuring of a handful of executives that was needed to promote the FSC in British Columbia and Sweden.

Similarly, with few exceptions, the existence of large, concentrated industrial forest companies in the US did not facilitate FSC efforts. Instead, market-based converting efforts by the FSC in the US had the effect of industrial forest companies supporting more vigorously the FSC competitor program, the Sustainable Forestry Institute (SFI). Of course, since we identify seven factors with direct effects that push in different directions, it is logical that not all will be able to strongly influence the dependent variable (forest company and forest owner choices to support the FSC). What is important is to understand better which factors "trump" other factors and under what conditions the hypothesized direct effects may intersect with other factors to create unpredicted outcomes.

The cases that did not have concentrated forest companies are also consistent with Hypothesis 3. That is, the absence of large, concentrated industrial forest companies in the UK and Germany made these regions much less fertile for direct targeting market campaigns by environmental activists. Instead, activists

---

4   As a result some officials assert that "large-scale industrial forestry doesn't exist in Finland." Personal interview, senior official, MTK.

relied on targeting bigger companies down the supply chain, but this was a second best option and, owing to the fact that most of these companies were in the same country (place in global economy), was made more difficult by charges that the FSC was a "foreign" organization and hence inappropriate for domestic forestry.

The research in all cases reveals the importance and influence of Hypothesis 4. Indeed, the cases revealed that the majority of non-industrial private forest owners in all of the cases under review either saw no need for forest certification at all, or worked to develop an alternative to the FSC where sustained pressure existed. In no country did the majority of non-industrial private forest owners accept, adopt, or express interest in, the FSC. Indeed, in all six cases under review revealed a common pattern: though there was divergence in industrial support for the FSC there was much stronger hostility from private forest owners towards the FSC—which Cashore, Auld and Newsom (2004, ch. 8) hypothesize may be partly explained by a sense of "independence" private forest owners have, qua owners, that profit-maximizing firms do not. Research on US forest owner attitudes supports these conclusions, revealing that many forest landowners in this region are ideologically opposed to FSC-style certification (Newsom et al. 2002).

Recognition of this highlights the need to understand the role of non-industrial private owners in facilitating or debilitating FSC efforts. In the case of Sweden the importance of non-industrial private owners and their eventual opposition to the FSC worked to limit slightly industry's commitment to the FSC. Given the non-industrial private forest owners, Swedish forest companies first looked to the FSC to *conform,* by pushing the program to alter its percentage-based claims approach. Later they went so far as to urge the FSC to reach out to the landowner-initiated program, the Pan European Forest Certification (PEFC), in an effort to develop a "made in Sweden" system that Swedish forest owners could deem appropriate. The UK case is a good example of how the effects of fragmented non-industrial private ownership were very real, but concentrated government owned lands and highly strategic maneuvering by FSC officials worked to downplay their significance. In the UK FSC strategists carefully read conditions in this country, allowing them to gain indirect support from most landowners in the UK

The US and Finnish cases reveal the importance of private forest owners resolutely opposing the FSC in influencing the support and evaluations of others along the supply chain, particularly industrial manufacturers. In both regions the opposition of non-industrial forestland owners to prescriptive focused certification in general, and the FSC in particular, trumped other effects such as industrial forest company concentration, greatly reducing FSC strategists' converting efforts. In both countries, wood processors require a continuous fiber supply in order to feed their highly specialized, capital-intensive mills. In the absence of non-industrial support, industrial forest companies were influenced by the sheer logistical problems associated with the FSC chain of custody requirements. Hence, the higher the importance of non-industrial private forest owners as a source of fiber means that everything else being equal, companies who might otherwise have been open to supporting the FSC will be much less likely to do so. When this occurs the "large, concentrated" hypothesis gets reversed—pulling in the opposite direction that we argue would have been the case had it operated by itself. Instead, and because of intersecting effects, industrial forest company concentration actually will work to hasten opposition to the FSC and increase support for FSC competitors such as the SFI in the US context and the PEFC in the Finnish example.

By the same token, our research reveals that when a region was characterized by a small number of large landowners, rather than thousands of small ones, the region was more easily converted to the FSC (largely owing to economies of scale in the costs of implementing FSC certification). This helps to explain why FSC strategists in the UK, by using a mix of conforming and converting efforts, were able to achieve strong support in their sector.

Finally, the hypothesized direct effects of a cohesive associational system (Hypothesis 5) did influence as predicted forest company and forest owner choices in BC and the UK (low associational system cohesion aided FSC converting strategies) and Germany and the US (high associational system cohesion limited FSC converting strategies), where well-developed associational systems helped companies and landowners to develop strategic alternatives to the FSC. We observed mixed results for the hypothesized direct effects in Sweden. The associational system cohesion for non-industrial private landowners was high, allowing them to vigorously create and defend the FSC competitor, the PEFC. However, the choice of Swedish industrial forest companies was not as expected. Indeed, the Swedish forest industry's associational system cohesion ultimately worked to enhance support for the FSC, rather than limit it. Again, the explanation for this has to do with understanding the role of intersecting effects during the early stages of FSC efforts to gain support. The effects of Swedish companies being highly exposed to foreign markets and also being large and concentrated ended up, ultimately, trumping the direct effects of associational structure. Once the association reversed its opposition toward the FSC, however controversial it may have been within the association, the associational structure ended up facilitating support for the FSC since once the association made a choice, all of its major forest industrial company members acted consistently with that choice. It was the specific timing and sequence of FSC's early efforts to gain support by focusing on foreign market pressure, followed by company decisions to support the FSC, that saw the associational system solidify its support for the FSC rather than work to create an industrial alternative.

The historically well developed associational systems in Finland were also critical in understanding the emergence of PEFC certification in Finland. Almost all private landowners with holdings greater than five hectares are members of one of over 200 local Forest Management Associations (Finnish

Forest Certification Council 1999; MTK 2001), who are in turn members of "Regional Unions of Forest Management Associations" who are themselves the united national Forestry Council of MTK—an influential and instrumental group in shaping national forest policy (MTK 2001). This longstanding cohesive associational system facilitated clearly facilitated Finland's forest owners efforts to craft a strong, made in Finland solution for staving off the pressures for FSC style certification, reviewed above. The association had immediate access to scientific information, communications budgets, and policy experts with which to develop their own strategic responses and convey them quickly and efficiently to the international market place (Cashore et al. 2007).

C. History of Forestry on the Public Policy Agenda

Hypothesis 6: (Public Dissatisfaction)   Forest companies and non-industrial forest owners in a country/region with sustained and extensive environmental group and public dissatisfaction with forestry practices are more likely to be convinced to support the FSC than those in a country/region with less dissatisfaction.

Hypothesis 7: (Open Forest Policy Processes)   Forest companies and non-industrial forest owners in a country/region where access to state forestry agencies is shared with non-business interests are more likely to be convinced to support the FSC than those in a country/region where forest companies and non-industrial forest owners enjoy relatively close relations with state forestry agencies vis-à-vis non-business interests.

Rationale   The rationale for Hypothesis 6 is that forest owners operating in regions where longstanding criticisms remain are more likely to support the FSC as a "shield" against being targeted in the present or the future since the FSC offers a set of standards endorsed by both domestic and international environmental NGOs. The rationale for Hypothesis 7 is that when the forest industry and/or non-industrial private forest owners enjoy close relations with governmental agencies (i.e. the subsystem is categorized as "clientelist" or "agency captured," forest companies and landowners are less likely to support a FSC-style certification program because it represents a fundamentally different approach in which business cannot dominate forest policy development. On the other hand, if the policy subsystem had already opened up to include an array of interests groups in which business is one of many, then, everything else being equal, business is more likely to support FSC-style certification since it does not represent a change in the status quo.

Applicability to cases   In each of the cases under review, the traditional public policy approach to forest management was identified as being key to understanding support for FSC-style certification, as it influenced whether forest companies and landowners viewed FSC certification as a threat or an opportunity. All of our cases showed support for our predicted relationship regarding sustained conflict and closed public policy networks.

The lack of business-government dominated public policy processes and the experimentation of a range of multi-stakeholder processes in BC during the early to mid-1990s meant industry was less threatened by the FSC multi-stakeholder, tripartite approach. The industry recognized that the closed processes dominant in the 1970s and 1980s would be difficult to reconstruct, even with the election of a more sympathetic administration in 2000. And the sustained scrutiny on BC forest practices both domestically and internationally meant that industry was more amenable to market solutions provided by the FSC approach.

Likewise, in Sweden increasing and sustained societal criticism of Swedish forestry practices, also from both domestic and international sources, meant that its industry was open to alternative solutions. And while lower level implementing networks were still closed, the increasing use of multi-stakeholder processes at the national level, with clear goals governing environmental stewardship, helped enhance support for the FSC as the way of addressing these goals. Similarly, in the UK increasing concern about domestic forestry practices helped the FSC in its efforts to convert forest owners to support the FSC. Relatively closed government-business networks mitigated against forest owner support, but once government decided to help facilitate certification discussions, these closed networks ended up, indirectly, supporting FSC efforts. This is because business interests entered into these discussions only because it was government, and *not* the FSC, with its highly disputed decision-making procedures, that was convening the process. And yet it was the FSC that was able to capitalize on this agreement by positioning itself as the dominant *certifier* of this negotiated standard.

The absence of such features on the German and the US public policy agendas worked to limit efforts to converting private forest owners to supporting the FSC. In Germany there was simply no discernible widespread society critique of domestic German forest practices and, hence, no strong rallying cry that FSC-style certification was needed to address a policy problem in this country. And unlike most other countries' domestic forest policy processes, Germany continues to maintain close relations between its state forestry agencies and its landowner clientele—at all levels of the policy process. In the German case the FSC multi-stakeholder approaches posed a radical departure in the way regulations would be made, one that found disfavor among most forest owners.

The role of the public policy process in the US was similar to Germany. National forest policy was, for the most part, off limits to FSC certification, which meant that core FSC environmental supporters would come to see the FSC as most relevant for privately owned commercial forestlands. Unlike the long tradition of sustained conflict on national forestlands, US private land regulation, for the most part, has not received high degrees of sustained and extensive scrutiny. In addition, forest industry and NIPF landowners are relatively more successful at influencing public policy networks at these levels. Both of these features worked against FSC efforts to gain support: in contrast to the educational and voluntary approaches encouraged by most state forest management agencies, companies and forest owners felt they had much to lose with FSC-style certification, in terms of both reduced access to the policy process and the prescriptive and perceived "stringent" approach of the FSC.

There is no question that, following the "forestry wars" in the 1990s, public attention was placed on the forest sector. Environmental group campaigning, frequent on-site protests, and physical attempts by ENGO supporters to stop logging activities, and resultant arrests, resulted in significant media coverage (Hellström 2001). While protests and campaigns initially focused on state lands, by 1994 they moved to include private forests. However, the Finnish governmental forest policy reforms served to significantly address, and minimize, widespread criticisms. This is in part owing to the Finnish government's leadership role in the Helsinki Process, and its June 1993 signing of the Helsinki resolutions which called for ecological sustainable development and biological diversity as an essential element of forest management.[5] And following its national forest policy reforms that concluded in 1997 the Finnish government now responded to societal scrutiny by asserting that all Finnish forest legislation was completely reformed with a new focus of promoting economically, socially, and ecologically sustainable forest management (Mikkelä et al. 2001a).[6]

Despite these proactive efforts to change and develop Finnish forest policy, public dissatisfaction with forestry practices in Finland never reached the level of dissatisfaction found in BC, the US, or even Sweden. In fact, some analysts assert that environmental groups' campaigns over old-growth forests and protected areas actually represented a conflict between environmental interests and the general *public*, who either directly or indirectly (through a member of their family) owned forestland. And governmental efforts to reform forest policy, precipitated by "changes in the international and societal environment of forestry, pressures for reducing the costs of forestry operations, and the active public debate on the sustainability of forestry" (Mikkelä et al. 2001a), appears to have satisfied the general public in ways that did not occur in other cases reviewed by Cashore, Auld and Newsom.

Summary

Table 12.5 reviews each of our cases for whether the particular feature exists that we hypothesized would influence the ability of the FSC to use converting

---

5   Personal interview, senior official UNFF.

6   The revised Forest Act and the new Act on the Financing of Sustainable Forestry provided a compensation incentive whereby small private landowners would be subsidized for safeguarding biodiversity and setting aside protective areas or special habitats.

efforts to gain support. We must emphasize as we discuss above, that each factor has equal causal weight, and that intersecting effects must be carefully analyzed. Nonetheless, Table 12.5 presents important overall trends that should be of interest to scholars attempting to theorize about the emergence of global NSMD governance in different countries, the particular flavor or approach it may take domestically; as well as to practitioners involved in shaping the emergence of NSMD. What is clear is that the three structural features we identified influence the choice, and impact of, strategic choices made by environmental groups, industry and forest owner associations, and others involved in the NSMD supply chain.

Table 12.5   **FSC ability to alter evaluation by hypothesis and case***

Case	Place in the global economy	Structure of the domestic forest sector			History of forestry on public policy agenda		
	H1–H2	H3	H4	H5	H6	H7	✓/6
BC	✓	✓	✓	✓	✓	✓	6
UK	✓	✗	½*	✓	✓	✗	3.5
Sweden	✓	½	✗	✗	✓	½	3
Finland	✓	✓	✗	✗	✗	✗	2
Germany	✓	✗	✗	✗	✗	✗	1
US	✗	½	✗	✗	✗	✗	.5

Sources: Cashore et al. (2004) and Cashore et al. (2007).

Notes: H1: high dependence on foreign markets for exports; H2: high dependence on imports; H3: concentration of forest industry; H4: low level of non-industrial forest fragmentation; H5: fragmented forestry associations; H6: long history of unresolved forestry conflict; H7: industry shares access with non-business interests.

*   The factor's effects described by each hypothesis do not have equal weight as we elaborate and explain in our case studies. We use the numbers simply to synthesize and present a general guide to understand the cumulative effects of each of the individual effects described by our hypotheses.

**   H4 Non-industrial forest land in the UK is distinguished from concentrated government ownerships, and fragmented private forest owners.

Conclusion

The prevailing consensus among most environmental policy scholars is that, for at least the last decade, Europe is a hare—becoming the place of advanced and innovative environmental policy development and either "catching up" to (Vogel 2003b), or forging ahead of (Speth 2004), the US. But are such stark generalizations across all efforts to address environmental deterioration accurate? How do we begin to assess complex and diverse policy innovations that vary within and across sectors, let alone countries, and whose short and long-term impacts are uncertain? Moreover, what can we say about the increasing use of *non-governmental* policy instruments in North America and Europe, including voluntary, self-regulation, reporting, and consumer-oriented labeling approaches? This chapter has shed light on these broader questions by examining what is arguably, of all recent private sector policy innovations, the furthest away from government control than any other: "non-state market driven" (NSMD) governance systems that turn to the market-place for policy making authority. Unlike public environmental policy initiatives in the EU which were, according to Vogel (2003b), pushed upward owing to *governmental* decisions that harmonization was instrumental in developing its single market,[7] support for, and implementation of, NSMD systems have diverged significantly *within* Europe, as well as North America

Two important conclusions emerge from our analysis of the emergence of private authority as a means to addressing environmental forestry concerns across the Atlantic. First, more attention must be placed on understanding how market-based systems that rely on economic demand intersect with longstanding economic and state-based structural factors. While scholars have long studied the role of market integration, including EU and US variations (Vogel 2003b) on public policy convergence and divergence limited attention has been placed on systematically understanding the emergence of private authority. This omission is problematic as market and domestic pressures appear to intersect in very different ways than they do in *public* policy processes. For example, in the EU's institutionalized single-market where mandatory environmental directives can, and do, provide upward pressure on EU-wide environmental policy (Vogel 2003b), the relative strength of international and domestic pressures in shaping policy innovation becomes much more important in the private sphere—where the coercive power of the state is absent and authority derives from support in market transactions.

---

7   The EU, according to Vogel (2003b) faces considerable pressure to promulgate higher EU-wide environmental standards for three reasons: through the increased concern of consumers about the regulatory policies safeguarding health and the environment in other EU states; to avoid a tarnished collective reputation as a consequence of a single bad apple; and because the EU roots its legitimacy on representation of broad EU interests and concerns.

Second, analysis also calls for greater attention to understanding not only the influence of existing public policy approaches to the emergence of private authority, but also on how the broad suite or basket (Gunningham et al. 1998), of policy innovations of the "next generation" of environmental policy (Chertow and Esty 1997) interact to produce different domestic choices. A major contribution in this regard has been made recently by Howlett and Rayner (2006) who have found that the divergence of national forest programs within Europe, especially regarding their content and character, can be explained, in part, by the role, and support of, domestic support in *forest certification* across these countries. That is, not only do public policy choices and public policy networks influence the emergence of non-state authority, but it is now increasingly clear that private authority is influencing the emergence of new public policy initiatives, including their content and instrument design.

Certification programs have presented the world of policy analysis with one of the most provocative and startling institutional designs since governments the world over first began addressing the impacts of human activity on the natural environment. Whether the forces emphasizing global standardization, or national sovereignty, will ultimately dominate, or whether intra EU and North American differences will remain, is arguably one of the more important questions facing students of transatlantic relations, and global environmental governance. For these reasons public policy scholars can no longer afford to ignore the emergence of private authority in the sectors they examine; and likewise, scholars of private authority must carefully assess how their emergence might influence domestic public policy processes.

Chapter 13

Mad Cows and Ailing Hens: The Transatlantic Relationship and Livestock Diseases

Kate O'Neill

This chapter addresses the transatlantic dimensions and impacts of outbreaks of two diseases affecting both animal and human health: bovine spongiform encephalopathy (BSE, or mad cow disease), and avian influenza (AI), during the 1990s and the early years of the twenty-first century. These diseases represent a transboundary risk that has become particularly high-profile in recent years, given the extent of globalization in the trade of animals and animal products. Neither the US nor the EU, nor trade between the two economic superpowers, has been immune to either.

Three questions are explored here. First, and especially given the contentious nature of relations between the EU and the US over food safety related issues such as GMOs or beef hormones, I ask how these diseases have affected the transatlantic relationship. When animal health or food safety in one country is threatened by a disease outbreak in a trading partner, the first line of defense is almost without exception a trade embargo. Outbreaks of BSE and avian influenza in recent years have been no exception. Although not entirely unproblematic, these trade bans between the US and EU have not caused anything like the same degree of contention that have surrounded GMOs or beef hormones. Nonetheless, and despite a history of dealing with animal diseases on both sides of the Atlantic that dates back to the nineteenth century, there is little evidence that the US and EU are working effectively together to minimize disruption to trade in the event of future outbreaks.

Second, comparing the responses of the US and the EU to BSE and avian flu yields some interesting conclusions about the governance of food safety and public health in the US and EU. In some ways, these two economic powers are not as different in their response to these sorts of diseases as one might expect: both are quite precautionary in their responses, not only at the initial, "outbreak" stage but also over the longer term, contra conventional wisdom on the use of the precautionary principle in the US versus the EU. But the differences that exist reflect different institutional structures and stages of development. Notably, one of the prime motivations for EU activities in this area has been to build and expand its authority as a new, supranational form of governance. The US, on the other hand, has not responded to BSE or AI with institutional change and reform at this level, continuing to rely on its existing network of agencies to coordinate its response. International organizations such as the World Health Organization (WHO) or the International Organization for Animal Health (Office International des Epizooties, or OIE) focus their aid and capacity-building efforts on poorer countries. Nonetheless, rules and standards established by these international organizations do influence US and EU responses, and provide guidelines to which all Member States should conform.

Finally, to what extent can we identify transatlantic policy diffusion or convergence between the US and the EU in the arenas of food safety and public health policy? To what extent do they demonstrate that they have learned from each other's experience? In developing two new agencies—the European Food Safety Authority (EFSA) and the European Center for Disease Prevention and Control (ECDC)—the EU clearly modeled their structures and functions on their US equivalents—the Food and Drug Administration (FDA) and the US Centers for Disease Control and Prevention (CDC), but adapted and changed these basic models to better fit the realities of EU politics.

This chapter draws on the respective experiences of and responses by the US and the EU to BSE (1988 to present) and avian influenza (2004 to present) focusing on:

- the EU's response to the emergence of BSE in the UK in the late 1980s, then to its spread across the EU in the 1990s;
- the US's response to BSE in the EU (and other countries) prior to 2003—and its response after its first case was diagnosed in December 2003;
- the EU's response to a minor outbreak of highly pathogenic AI (the H5N2 strain) in Texas in 2004;
- the emerging response of the US and the EU to the global spread of AI (H5N1) that led to the death or culling of over 200 million birds worldwide, and causing over 100 human fatalities.

These cases can be categorized according to whether authorities in either region perceived the threat as internal (within its borders) or external (coming from abroad), and its source as specific (a particular, identifiable point of origin) or general (origin unknown or multiple origins). These perceptions help condition authorities' choice of response, both to new threats and to diseases as they shift categories—from external to internal, for example, or from specific to generalized. Figure 13.1 illustrates these categories.

	Internal Threat	External Threat
Specific Origin	EU's response to BSE in UK	US response to BSE in EU EU response to AI in US, 2004
General Origin	Spread of BSE across the EU Emergence of BSE in the US	Global spread of AI (H5N1), 2004- 2006

Figure 13.1   Threats posed by BSE and Avian Influenza

Responding to Transnational Epidemics

BSE and AI are both zoonoses: animal diseases that can jump the species barrier to humans. Generating an effective response requires mobilizing authorities responsible for animal health, food safety and human health. Both poultry and cattle are heavily traded commodities, forming the bulk of the world's consumption of meat. Any infection or disease—be it BSE or AI, or diseases with fewer implications for human health, such as foot and mouth disease—thus can be rapidly spread around the world. Their emergence and spread is exacerbated by practices of industrialized agriculture—animals kept in crowded, enclosed conditions, and fed "unnatural" feedstuffs are more prone to diseases, and epidemics spread more rapidly under these conditions.

In general, countries adopt certain patterns of national responses to diseases like BSE or AI at a number of levels over time. Although later sections will elaborate the differences in how the EU and US combat both diseases, broad similarities exist across most countries in terms of short and longer term responses. We identify these now.

First Lines of Defense: Emergency Measures and Trade Embargoes

At the onset of an outbreak, authorities in the afflicted country or region undertake emergency measures. These include immediate slaughter of infected or potentially infected animals or birds, their herds and flocks, and quarantine of infected areas. Usually, these measures are accompanied by compensation for affected farmers (although the impact on farmers and level of compensation is frequently a focal point for social conflict in these cases).

For unaffected countries, their immediate concern is usually to keep the disease out of their domestic animal populations and out of the human food chain. Trade embargoes in the event of animal disease are by no means new. In 1878, the British government adopted legislation requiring strict health examinations for cattle from other countries—and thus, in March 1879, US cattle exported to Britain became subject to an immediate slaughter order due to an outbreak of pleuro-pneumonia in Virginia and New York. This impasse was only broken when the US government, at the urging of its domestic producers, introduced its own animal and meat inspection plan, thus restoring confidence in its exports (Kastner and Powell 2002, 284–5).

The main difference now is that these bans are subject to more international scrutiny, and must be justified according to internationally established standards, or risk being labeled protectionist. While the WTO's 1995 Sanitary and Phytosanitary (SPS) Agreement accepts the right of countries to adopt controls to protect public health, as long as those controls "do not represent arbitrary, discriminatory or scientifically unjustifiable restrictions on international trade" (Kastner and Powell 2002, 289), other organizations have expressed concern at the speed and extent of trade embargoes in the event of even minor disease outbreaks. In March 2004, the UN's Food and Agriculture Organization (FAO) warned that reactions to animal diseases were, at that point in time, affecting approximately one-third of global meat exports, potentially wiping $10 bn from an annual market worth $33bn. Earlier in 2004, the OIE, whose responsibility is oversight and information provision on animal diseases worldwide, issued a press release indicating concern over swift trade embargoes in the absence of scientific risk assessment as to the actual threat posed by the outbreak, and in violation, or misinterpretation of its standards (OIE 2004).

Institutionalizing the Response: On the Ground (or in the Barnyard)

After the immediate crisis has passed, authorities take steps to institutionalize internal reforms designed to prevent future outbreaks of the disease, and/or to minimize the impact of a chronic epidemic. These reforms may be in response to scientific advice, the results of official commissions or inquiries, consumer pressure, or the demands of trading partners. As authorities realize the need to build their capacity to deal with similar sorts of threats in the future, they may shift from emergency measures to a routine control model, with long term policies and programs in place to detect, control and/or prevent outbreaks of disease.

In the case of BSE, key points of intervention occur along the beef supply chain, targeting what cattle are fed, how they are slaughtered and ensuring that only healthy animals enter the food chain. Testing, monitoring and surveillance are also critical components of these policies. Testing programs and protocols show extensive cross-national variation—from Japan's (now discontinued) policy of testing all cattle sent for slaughter, to the smaller, primarily high-risk samples used by the US prior to 2004. Control programs include monitoring human health, and undertaking reforms to prevent human to human transmission of vCJD. Notably, this led to additional restrictions on blood donations in many countries (O'Neill 2003).

Reforming Health and Food Safety Governance

Broader institutional reforms are not unusual in response to a new disease threat. These may involve building new agencies and institutions, fundamentally reshaping existing agencies, or working towards coordination between agencies that previously had not worked so closely together. Ideally, these new institutions, or institutional configurations, address not only BSE, but also to ward off the worst impacts of other emerging disease or food safety threats. The emergence of BSE in the UK and its subsequent spread across Europe led to a fundamental shake-up of regulatory institutions for governing food safety in the UK and the EU (Shears et al. 2001). While not unprecedented, the twin threat to human and animal health posed by BSE led to a reorientation and coordination of EU food safety policy. Likewise, the threat of a human epidemic from AI was one of several driving forces in creation of a new EU Center for Disease Prevention and Cure. The US, by comparison, has chosen to work within existing institutional structures in order to address the BSE and AI threats, focusing its energy instead on coordination of existing agencies and advisory bodies.

BSE in the EU and the US, 1988–2007

Emergence and Spread of BSE

BSE was first reported in the UK in 1986, and soon became epidemic among British cattle. In 1992 and 1993—the peak years of the epidemic—over 70,000 UK cattle were found to have the disease (OIE 2008a). Most scientists believe that cattle were infected through being fed ruminant-derived meat-and-bone meal (MBM), a practice that resulted from the need to find a cheap source of protein for mass-produced beef. BSE is a form of transmissible spongiform encephalopathy (TSE), diseases that destroy brain tissue, causing disorientation, loss of motor and cognitive skills, coma, and, quite rapidly, death. At the moment, there is no vaccine, cure or reliable ante-mortem test for BSE or its human form. Although there are several known human TSEs, the UK BSE epidemic represented the first time that the infectious prion causing the disease actually jumped the species barrier. It took nearly eight years from the onset of the BSE outbreak for the British government to acknowledge this fact, even as the media and scientists publicized a link between BSE and variant Creutzfeldt-Jakob Disease (vCJD), the new human form of TSE (Powell and Leiss, 1997).

In Britain, the crisis led to the slaughter of millions of cattle, long-standing trade embargoes, and severe loss of public confidence in the governance of food safety (Jasanoff 1997; Powell and Leiss 1997). BSE has subsequently been reported in 25 other countries. At first many of these cases were in cattle born in

Britain. Its incidence in indigenous cattle has now overtaken the imported cases (OIE 2008a). In 2004, 536 cases of BSE were reported worldwide in 17 countries, not including the UK (OIE 2008b). By the end of 2007, worldwide incidence of BSE had dropped precipitously, with only 126 cases reported that year (OIE 2008b, a). As of early 2008, 163 deaths from vCJD (confirmed and probable) had been reported in the UK, at a median age of 28 (UK CJD Surveillance Unit 2008). Although the human death toll in the UK is nowhere close to meeting the most dire predictions, the nature of BSE/vCJD has significantly amplified risk perceptions around TSEs, and even a handful of cases necessitate strong policy responses on the part of afflicted countries (O'Neill 2005).

The European Union Response to BSE, 1988–2007

On the outbreak of BSE in the UK, the nascent EU faced a series of difficult challenges, which it eventually turned into opportunities to create EU-wide competency in food safety regulation. In particular, it took advantage of the crises in consumer confidence across the Member States caused by the disease, and the disastrous mishandling of the outbreak by the British government. However, to date it has not restored global confidence in the safety of its cattle and beef products.

At first, when BSE was apparently confined to the UK, the EU (then the European Community) acted to ban all movement of British beef and beef products throughout the Community, while the UK government undertook its own emergency measures. But, given the relatively slow incubation of the disease, such measures were not successful: BSE started showing up in cattle imported from Britain in other countries in 1989, and by 2001, BSE had been diagnosed in native-born cattle in 16 European nations (including Member States, future members and non-members) (OIE 2008b).

BSE also took a terrible toll on the production and export of EU beef (Pickelsimer and Wahl 2002). In 1995, the UK exported 77,000 metric tons of beef and veal; in 2000, it exported less than 2000 metric tons. The EU exported 934,000 metric tons of beef and veal in 1995, but only 640,000 in 2000 (Pickelsimer and Wahl 2002, http://www.fas.usda.gov/). In 2004, exports of EU beef and veal reached 363,000 metric tons, dropping in 2007 to 139,000 tons (USDA 2008).

The EC first banned export of UK cattle born before July 1988 (Decision 89/469/EEC), following it up with progressively stricter bans, culminating in March 1996 with the ban of all UK beef exports worldwide (Decision 96/239/EC). These moves both angered the British, and came under fire from member governments and the media for being based on the disingenuous assumption that BSE would remain confined to the UK, and for a fragmented and opaque approach to policymaking (Vos 2000; Buonanno et al. 2001; Vincent 2004).

These reports helped set the stage for a multi-phased shake up of existing EC policy regarding consumer health and food safety over the following five years. The major planks of the Commission response are laid out in its Communication on Consumer Health and Food Safety in April 1997, and in a January 2000 White Paper on Food Safety. In 1997, all scientific committees whose work concerned consumer interests were transferred to DG XXIV, which in 1999 became the DG of Health and Consumer Protection (DG SANCO) (Bergaud-Blackler 2004). This followed the establishment in 1996 of the Scientific Steering Committee (SSC), whose role would be to provide "the best possible" scientific advice across issues involving consumer health and safety. The 2000 White Paper went even further, calling for the establishment of a European food safety agency, whose task would be to provide scientific advice on all aspects of food safety to both EU-wide agencies and to Member States, and provide the central node in a network of cooperation between all relevant actors within the EU. Thus in 2003, the EU launched the European Food Safety Agency, under DG SANCO, whose "farm to fork" philosophy is designed to guide the EU through any and all food safety-related crises into the future. While the EFSA has no legislative function, it is the filter through which all EU food safety decisions must pass. It has a powerful agenda-setting role within the EU, and its emphasis on developing authoritative scientific bases for political action—based on risk analysis and the precautionary principle—likely sets the stage for future reforms of EU governance (Vincent 2004, 517). Further, it very much represents a role the EU itself sees as critical to its identity as an effective supranational governance institution.

Against the background of governance reform, the EU policy regime for handling BSE has itself gone through a number of different phases. First, the EU has shifted from viewing the epidemic as a threat contained to the UK to one that has afflicted virtually all Member States to greater or lesser degrees. Subsequently, from 2001, the EU , in essence, adopted a routine control and prevention regime for BSE, requiring testing of all symptomatic animals, and all animals over 30 months sent for slaughter, and banning feeding of all mammalian proteins to all farm animals (SSC 2001). Countries finding BSE are immediately subject to short-term intra-EU trade embargoes, and to subsequent careful monitoring and surveillance by the European Food Safety Authority.

In July 2005, the Commission released a discussion document, "The TSE Roadmap" (COM(2005) 322 Final), which envisages how TSE control policy is likely to evolve over the medium (2005–2009) to long term (2009–2014) should current positive trends in overall TSE reduction in farm animals continue (European Commission 2005d). More recently, the EU announced the lifting of final controls on UK cattle and beef exports to the rest of the Union ("EU Lifts British Beef Ban" 2006). They admit the eradication scenario faces potential challenges, but these discussions certainly represent confidence in the EU's ability to control and eradicate these diseases. Yet, these actions have, so far, done little to convince the rest of the world to lift its embargoes on EU beef.

The US Response to BSE, 1988–2007

The US response to BSE's emergence in the UK and the EU makes an interesting comparison to that of the EU governing authorities. First, the US clearly defined BSE as an external threat, and, until December 2003, when its first case was diagnosed, based its policy regime primarily on keeping the disease out of the country, with correspondingly weaker internal controls. Second, US authorities, led by the US Department of Agriculture (USDA), were concerned from the outset to avoid the mistakes made by their counterparts in the UK and EU, particularly in downplaying the risk to public health. Finally, unlike the EU, the US policy response involved several well-established federal agencies (there is next to no state-level deviation from federal rules and standards laid out by the FDA and USDA). Although coordination remains an issue, there have been few calls for the consolidation of food safety authority or fundamental reorganization that happened in the EU. In fact, domestic consumer response to BSE has been remarkably muted, even non-existent, especially in comparison with the EU public. The major impetus for policy change in the US has, instead, come from federal authorities and from US trading partners—notably Japan—who led calls for changes in basic US BSE control policies.

Beginning in the late 1980s, the US took a three-step approach to controlling BSE and vCJD (O'Neill 2005): first, prevent introduction; second, should introduction occur, prevent infection of large numbers of cattle; and third, prevent vCJD emerging in the human population. This led to the so-called "triple firewall" of import bans, surveillance and testing, and a ban on feeding most mammalian proteins to ruminants (USDA 2003). As early as 1988, the USDA set up an inter-agency BSE working group. In 1989, the US banned the import of live cattle, feed and beef products from the UK (and any other country in which BSE might be found). In 1997 such imports were banned from all of Europe. Finally, the US acted within hours of the May 20 announcement by the Canadian government of the single BSE case to ban cattle and beef products from Canada.

US policy toward BSE and beef imports from the EU appears to be based more on risk assessments of its vulnerability to BSE entering from outside the country than on a desire to protect its beef markets. For one, the US imports relatively little beef, and has always relied more on production within North America (including Canada). The study providing the cornerstone of US BSE policy was commissioned by the FDA in 1998 and authored by the Harvard University Center for Risk Analysis (Harvard Center for Risk Analysis and Harvard School of Public Health 2001). Published in December 2001, this study was highly influential in shaping official policy debates around BSE. It took the position that while the appearance of BSE in the US cattle population was quite likely on a minor scale, a full-scale epidemic was next to impossible. Its recommendations focused primarily on technological and epidemiological factors, and getting the right policies and practices in place, rather than on potential failures in implementation and compliance. As a result, US policy towards BSE has been heavily criticized by stakeholder groups, most especially consumer activist groups, who saw internal measures as weak, fraught with loopholes, and subject to serious implementation problems (Rampton and Stauber 1997).

This situation changed in December 2003, when a dairy cow in Washington State tested positive for BSE. USDA and FDA immediately began speeding up a process of policy reform already under discussion in the months since May 2003, when Canada announced its first case. These included measures to remove high-risk cattle and meat by-products from the food chain, to tighten feed rules, and significantly expand the cattle-testing program, using the rapid testing techniques favored by the EU (O'Neill 2005).

Many of these new measures were strongly disputed by the US beef industry, much more politically powerful than its EU counterpart. However, their implementation—albeit slow—proceeded apace, at least for a couple of years. By early 2006, the US had restored trade relations with most of its erstwhile partners, and domestic consumption of conventionally-raised beef remained stable. BSE has, in effect, been far less of a crisis in the US than the EU, but has still triggered significant reforms in how beef is raised and processed. By late 2007, however, it had become clear that US authorities were taking an optimistic view: testing programs enacted in 2003 and 2004 had been rolled back, and enforcement of existing rules remained a problem. Nonetheless, the discovery of abuses at a southern California slaughterhouse in early 2008 reminded the general public of potential problems with beef production, and in May 2008, authorities enacted a ban on sick animals entering the food chain. Although so far the US has escaped a BSE epidemic, it is not yet certain what the future trajectory of identified cases of BSE is going to be, and what the impact of finding tens, or hundreds of cases could be on the overall political landscape of US food safety politics, which remain a work in progress.

Avian Influenza, 2004–2007

History and Recent Outbreaks

If the 1990s were marked by the emergence of BSE as a major global threat to human and animal health, leaving authorities scrambling to generate an adequate response, than the early years of the twenty-first century belong to avian influenza (AI, or bird flu). The lethal H5N1 strain of highly pathogenic AI, which emerged in South East Asia in 2004 and subsequently spread to the rest of Asia, Europe and Africa, generated fears of a human flu pandemic which could potentially kill hundreds of millions of people.

Like BSE, AI can jump from animals to humans. Its spread and impact, are also magnified by practices of industrialized agriculture, under which animals are raised and kept in crowded, enclosed conditions, making them far more vulnerable to disease epidemics. Yet, AI is not a new disease. Its incubation periods are far shorter, and while it can be spread through trade in live or recently killed birds, the main vector for the disease is thought to be migrating wild birds. Further, AI cannot be transmitted through eating cooked chicken. To date, the only humans infected by the disease are those involved in handling live birds.

Avian influenza (AI) infects many bird species commonly raised for human consumption (for information on avian flu, see USDA 2002; World Health Organization 2002; USDA 2004). Avian flu virus strains are categorized into two main forms: low pathogenic (LPAI) and highly pathogenic (HPAI). LPAI is by far the less dangerous of the two, rarely fatal to infected birds. It does not infect humans. However, it can rapidly mutate, especially if left unchecked, into HPAI. Highly pathogenic AI is, by contrast, extremely dangerous to birds and to humans. Human infection from HPAI is relatively recent, suggesting that the virus has mutated. In 1997, 18 people in Hong Kong were infected in the course of an HPAI outbreak, of whom six died—the first known instance of human deaths from HPAI (World Health Organization 2002). Hong Kong's entire poultry population, around 1.5 million birds, was culled in three days.

Between 1959 and 2003, there were only 21 recorded outbreaks worldwide of HPAI, mainly in the Americas and Europe (World Health Organization 2004). In the case of one outbreak, in the US in 1983–84, the virus began with relatively low mortality, but within 6 months was killing nearly 90 percent of infected birds. Seventeen million birds were culled at a cost of nearly $65 million. Another, in Mexico, broke out in 1992 and, due to lack of prompt control measures, lasted until 1995 (World Health Organization 2002). The most recent outbreaks, which began in Asia in 2004 proved even more deadly (mortality rates of 60 and 70 percent, cf. 30 percent in Hong Kong in 1997), alarming officials at the speed of travel of the virus, and its possible mutations (Parry 2004).

Early 2004 saw one major and one minor outbreak of HPAI around the world. The major outbreak of HPAI (H5N1 strain) occurred in East and South East Asia. First reported in South Korea in December 2003, cases were subsequently confirmed across East and Southeast Asia. International and national officials swung into action. The WHO Global Influenza Surveillance Network, the FAO, and ASEAN all took on major roles in helping stem the spread of the disease. However, the WHO noted that "the present outbreaks in poultry are historically unprecedented in their geographical scope, international spread, and economic consequences for the agricultural sector" (World Health Organization 2004, 2).

By March 2006, the H5N1 virus had been found in birds across Asia, Africa and Europe. The first outbreaks in the EU were identified in wild swans in Italy, Greece, Germany and Austria; cases have subsequently been found in France and across Central Europe (BBC News 2006). Of the 186 human cases diagnosed in eight countries as of late March 2006, 105 patients have died

(World Health Organization). An estimated 200 million birds have died as a result of the outbreak (Altman 2006).

H5N2 Avian Influenza in North America, 2004: The EU Responds to a Specific Threat

The second, more minor outbreaks of HPAI in 2004 occurred in North America: in Texas and in British Columbia, both reported in late February 2004. On February 20, a case of HPAI turned up on a small farm in Texas—the US' first case since 1984 (Murphy 2004). A different strain of HPAI than the one that appeared in Asia (H5N2 rather than H5N1), the chickens did not exhibit the usual symptoms of HPAI, and only further testing revealed the nature of the strain. Immediately, the Texas flock was culled and quarantine established. While it was possible that the chickens were infected by wildfowl, it was also possible they had been infected at a live market in Houston (Murphy 2004).

Immediately, the EU and many other countries suspended imports of eggs, poultry and poultry products from the US (Commission Decision 2004/187/ EC, 'International Roundup' 2004). The meaningful part of this ban is that on egg imports, which accounted for about one-quarter of total EU egg imports, or about 9 million eggs, worth $20 million annually, and on live chicks. As *Food Chemical News* points out, "the poultry meat ban is academic, because the EU already refuses exports on the grounds that American processors use chlorine to sanitize carcasses" ("International Roundup" 2004). Although the EU recognized that the Texas strain was likely not as dangerous as the Asian version, the EU justified its decision on the basis of risk posed to European flocks. The suspension was set for an initial period of 30 days, subject to review. On March 9, 2004, authorities found HPAI among avian flu cases in the Canadian province of British Columbia, and the EU imposed a similar ban on Canadian exports of poultry.

One month later, the Commission's Standing Committee on the Food Chain and Animal Health met to consider the North American ban, which was partially lifted (European Commission 2004a, 2005a). The EU left in place a limited ban, on areas of each country in which the HPAI outbreaks occurred, namely the entire state of Texas, and large parts of British Columbia. As 15 percent of the poultry in Texas is exported, this imposed some significant costs on its poultry industry (Hart 2004). On August 23, 2004 all restrictions were lifted (APHIS 2004). In announcing this decision, European Health Commissioner David Byrne was quoted as saying, "this demonstrates the proportionality and flexibility of the EU's decision-making capacity based on risk analysis" (Clapp and Lewis 2004).

H5N1 Avian Influenza: The EU and US Respond to a Generalized Threat

As the H5N1 strain of AI spread around the world in 2004–2006, governments and international organizations began scrambling to formulate responses to a twin threat: to commercial and domestic poultry operation, and to human health (both poultry workers and the general population). From the outset, this outbreak of AI has been treated as an emergency. Actions taken included the culling of infected flocks, and measures to prevent exposure in healthy flocks (e.g. keeping them indoors; vaccination; preventing export from infected areas and countries). Many countries have, following WHO and OIE guidelines, established educational and outreach programs to help citizens recognize and report signs of the disease. In terms of human health, the race is on to find appropriate treatments for human victims, and to identify, manufacture and stockpile appropriate vaccines and treatments for a possible outbreak. The level of alarm in the popular press about a possible pandemic was high, but the health establishment is split about the probability of its occurrence (*New York Times*, March 28, 2006). All experts, however, agree that the speed and distance of travel of such a virus around this globalized world would both be far higher than in past epidemics.

In both the US and the EU addressing the threat of H5N1 has engaged agencies across human health, agriculture and wildlife agencies. In the US, the USDA (APHIS), the CDC, the National Wildlife Service, and the FDA are all engaged in the effort to prevent, or minimize, an AI outbreak. In the EU, in addition to the EFSA, the AI epidemic is the first major test for its new European Center for Disease Prevention and Control.

The ECDC, based in Sweden, near Stockholm, opened for business in May 2005, with an initial staff of 10 (Clapp 2004). Its mandate is to monitor and control the spread of infectious disease across the EU and neighboring states, in essence coordinating and expanding an existing network of surveillance and informational agencies across the Member States (MacLehose et al. 2002). It will take over the EU's Early Warning and Response System on infectious disease, and provide scientific advice and training to Member States. By 2010, the ECDC is expected to employ 300 staff, with an annual budget of 90m Euros (Jack 2005). As an agency, it has been compared to the US CDC (EPHA 2004; Wigzell 2005). However, it differs from its original model in significant ways. First, it does not have anything like the same size or powers as the CDC. Instead it is designed (rather like the EFSA) to be a central point in a network of related agencies and organizations, particularly the national bodies of the Member States. It thus takes a more decentralized approach than its US equivalent (Jack 2005), providing advice and information rather than taking control of disease situations. It also does not (at present) have plans for its own laboratories, another function to remain in the domain of the Member States. In May 2004, when the ECDC was first proposed, authorities did not expect it would be tested so soon. It remains to be seen how well it will cope with a potential flu outbreak among the EU's human population. However, the founding of the ECDC, along with the EFSA reflect the EU's current philosophy of supranational governance when it comes to issues of scientific knowledge and advice across all Member States.

Analysis

The transatlantic dimensions of global disease outbreaks should be seen in the context of large-scale economic integration and the expansion of international trade. The problem of cross-border transfer of disease in the face of differing vulnerabilities, threats, and standards is exacerbated by trade liberalization, even as that liberalization brings economic opportunity. This section analyzes how the world's two largest economies—the US and the EU—handle these risks. While each economic superpower has strengthened its borders and improved or reformed domestic institutional capacity, there is less evidence that the US and the EU are engaged in joint risk management or harmonization strategies. Still, clear differences in longer-term risk mitigation strategies on the part of each demonstrate some interesting contrasts between the EU—a relatively new supranational authority– and the US, a more mature federal governing institution.

The Transatlantic Relationship: Trade Embargoes and Risk Mitigation

Transatlantic trade in live animals, animal products, and other agricultural commodities has been fraught with conflict for a long time now (Keilbach, this volume). However, trade embargoes in response to infectious livestock diseases have caused little inter-governmental conflict, although they are not cost-free.

Trade embargoes in response to animal diseases in foreign countries may perform any of three functions. First, they can serve as a precautionary measure to protect domestic livestock and public health. Second, they can be a protectionist measure, put in place to protect domestic markets from foreign encroachment. Third, they can be a tool to get another country to change its policies in handling a given disease. In this way, they are used to mitigate negative effects (and risks) of economic interdependence over the longer term.

In the context of transatlantic trade relations, the US response to BSE in the EU, and the EU response to HPAI in the US in 2004 are most pertinent. Each of these outbreaks generated different trade embargoes. The US responded to BSE in Europe with a complete ban on imports of beef and beef products from the UK (1989) and the EU (1997). These bans remain in force to this day. The EU ban on poultry products and eggs from the US was far more targeted. Designed to be temporary (based on monthly reviews), in April 2004, the ban was limited only to the state of Texas, where the outbreak of HPAI occurred.

There is some dispute over the ultimate purpose of trade bans in response to disease outbreaks. Conventional wisdom has it that trade bans are blunt and defensive instruments, especially over the longer term. Critics argue, too, that such bans tend to benefit domestic producers more than they benefit public and animal health, and may constitute non-tariff barriers to trade. This discussion disputes both these points. First, I argue that the US reaction to BSE in Europe is more accurately seen as precautionary (protecting human and animal health) rather than protectionist. Second, the case of the EU's selective, and targeted embargo of US poultry exports in response to avian flu may demonstrate a particular use of trade embargoes: to generate positive incentives for policy change in the target country under particular conditions—for example, when domestic consumer pressure for such change is not present. Although not directly relevant to this chapter, Japan's ban on American beef following the US' first case of BSE presents another example of this sort of targeted, conditional trade embargo.

In general, the US is not noted for taking a "precautionary" approach—in the sense of acting either in the context of significant uncertainties, or going above and beyond what prevailing evidence might suggest—in leading environmental and food safety issues. However, even countries not generally known for their advocacy of the precautionary principle, tend towards precautionary action when human health is at stake (Wiener and Rogers 2002). Further evidence in favor of US precaution, rather than protectionism, lies in the lack of international dispute of its actions—neither international organizations nor the EU have disputed this embargo. Second, US authorities justified their decisions strictly, and effectively, in terms of threats to human and animal health.

Certainly, the US experienced some benefits from the worldwide embargo on EU beef products. However, a close examination of the debate in the US policy establishment, where policies were fairly clearly justified on the basis of scientific risk assessments, suggests that these measures were designed primarily to protect human and animal health against the emergence of BSE and vCJD within the US. One may argue that the framing of BSE as a purely external threat does, on the other hand, protect the US beef industry. By denying possible internal sources of infectivity, and not addressing existing loopholes in policy at feedlots and slaughterhouses, consumer advocates argued that USDA was effectively denying multiple sources of risk.

A further argument that these trade embargoes are more precautionary than protectionist is that the main beneficiaries are frequently third parties who are unaffected by the disease. Latin America has been the big winner in both the recent BSE and avian flu outbreaks, while the Australian beef industry has benefited immensely from the closure of Canadian and US beef exports (Lewis 2004). Losers, on the other hand, include farming communities in the embargoed countries (who often face bitter battles with the government over compensation) and consumers in countries imposing the embargo, who often have to pay higher prices for beef or poultry products.

Given the blanket nature and permanent status of the US trade ban on the EU because of BSE, it was clearly not designed to change EU behavior. This can be seen by comparing the more targeted and review-based ban on US poultry products which the EU put in place in 2004, which appeared to be targeted at changing US behavior and policy. It was clearly based on sets of conditions the US had to meet in order for it to be lifted, and subject to periodic review. One possible motive behind this set of actions on the part of the EU was a fear that US consumers lacked the concern to push for change, as, unlike in the EU, US consumer confidence in beef barely dipped (Hallman et al. 2004). The EU, too, may have learned from its experiences with its Member States and BSE.

The use of trade embargoes as a short-term, targeted tool to change specific policies and practices in, or within, the target country has several advantages. First, it avoids large-scale costs, and therefore the likelihood of worsened relations over the longer term between the parties involved. Second, as a practice it is generally more in harmony with international trade and food safety regulations. Third, and perhaps more controversially, it is a way for one country, or group of countries to influence domestic politics and practices of their trading partners. This is seen as a particularly useful tool for trading partners of the US—including the EU and Japan—in cases where they are concerned about the relative absence of concern on the part of most US consumers. Further, this proved relatively successful in recent cases of BSE and avian flu, and appears to have inflicted little damage on transatlantic relations more generally. Whether or not these sorts of trade embargoes can be over-exploited is yet to be seen.

Beyond trade embargoes, while steps on both sides have been taken to reinforce domestic defenses and coordination in the event of an outbreak, less has been done to develop cooperative mechanisms that would allow the US and EU to ward off potentially damaging trade embargoes in advance. The two economic superpowers are following "separate but parallel" tracks in terms of regulatory policy, and paying less attention to developing joint, long-term solutions to preventing the transmission of livestock diseases across national borders. In July 1999, the EU and the US signed a veterinary agreement, whose aim was, according to a Commission press release, to "facilitate trade in live animals and animal products between the EU and the US by establishing a mechanism for the recognition of equivalence of sanitary measures operating in the two regions" (Freeman 2002, 366). As Freeman notes, this agreement essentially validated the existing status quo, and has apparently generated little analysis or excitement. No mention of its existence has been made in recent analyses of BSE or avian flu outbreaks.

The Comparative Story

Examining, particularly, their respective responses to BSE, but also their more nascent responses to the H5N1 strain of bird flu yields some interesting differences and similarities between the EU and the US, both in the development of "on-the-ground" strategies to combat the disease, and in the development or reform of related governance institutions.

First, countries tend to act in remarkably similar fashions in an initial outbreak crisis. It is in how countries react to a potential outbreak over the longer term, learn to manage a disease that has become endemic, or undertake deeper government or institutional reforms where key differences emerge.

To return to Figure 13.1, and using BSE as an example, the US and EU of course differed in terms of how and when they experienced "mad cow" disease. In the EU, the epidemic began within its borders, and spread across the continent over the course of a few years. While the US began taking defensive measures in the late 1980s, it was not forced to reexamine its BSE control regime until late 2003, despite heavy criticism from several consumer groups and the Government Accountability Office (Rampton and Stauber 1997; United States General Accounting Office 2002). As of May 2008, only three cases of BSE have been officially diagnosed in the US (the first of which is classified as Canadian), compared to many thousands across Europe. To some extent, the perception held by the US policy establishment that BSE was a minor, and external threat (a view backed by important scientific studies) helps explain what might be termed a "selectively precautionary" response, focused primarily on keeping the disease out, and minimizing the possibility of an "internal" outbreak. However, until 2003 (and even to some extent afterwards), the US government avoided the widespread criticism leveled at the EU after its first and inadequate response to BSE (Vincent 2004). Since then, the EU and its neighbors have moved to a strong routine control model to which Member States adhere (though not without many initial compliance problems). It is currently contemplating how it can move to eradicate the disease, and thus ramp down its control regime over time.

A focus on the timing of, and vulnerability to diseases such as BSE and AI, does not, however, capture some of the important points of comparison. One must also look to the institutions and actors involved in driving and formulating particular responses, and at the longer term institutional outcomes. While there is not space here for an in-depth study of comparative regulatory structures, styles and participation in these two regions, a few observations are in order.

First, many have pointed to the relative absence of consumer concern in the US in response to the emergence of BSE, especially in comparison to the response in Member States—notably Germany—after the discovery of only one or a few BSE cases (more generally on consumer responses in the EU and EU to food safety issues, see Vogel 2003b; O'Neill 2005). In the EU, it is possible to trace clear connections between consumer confidence and the emergence of new institutions (Shears et al. 2001; O'Neill 2005). US policy makers and lobby groups have not had the force of consumer outrage to drive fundamental reform of animal and public health governance institutions, while the beef industry has remained a politically powerful voice in this process (Sugarman 2004).

Second, governance institutions and powers at the "federal" level differ across the EU and US. The US centralizes control over food safety and animal and public health at the federal level. US states have very little ability to determine their own standards, or to allow their individual standards to exceed federal ones. These regulatory powers are spread out over a wide array of federal agencies and subagencies (USDA, FDA, CDC, Wildlife, APHIS). Under these conditions, a central problems faced by the US government is one of inter-agency coordination and cooperation, rather than one of institution-building (despite some calls for change—see Taylor 2004). The EU, by way of comparison, is far more decentralized. Member State governments bear the ultimate responsibility for ensuring the health and safety of their citizens (and livestock). But the EU has long struggled with developing the content and form of a Union-wide public health establishment (Cucic 2000; Randall 2000), as well as a coordinated response to biosecurity threats (Sundelius and Grönvall 2004). EFSA and ECDC reflect this context. Both agencies were established specifically as providers of authoritative knowledge, not as rule-making, political authorities designed to supplant the authority of the Member States, and both integrate broad policy areas. To that extent, it seems that EU authorities are seeking to avoid the coordination problems they see in the US.

Third, Brussels' response to BSE and other transnational epidemics cannot be separated from its status as an emerging form of supranational government, endeavoring to establish its powers and legitimacy to its members and to the outside world. Such efforts in the context of areas of deeper integration—the establishment of a single currency, constitution-building—have met with differing levels of success. The establishment of EU-wide information and coordination agencies around issues of health and food safety reflect yet another facet of the EU's "state-building" process.

Transatlantic Policy Diffusion and Learning

The final aspect of this story concerns the extent to which any transatlantic policy diffusion or learning is evident in how the US and the EU have responded to AI and BSE. Have particular policies or institutional structures crossed the Atlantic? Has either party learned from the other's experience?

The US and EU are far from achieving convergence in responding to BSE and AI, beyond short-term emergency measures, despite the possibility that harmonized structures of risk regulation (and mutual recognition of such) might help avoid, at least, the economic disruption of a disease outbreak in one country or the other. As is the case with general security problems or concerns, states remain committed to protecting their own populations when under threat from a disease like BSE or AI.

Nonetheless, it is possible to identify learning and policy diffusion across the Atlantic, at least in two respects. First, the US authorities learned from the rather disastrous mistakes made by the UK government, and by European authorities on the initial outbreak of BSE. Their policies addressing BSE have been transparent from the outset, and they recognized it as a threat to both cattle and human health early on. US authorities have also done a better job of communicating their policies and programs to stakeholders and to the public at large. These levels of transparency and communication have, of course, not satisfied critics of policies pre- and post-December 2003, but the American public seems assured that their government is not letting them down (O'Neill 2005).

Second, the EFSA and ECDC are fairly clear examples of transatlantic policy diffusion. Modeled on their US equivalent agencies—the FDA and CDC—they nonetheless demonstrate some clear differences. Their powers are less extensive, designed to function in a system where a good deal of power and governing capacity remains concentrated at the Member State level. Yet, their commitment to providing objective scientific advice (authoritative knowledge) in the event of threats to animal or human health, and to food safety mirrors the missions of their US partners. It is still too early to tell, however, how well these agencies will perform their designated functions, and how they will evolve in the future.

Conclusion

Outbreaks of livestock diseases—even those with potentially severe impacts on human health—have not been a major source of tension in the transatlantic relationship. While they have caused serious economic and other costs for both partners, as well as major trade disruptions, the two partners have not seriously disagreed over their respective approaches (and trade bans placed on one by the other) when either has experienced an outbreak of BSE or avian influenza.

Both take such outbreaks seriously. Even the US approach to BSE could be characterized as precautionary in several respects, and both the US and EU are putting strong measures in place to counteract a potential global human flu pandemic. Still, it is probably most appropriate, from the evidence presented in the BSE and AI cases, to argue that they are each following "separate but parallel" tracks in managing the risks of transboundary livestock or human diseases. Further, their particular responses, while partially explained by their individual experiences of the diseases, have also been conditioned by wider institutional priorities—notably, the EU's "state-building" project. Mutual coordination of risk management strategies is rare. This perhaps reflects the highly "nationalist" approach the European countries and the US have historically taken when faced with a direct threat to health security within their borders. It remains to be seen whether these practices will change, or whether we will see any further convergence in policy institutions if such threats of transboundary disease intensify over the coming decades.

Conclusion

Chapter 14

Transatlantic Environmental Relations: Implications for the Global Community

Miranda A. Schreurs, Henrik Selin, and Stacy D. VanDeveer

Environmental and energy related politics have become increasingly central and contested areas of transatlantic relations. These issues are critical to the transatlantic relationship, involving public officials and organizations and various levels of governance as well as large numbers of private sector and civil society actors on both sides of the Atlantic. As the Atlantic Council, a non-governmental organization for the promotion of transatlantic cooperation, assessed the situation in 2002: "[T]ensions over environment and food safety issues are just below the surface and—if not addressed—will have enduring corrosive and divisive effects. Indeed, the recent acrimony over these issues has contributed to concern about an erosion in shared transatlantic values and a deterioration in US-European relations more generally ... Unless they now find a way to reconcile their different perspectives and approaches, the United States and the European Union will miss real opportunities to work together in addressing global environmental and public health issues" (Aaron and Gray 2002, 1).

The politics of climate change is one area of notable transatlantic tensions. At the G8 meeting in Heiligendamm, Germany in June 2007, clear tensions remained between the US, on the one hand, and Germany and other European states on the other, over what approach should govern international climate change policies in the post-Kyoto Protocol (i.e. post-2012) period. These differences largely persisted through the December 2007 UNFCCC negotiations in Bali and the July 2008 G8 Summit in Japan. Whereas the EU was calling for a significant cut in GHG emissions relative to 1990 levels by 2020 for industrialized countries, the US continued to push voluntary measures and technology research and development.

There are a substantial number of other environmental and energy issues where the EU and the US have pursued different regulatory approaches. Fears about the health implications of the thousands of chemicals used in our daily lives, concerns about what to do with discarded electronic products, worries about tainted foods and communicable diseases, have furthermore put environmental and energy issues high on transatlantic political agendas. On many of these issues, the EU and the US have failed to see eye to eye on how best to approach sustainability issues. This is the case with chemical regulations, the banning of asbestos, product standards, GMOs, renewable energy, and climate change. Although the respective regulatory responses of the EU and the US have been formulated primarily in response to domestic interests and demands, the different policy outcomes have impacted transatlantic trade and foreign policy relations in significant ways.

These issues have injected considerable tension into transatlantic relations and complicated efforts between the EU and the US to work cooperatively. This book was motivated by a desire to understand how these tensions emerged in transatlantic relations, how deep and wide the transatlantic divide really is, and what the potential for an improvement in transatlantic environmental relations might be in the coming years. It has examined the significance of environmental and energy issues in this relationship, the causes of frequent discordance, the prospects for cooperative problem solving across the Atlantic, the role of EU expansion in the transatlantic relationship and the significance of these relations for global politics. Together, the volume's chapters have much to say about these concerns.

Part of what makes the tensions in transatlantic environmental relations so surprising is that Europe and the US have a long history of transatlantic environmental learning and cooperation. The establishment of the US Environmental Protection Agency in 1970, for example, followed the establishment of the first environmental protection administration in Sweden in 1967. In the 1970s, the US took the lead by establishing a wide range of environmental organizations and regulations, many of which subsequently influenced regulatory developments in Europe. European countries and the US also cooperated on the development and enforcement of a number of multilateral environmental agreements including the 1973 Convention on International Trade in Endangered Species and the 1987 Montreal Protocol on Substances that Deplete the Ozone Layer.

The EU and the US continue to cooperate on a wide range of environmental matters and there is still much cross-societal learning between them—a process some analysts have called "hybridization" (Vig and Faure 2004b). For example, Kate O'Neill's chapter (this volume) illustrates how the EU and the US learned from each other in terms of responding to the outbreaks of mad cow disease and avian influenza. Of particular interest is the fact that EU officials decided to develop new food safety and disease prevention agencies modelled on similar US institutions. Another example is the EP's decision to establish a Climate Change Committee shortly after the newly-Democratically-controlled US House of Representatives established a House Select Committee on Energy Independence and Global Warming (Euractiv 2007b). Yet, the EU and the US have clearly pursued different approaches towards many pressing environmental problems. This has led to strains in transatlantic trade relations and a failure to cooperate on a number of international environmental agreements.

In response to growing concerns about the potential link between the tens of thousands of chemicals used in modern society with asthma, allergies, edocrine-disrupters, skin diseases, cancer, birth defects and other health problems, in June 2007, the EU put the REACH (Registration, Evaluation, Authorization, and Restriction of Chemicals) regulation into force. REACH will require the registration of thousands of chemicals produced or imported into the EU. Industries will be required to test chemicals for safety, report information about the chemicals they are using, and shift towards less hazardous chemicals when substitutes are available. The US chemical industry and the George W. Bush administration opposed REACH, instead arguing for a US-style risk assessment approach and no mandatory testing of existing chemicals. The US has taken the position that it makes little sense to test chemicals that are already widely in use and for which there is no clear evidence of any health or environmental problem (Selin, this volume).

Similar differences exist between the EU and the US in relation to the use of asbestos. Whereas the EU has banned all uses of asbestos, the US and Canada still permit some, and Canada continues to mine it (Carson, this volume). The EU and the US also differ in their regulation of hazardous substances in electronics. The EU has gone much farther than the US in requiring manufacturers to change production processes to reduce human and environmental risks and to facilitate recycling. Since July 2006, the EU has required electronics manufacturers to remove lead, mercury, cadmium and brominated flame retardants from their products. US manufacturers, in contrast, have no obligation to remove these toxic chemicals from their products; they can voluntarily take their products back, but few companies have schemes to do so. Here too, the EU has taken a more precautionary approach than the US (Iles, this volume).

Another sharp divide between the EU and the US has been in their attitudes towards GMOs (Keilbach, this volume). Responding to consumer concerns about food safety and the potential deleterious effects on natural ecological systems if GM products "get loose," the EU established a de facto ban on commercial introduction of new GM crops (and some EU countries banned previously approved GM crops). The EU continues to pursue a relatively restrictive approach towards GMOs. The US, Argentina and Canada challenged the EU's de facto ban on GM products in the WTO, arguing that it was costly and unjustified given that there was no evidence that the GM products are unsafe, and contributed to world hunger as GM products could produce higher yields. The EU's revised GMO policy was developed in part as a response to this challenge.

There have been sharp differences in EU and US responses to climate change as well as to means to promote renewable energy use. The EU has embraced the Kyoto Protocol, which requires the EU as a whole to reduce its GHG emissions by 8 percent of 1990 levels by 2008–2012. In early 2007, the Council further agreed to work toward a 20 percent cut in 1990 emission levels by

2020. In contrast, the US federal government has rejected the Kyoto protocol's mandatory emissions cuts and instead pursued programs that encourage voluntary GHG reduction measures and technology research and development (Schreurs, Selin and VanDeveer, this volume). In 2001, the EU introduced a directive with a target to obtain 12 percent of its energy from renewables by 2010. In 2007, this goal was strengthened to 20 percent by 2020. In contrast, while the US Energy Policy Act of 2005 encourages renewable energy development through tax incentives, no explicit targets and no federal renewable energy portfolio standard have been required (Rowlands, this volume).

The volume's contributors found that several important factors lie behind the relatively stronger EU environmental activism and policy programs of the 1990s and the 2000s. These include the growing power and influence of EU institutions, Europe's eagerness to wield foreign environmental policy influence, the stronger embrace in Europe of the precautionary principle, the neo-corporatist traditions found in Europe and the agenda-setting role played by a handful of individual Member States. In other words, institutions and their development lie at the center of explanations for the growth of European environmental and energy regulations.

In contrast, in the case of the US, the authors have argued that there is a stronger neoliberal economic model shaping approaches to environmental protection, there was a shift towards a more conservative politics during the 1990s (and some would argue even earlier), and there is a deeper embrace of cost-benefit analysis. All these developments have greatly shaped US domestic environmental politics as well as influenced transatlantic environmental and energy relations. In addition, the cases suggest that whereas the EU has moved toward a stronger supranational role in environmental politics in the past several decades, the US has devolved much environmental policymaking and enforcement responsibility from the federal to the state level.

The Transatlantic Environmental Divide: What's at Stake?

As environmental and energy issues have become increasingly significant within the transatlantic relationship, the stakes inherent in the transatlantic divide have also grown. From an environmental perspective, the failure of the US to participate in a wide range of multilateral environmental agreements has limited their effectiveness. With the US being among the world's largest consumer of energy, goods, and natural resources, its absence in the Biodiversity Convention, the Cartagena Protocol on Biosafety, the Kyoto Protocol, the Basel Convention, and numerous other agreements weakens their environmental impact and legitimacy (Andresen and Hey 2005). The US absence from these treaties leaves a glaring gap in their coverage and undermines the construction of global standards of conduct. The lack of US involvement limits the scope of the treaties. It also means that the industrialized north does not speak with a unified voice regarding either the seriousness of problems like biodiversity preservation, chemical safety, renewable energy, and climate change, or the means to address such problems.

While domestic US policy may in fact do quite a lot more in these issue areas than its lack of involvement in multilateral environmental agreements might suggest—this for example, is the case in relation to biodiversity conservation and increasingly, at the sub-federal level in relation to renewable energy and climate change—the lack of leadership from Washington DC at the federal and multilateral levels leads many in the international community to question the US commitment to sustainability as well as it commitment to cooperative problem solving, generally. Indeed, the US Congress prior to the 2006 Congressional elections, resisted the introduction of domestic regulatory measures requiring the introduction of renewable energy portfolio standards, CO_2 emissions controls, strengthening of vehicle fuel efficiency standards, and the banning of remaining uses of asbestos as has been done in the EU. Since 2006, there are some signs of change, but no vast sea-change in US federal environmental and energy policies occurred.

Furthermore, from a foreign policy perspective, the US has been displaced as the global leader by the EU in the environmental and renewable energy areas. Whereas in the 1960s and 1970s, the US was leading the world in environmental scientific research, the introduction of environmental regulations and institutions, the deployment of renewable energies, and innovative policy ideas, this position has shifted in large part to Europe. While there are still a handful of environmental issues where the US takes a policy lead, as with the development of regulations governing export credit agencies (Schaper, this volume), on the whole, it has been Europe and not the US that has most forcefully and effectively pursued multilateral environmental agreements and domestic regulatory change since the 1980s. The EU is wielding soft power in order to shape global sustainability norms, having won itself the mantra of global environmental leader due in large part to Washington's retreat.

The differences across the Atlantic also have taken on added significance due to the growing size and unity of the European market place. The EU's population of about 450 million substantially exceeds the roughly 300 million living in the US. While the EU's recent expansion means that it now includes many Member States that are not as economically well-off as Western Europe was, the total size of the EU economy rivals that of the US. Thus, the European market is simply too big for business actors to ignore. When Europe regulates, international businesses have little choice but to comply with EU rules if they wish to have access to the European market. Their only other option is to convince their governments to challenge EU regulations in the WTO, as has occurred with the EU-moratorium on importing agricultural products with GMOs (chapters by Keilbach, and Bernauer and Aerni).

Many US pundits and policymakers express fascination with the weaknesses and challenges of EU institutions and politics. Yet, there is much more widespread, and growing, European influence on US policymaking and regulatory standards than many—probably most—in the US appreciate. Increasingly, it is the EU (often together with Japan) that dominates environmental markets (as with wind and solar energy technologies) and determines global product standards. It is now Europe that sets environmental standards for international commerce, "forcing changes in how industries around the world make plastic, electronics, toys, cosmetics and furniture" (Cone 2005). EU officials and firms have discovered the global power of high standards (Selin and VanDeveer 2006b).

The EU directive restricting the use of hazardous substances in electrical and electronic equipment prohibits the sale of equipment containing lead, cadmium, mercury, and other toxic substances. US manufacturers must meet these standards if they wish to sell their products in Europe. The EU Directive on Waste Electrical and Electronic Equipment, which requires electronics producers to design products in such a way that they can easily be recycled at the end of their life cycle, affects US manufacturers as well. As Iles' chapter notes, manufacturers must change how they design their products if they wish to continue to sell them in the European market, which as a result of the EU's expansion now covers 27 countries. The EU has become too big for manufacturers to ignore. For the vast majority of international firms, the costs of abiding by new regulations are dwarfed by the costs of forgoing doing business in Europe (Selin and VanDeveer 2006b). So, they implement changes and seek to comply with new EU regulations.

The competition between the EU and the US over the setting of standards and regulations also has implications that stretch beyond the transatlantic relationship to the global level. As commented on by Thomas Fuller, "European environmental and safety rules conceived in Brussels are increasingly becoming de facto Asian standards on the factory floors [in China] that churn out the televisions, clothing and furniture that fill most homes in the West" (Fuller 2006) This shift toward use of European as opposed to US standards is because European standards tend to be the higher ones; if a factory interested in exporting products can meet the European standards, it is most likely to be in compliance with most other countries' standards as well. In such cases, international competition and trade help to push standards up, toward those enacted by the highest standard jurisdiction—what David Vogel (1995) calls the "California effect." Another reason EU standards take on a particular significance is that they apply to many national markets simultaneously. Beyond this, it is easier and cheaper for firms to meet a single standard than to have to try to adjust products to different national standards.

These dynamics are evident in chemicals regulation as the chemical industry is a multi-billion dollar industry with a global reach (Selin, this volume). The EU directives restricting the use of specified hazardous substances in electronics products and requiring electronic product recyclability also forced Chinese manufacturers to choose between meeting Europe's higher standards or risk being shut out of the European market (*China Daily* 2006; Lee 2006). The EU-

US battle over GMOs has also been extended to developing countries. In recent years, the EU has been more successful in exporting its anti-GMO preferences and regulatory approaches to developing countries than have pro-GMO forces, including the US. The competition to win the public trust of developing country publics is likely to continue into the foreseeable future. Bernauer and Aerni (this volume) predict, however, that the tide may turn in the pro-GMO direction as more pragmatic attitudes towards GMOs begin to take root in developing countries and they develop their own agri-biotech industries.

Domestic Politics and the Transatlantic Divide

Policy diffusion can be a powerful force promoting policy convergence across jurisdictions (Lopes and Durfee 1999; Tews et al. 2003; Rabe 2004; Levi-Faur and Jordana 2005), and pathways of transatlantic politics can serve as channels of norm diffusion, learning, and strategic action (O'Neill et al. 2004; Slaughter 2004; Vogel 2005; Selin and VanDeveer 2007). The number of channels through which information, norms and ideas move across the Atlantic has grown substantially. These transatlantic pathways have become increasingly institutionalized over time, with the construction and stabilization of transatlantic firms, professional associations (for public and private sector professionals) and activist organizations all seeking to exercise political and economic influence. As a result, the number of networks competing with each other and with state actors to influence policy development and market forces across the Atlantic has expanded.

Given that the forces of economic globalization are alleged to be so strong and bringing countries and markets closer together, why in the past decade or more have the EU and the US diverged on so many important policy issues? Why, when transatlantic economies are increasingly integrated, are the accompanying politics so often discordant?

At the cultural level, Europeans have embraced precautionary thinking and sustainability concerns more deeply than have many North Americans (Bomberg, this volume). In densely populated Europe, where environmental pressures are felt quite acutely, the precautionary principle developed relatively strong roots (O'Riordan and Cameron 1994; Wiener and Rogers 2002; Wiener 2004), even though the EU still struggles with implementation of its precautionary regulations (Eckley and Selin 2004). The strong European embrace of the precautionary principle and sustainability norms may also be both a reflection of the rise of the green movement in Europe and a reason behind the movement's electoral success both in national governments and parliaments and within the EP. Certainly in terms of its political economic culture, the EU embraces social welfare and equity concerns more deeply in its policies than does the US.

In the US, a more laissez-faire approach to the market prevails. Thus, whereas Europe has introduced a growing number of environmental regulations affecting economic interests in recent years, US policymakers have avoided environmental regulations and instead tended toward voluntary measures and support of technology research and development (Wälti, this volume). There have also been major institutional changes in US and European politics that have affected the abilities of different groups to influence political outcomes (Hall 1986; Steinmo et al. 1992; Baumgartner and Jones 1993). During the 1970s in the US, there was strong public demand for a governmental role in environmental protection. The influence of environmental advocacy groups was strong, and Congress was relatively receptive to environmental regulatory reform (Rosenbaum 2005). In this period, the US often championed the formation of international environmental agreements. In contrast, since the 1980s, there has been a growing push in the US by supporters of free market economics to reduce the role of the government in economic activities (Bryner 1993).

The neoconservative revolution that began under the Reagan administration changed political institutions in directions favoring corporate over societal interests. Strong support for neoconservative economics in many parts of the country were behind the rise of the Republican Party at all political levels during much of the 1990s and the first half of the 2000s. The Republican Party has traditionally favored small government, supported business interests and put a lower priority on environmental protection than its Democratic counterpart. In the relatively corporate-friendly climate that has prevailed in the US since Ronald Reagan entered office in January 1981, corporate interests have been relatively successful in spreading liberal economic norms and challenging the introduction of new environmental regulations.

Environmental regulations have been a prime target during the anti-government backlash (Layzer 2006; Vig and Kraft 2006). Compared with the 1970s, there is far less bi-partisanship in the US Congress today, in part a reflection of the loss in influence of moderates on both sides of the aisle. Congressional stalemate has stymied efforts at regulatory change. Added to these developments is the institutional division of authority in the US federal system between the executive and congressional branches and between the two legislative chambers. This makes it much easier to stop new proposals, than it is to pass them. Private sector actors have used this institutional reality to their advantage as they work to kill or weaken many proposed environmental regulatory proposals.

The domestic changes in the US, together with the institutional structure of the federal government and the need for a two-thirds majority in the US senate for treaty ratification, have also altered the direction of US foreign environmental policy (Harris 2000b; Jacobson 2002; DeSombre 2005). Jutta Brunnée (2004) suggests that, because many of the international agreements negotiated since the early 1990s are more complex and seek to restrict a greater range of activities than earlier treaties, they have been more likely to meet with opposition by a variety of US domestic interest groups opposed to treaty provisions. Thus, even in instances where the executive has supported an international environmental agreement, as suggested by President Clinton's signing of the Biodiversity Convention and the Kyoto Protocol, domestic opponents can block Senate approval. The US has also sought to restrict the inclusion in international agreements of references to principles not fully embraced in US legal tradition and/or among US politicians, including the precautionary principle and the concept of common but differentiated responsibilities (Christoforou 2004; Durant 2004; Eckley and Selin 2004).

European trends since the 1980s show a transition in the opposite direction. Public demand for environmental regulatory change in Europe became stronger over the course of the 1970s and 1980s, especially in the northern and central states of Western Europe. This was linked to mounting concerns about environmental conditions, and in particular the effects of acid rain on European ecosystems and the safety of nuclear energy after the Chernobyl nuclear disaster. Green parties and environmental movements were able to take advantage of Europe's parliamentary systems, most of which employ some form of proportional representation. In numerous European governments during the 1990s and 2000s, green parties were either in coalition governments or well-represented in national politics. They also became one of the largest groupings in the EP (Papadakis and Schreurs 2007).

Other European political parties also developed relatively strong environmental platforms in the 1980s and the 1990s including the Social Democratic parties in northern Europe, which have continued to wield strong influence in European national and regional politics. In addition, many of Europe's liberal and conservative parties have taken relatively progressive stances on environmental matters at least when compared with the Republican Party in the US. Examples include the Christian Democratic Union in Germany, which has at a number of occasions pursued a leadership role internationally on climate change, and more recently, and the French conservatives under Jacques Chirac and more recently Nikolas Sarkozy. All these European political developments have given environmental advocates in Europe many different opportunities and channels to bring about environmental policy changes that the US environmental movement lacked (Schreurs and Tiberghien 2007).

As environmental awareness was growing within individual European states, the process of European integration and institutional strengthening was progressing apace (Carmin and VanDeveer 2004). The 1992 Treaty on the European Union (the Maastricht Treaty) established three main pillars (or areas) of activity. The first pillar, the European Communities, is the most developed. Over time, this has been expanded from free trade issues to deal with social and environmental issues as well. In these areas EU policymaking is most supranational in character. Excepting taxation policy and monetary integration, decisions generally can be reached by Member States through a weighted majority in collaboration with the EP and the Commission. The other two pillars—the Common Foreign and Security Policy (second pillar) and Cooperation in Justice and Home Affairs (third pillar)—are still primarily inter-governmental in character. This means that environmental issues are one of the few areas where Europe can more easily be in the vanguard, both at the regional and the global levels.

As Elizabeth Bomberg argued in Chapter 2, the EU has become an increasingly strong policy actor in its own right. EU harmonization of environmental standards has been guided by the institutions of the EU at the same time that these institutions have been strengthened through the processes of harmonization. Sustainability has become an important area for EU foreign policy. Much early EU environmental legislation was pioneered by "green" Member States that wanted to raise and harmonize environmental standards throughout Europe to improve environmental standards and assure that their own industries would not be disadvantaged by stricter national regulations (Liefferink and Andersen 1998; Jänicke 2005). As the powers of EU institutions in the environmental realm were strengthened, the Commission, the EP and the Council began to play more activist roles in pushing environmental issues and raising standards of protection both within the EU and at the international level (Grant et al. 2000; McCormick 2001; Axelrod et al. 2004; Jordan 2005; Selin 2007).

The development of EU institutions, moreover, has provided new avenues for societal interests to influence policy outcomes in ways that promote increases in European environmental (and other) regulatory standards. Environmental interests can work to affect change simultaneously through their national governments and through EU institutions. Tellingly, there is an increasingly large group of environmental advocacy groups working in Brussels. Many collaborate within the umbrella organization European Environmental Bureau and receive public financial support from national governments and the Commission. As Sonja Wälti shows in her chapter, European businesses also has had an interest in having environmental standards harmonized across the continent rather than having to deal with an array of differing standards in different European states. In the neo-corporatist arrangements that are relatively common in Europe, businesses could, moreover, more easily be persuaded of the necessity of pursuing precautionary measures. Thus, the opposition that US industries have voiced towards the imposition of new environmental, safety, energy, and product regulations has been less prevalent in Europe.

Another factor that has made EU environmental leadership possible has been the adoption of the burden sharing concept (Schreurs and Tiberghien 2007). Burden sharing allows for differentiated targets for Member States dependent upon their economic conditions. A burden sharing agreement covers the EU commitment under the Kyoto Protocol and is being worked out in relation to the 20 percent renewable energy target for 2020 adopted by the Council in March 2007 (Council of the European Union 2007b). The new accession states were given extended time frames to fulfil the EU's vast array of environmental regulations. The concept of burden sharing grows directly out of the EU's embrace of notions of social equity. In order to achieve this, Member States have agreed in certain circumstances to the adoption of differentiated targets based on national circumstances. Without this agreement, it is difficult to imagine that the EU could have established a unified position on renewable energy targets or climate change mitigation goals.

It is also important to remember that the EU is a collection of states with much political, economic, cultural and linguistic diversity. In general, northern European states have been stronger supporters of stringent environmental targets and regulations than have their southern neighbors, the combined impact of better economic situations and relatively strong influence of green movements. As Cashore, Auld, Newsom, and Egan show in their case study of forestry certification movements in Europe and North America, there is often considerable diversity within Member States on which policy approach to pursue. Patricia Keilbach and Thomas Bernauer and Phillip Aerni note in their respective chapters that there is considerable diversity within Europe on the desirability of GMO-bans. Whereas some Member States strongly oppose the introduction of GMOs, others are more interested in the possibilities GMO crops may provide.

Lastly, to be clear, the chapters do not suggest that Europe lacks opponents to increased environmental, food safety and energy regulation. Corporate actors worked hard to weaken new EU chemicals regulations (Selin 2007), many firms and several Member States have labored to weaken EU climate and energy policy and German firms sought to avoid applying US style regulation in the area of export credits (Schaper, this volume)—just to name a few. Opponents exist and they are active. In short, the institutional factors discussed above and throughout the chapters help to explain why opponents of environmental and energy regulation in Europe (not all of them Europeans) tend to lose their fights against regulatory proposals more often than their US counterparts.

More Nuanced Comparison Needed: US Leadership at the Local and State Levels

Although the transatlantic environmental divide, at the time of this writing, remains very real, there are numerous signs that the differences in the regulatory approaches of Washington and Brussels are only part of the story. In fact, there are numerous countervailing forces to those shaping the divergent federal environmental policies of Europe and the US, demonstrating the domestic politics and institutions continue to shape (and be shaped by) international environmental politics (Steinberg and VanDeveer forthcoming). These include powerful market influences, public pressure, and sub-federal level (states and municipalities) policy initiatives. When viewed at the sub-federal level or from the perspective of at least some key market players, the differences between the EU and the US are far less striking and suggest that greater transatlantic cooperation is both desirable and possible.

Since the 1980s, there have been few new federal environmental statutes introduced in the US. There are several reasons for this. First, as noted above, one of its main goals of the neoconservative forces that came to dominate Congress and the White House was to deregulate including in the environment area. Second, as a result of a growing partisan divide, Congress has often been incapable of reaching the bipartisan consensus necessary to pass major new environmental statues. Third, in those areas where the federal government is still involved, it has shifted its focus towards new environmental governance modes as opposed to regulations premised on command and control approaches. These include voluntary programs, market-based mechanisms, public-private partnerships, and community-based environmental protection (Kraft and Scheberle 1988; Durant et al. 2004).

The end result has been a devolution (or what some call an abrogation) of responsibility from the federal to the state and municipal levels in relation to a number of environmental and energy issues (Schreurs and Epstein 2007). In the US, it is also at the state and municipal levels where much of the most innovative environmental policy change is occurring (Rabe 2006a). On a range of critical sustainability issues, California, the Pacific Northwest, and the New England states have tended to assume environmental positions closer to those embraced by the EU than by federal policymakers in Washington DC. Thus, whereas the Brussels-Washington relationship has been mired in tension and disagreement over their divergent policy responses to sustainability issues, many counter-trends are visible at the sub-federal level.

For example, the lack of a strong federal presence in climate change politics has spawned much policy innovation at the state and municipal levels (Selin and VanDeveer 2009a). As of mid-2008, over two dozen states and hundreds of US municipalities had established their own plans and regulations addressing GHG emissions. The Regional Greenhouse Gas Initiative (RGGI), launched by states in the US Northeast, created a regional cap-and-trade scheme. RGGI covers CO_2 emissions from major power plants with the goal of stabilizing CO_2 emissions between 2009 and 2015, and a total 10 percent reduction by 2019 (Schreurs, Selin, VanDeveer, this volume). Furthermore, over half of all US states have some form of renewable energy portfolio standard and many have enacted green purchasing requirements (Rowlands, this volume).

In relation to product standards, several states have started to follow European approaches. Alastair Iles notes in his chapter that in July 2002, Maine became the first US state to prohibit the landfilling of mercury switches and that New Jersey, Arkansas, Texas, Washington State, Illinois, New Hampshire, California and Rhode Island have followed suit. And, in July 2003, California became the first state to require limited producer responsibility for recycling when it obligated manufacturers to develop, finance and implement an e-waste recovery system for reuse and recycling of computer monitors and television sets.

Even in the area of GMOs, we see some local communities opposing US federal policy direction. In June 2006, the Organic Consumers Association announced that Santa Cruz had become the fourth California county to ban GMOs.[1]

In addition, state and municipal officials in the US states interact with their counterparts—and with officials at various levels of governance—in Europe as they assess European policies and their effectiveness and decide what aspects of these to emulate or alter in their policymaking efforts (Schreurs, Selin and VanDeveer, this volume). These many sub-national developments suggest that there are more similarities between the EU and the US than simple comparisons of the US and the EU at the federal level suggest. In other words, the transatlantic relationship is far more complex than is often portrayed, indicating that there are avenues by which normative and regulatory convergence between the EU and the US may be promoted.

Looking to the Future: Competition and the Potential for Greater Transatlantic Policy Convergence

Whereas the EU and the US have been diverging on their approaches to environmental governance over the past two decades, there are signs that the future could bring more convergence—potentially at several levels. First, the growing pressures emanating from local and state governments in the US and rising public apprehension about global environmental degradation have the potential to push the US Congress into taking action. The 2006 Congressional election opened a window of opportunity for environmental policy change. The signs of change in the Congressional mood can be seen in multiple environmental areas. There are new legislative initiatives in the works that address many of the issues covered in this book: asbestos, renewable energy, and GHG emissions.

For example, in March 2007, California Senator Patty Murray reintroduced for the third time the Ban Asbestos Act, which would prohibit remaining uses of asbestos and enhance funding for research on asbestos-related diseases[2] (Carson, this volume). In April 2007, the House of Representatives passed the Clean Energy Act of 2007 with the intention of repealing money targeted to the oil and gas industries and steering it instead toward renewable energy in the process weaning the country off of its heavy dependence on imported oil while promoting cleaner forms of energy[3] (Rowlands, this volume). In December

---

1   Organic Consumers Association, "Fourth California County (Santa Cruz) Bans GMOs," June 20, 2006, http://www.organicconsumers.org/articles/article_858.cfm.

2   Senator Patty Murray, homepage, http://murray.senate.gov/news.cfm?id =270031.

3   HR 6 Clean Energy Act of 2007, http://www.govtrack.us/congress/bill.xpd? tab=main&bill=h110-6.

2007, Congress passed and President Bush signed into law the first increase in automobile CAFE standards since the 1970s, along with new regulations to raise the energy efficiency of numerous other products. There are many bills in Congress addressing GHG emissions.[4]

In spring 2007, Congress took up the issue of food safety and proposed strengthening the power of the Food and Drug Administration to inspect and recall products. There is also increased federal interest in strengthening hazardous chemicals policy (Selin, this volume). The Child, Worker and Consumer Safe Chemicals Act was introduced in 2005, and the Kid Safe Chemical Act was put forward in 2008. Both these proposals, which borrowed recent policy ideas and regulatory approaches from the EU, failed to pass. Nevertheless, such developments across several major environmental issue areas suggest that the transatlantic divide at the federal level may narrow in the coming years.

Beyond this, California has the potential to emerge as a formidable challenger to EU policy dominance in the environmental realm in the future. Given California's large economic size—if it were a country it would rank as the world's fifth or sixth largest economy—changes in its policies reverberate in other US states and internationally (Schmidt 2007). In recent years, California has followed the EU lead introducing its own renewable energy portfolio standards, mandating CO_2 emissions reductions from automobiles, and requiring manufacturers to take back electronic equipment for recycling. California lawmakers are currently considering a green chemistry policy, influenced by the EU, which could exceed the federal Toxic Substances Control Act.

California is trying to reap first-mover advantages. California's policy changes are being driven not only by the environmental concerns of its voters and elected officials, but also from its economic interests. A 2006 report prepared for the California Senate Environmental Quality Committee and the California Assembly Committee on Environmental Safety and Toxic Materials states: "Large 'sunk' investments by industry in existing chemical technologies will make it difficult to transition to an industrial system based on cleaner technology, including green chemistry; this transition, however, will have to be made if California is to respond proactively to developments in the E.U. and address a host of chemical problems affecting public and environmental health, business, government, and industry in the state" (Wilson et al. 2006, 14).

Many Europeans are watching the "California effect" with considerable interest. In fact, in some areas, California is now pushing ahead of Europe. California statute AB 32 requires a reduction in GHG emissions to 2000 levels by 2010, 1990 levels by 2020, and 80 percent below 1990 levels by 2050 (Farrell and Hanemann 2009). The 80 percent reduction goal exceeds the 60 percent reduction goal proposed in the United Kingdom in March 2007 (Cowell 2007).

---

4   Pew Center on Global Climate Change, http://www.pewclimate.org/docUploads/ Cap%2Dand%2Dtrade%20bills%2011th%5FFeb5%2Epdf.

In January 2007, California Governor Arnold Schwarzenegger signed an Executive Order establishing a Low Carbon Fuel Standard, the first in the world, establishing a state-wide goal of reducing the carbon intensity of California's transportation fuels by 10 percent by 2020 (Low Carbon Vehicle Partnership 2007). As policy leaders such as California and the EU learn from each others' policies, some 'hybridization' of policies can occur (Vig and Faure 2004b).

For much of the past two decades, global environmental protection has been pushed forward by the EU. It is not a given, however, that Europe can continue to play this pioneering role. As the EU expands to include a larger number of Member States, passing stringent environmental regulations may become more difficult. For many of the newest members of the EU, environmental protection is less high on the policy agenda than is economic development. So far, the Commission has been able to push forward with bold new environmental initiatives, but the ability and will of Member States to implement all environmental EU environmental legislation remain to be seen (Carmin and VanDeveer 2004). Even if a handful of the new Member States develop records like Italy, Portugal, Spain and Greece have in terms of their failure to implement a substantial number of EU environmental directives, EU leadership could be questioned.

It is not at all certain that an EU with 27 plus Member States will be able to muster the same kind of internal support for global environmental leadership as in the past. In a number of issue areas, the EU may find it hard to achieve the relatively demanding goals it has established. In relation to renewable energy expansion and GHG emissions reduction targets, for example, some of the least costly and demanding measures for expanding wind and solar energy and improving energy efficiency have already been taking. Future renewable energy expansion and GHG emission cuts could be even more demanding and costly. The EU is also not likely to have a second "windfall" in GHG emissions reductions as occurred after the collapse of the east bloc economies with the break up of the Soviet Union. While Germany may continue to be a leader in climate change policy formulation, it will not be able to carry the EU in the way it did in the 1998 burden-sharing agreement.

Beyond this, many of the more recent Member States are likely to be the ones that will experience the fastest growth in their economies over the next decade. The declines in GHG emissions most experienced due to the collapse of their Soviet-style economies are likely to reverse as their economic transformations progress—in part as a result of their EU membership. In addition, growing competition from the developing world may also strain the EU's willingness and ability to lead. If countries with whom the EU is in growing economic competition, including China, India, Brazil and Mexico, are not also constrained by similar environmental regulations, then eventually cost considerations and competitiveness concerns could gain the upper hand in EU policy discussions the way that they did in the US in the 1990s.

Still, despite all of these potential roadblocks, early indications are that an EU of 27, at least in the immediate term still intends to lead on environmental matters. Member States and organizations have shown this with their recent passages of regulations addressing hazardous chemicals, renewable energy, GHG emissions, product standards, recycling requirements, and the like. The real onus at this point in history is on the US to join the EU in leading the global community in addressing global environmental and health problems. This may, in fact, occur in the coming years as Washington finds itself pressured increasingly strongly to take on a leadership mantel together with the EU. Hopefully, growing transatlantic environmental policy cooperation to tackle the many vital environmental and energy issues threatening our planet will characterize the next decade.

References

Aaron, D.L. and Gray, C.B. (2002), "Risk and Reward: US-EU Regulatory Cooperation on Food Safety and the Environment," Policy Paper, November, Washington, DC: Atlantic Council of the United States.

Ackerman, F. and Massey, R. (2004), "The True Costs of Reach," *TeamNord 2004:557*, Copenhagen: Nordic Council of Ministers.

Ackerman, F., Stanton, E.A., Roach, B., and Andersson, A.-S. (2008), "Implications of Reach for Developing Countries," *European Environment* 18:1, 16–29.

Aerni, P. (2001), "Assessing Stakeholder Attitudes to Agricultural Biotechnology in Developing Countries," *The Biotechnology and Development Monitor* 47, 2–7.

Aerni, P. (2002), "Stakeholder Attitudes Towards the Risks and Benefits of Agricultural Biotechnology in Developing Countries: A Comparison between Mexico and the Philippines," *Risk Analysis* 22:6, 1123–37.

Aerni, P. (2003), "The Private Management of Public Trust: The Changing Nature of Political Protest," paper presented at the *Workshop of the European Consortium of Political Research*, Marburg.

Aerni, P. and Bernauer, T. (2006), "Stakeholder Attitudes Towards GMOs in the Philippines, Mexico and South Africa: The Issue of Public Trust," *World Development* 34:3, 557–75.

Ahmia, T. (2007), "Umweltschutz Wird Zur Verschlusssache—Die Bundesregierung Druckst Herum, Unter Welchen Bedingungen Sie Für Den Großstaudamm Ilisu Bürgen Will," *taz, die tageszeitung*, January 31, 8.

AI (1996), "Background Information on the Ban of Asbestos Products in France," *Asbestos Institute*, <http://www.asbestos-institute.ca/specialreports/specialindex.html>, accessed March 7, 2001.

AI (1997), "Politics, Not Science, Basis for France's Ban of Asbestos," *Asbestos Institute Newsletter*.

AI (2001), "About the Asbestos Institute," *Asbestos Institute,* <http://asbestos-institute.ca/asbinfo.html>, accessed March 7, 2001.

Aitken, D.W. (2002), "The American Re Transition," *Refocus* 3, 28–9.

Akerlof, G.A. and Kranton, R.E. (2000), "Economics and Identity," *Quarterly Journal of Economics* 19:1, 9–32.

Alden, E., Pilling, D., and Williamson, H. (2006), "Export Credit Agencies' Graft Crackdown Stalls—Germany and Japan Are Blocking the Introduction of International Guidelines Designed to Prevent Corruption," *Financial Times*, February 15, 8.

Altman, L.K. (2006), "Tackling the Animal-to-Human Link in Illness," *New York Times*, March 25.

Ambrose, W.W. (2003), "Green Utilities: Renewable Energy Utilities Forge Global Strategies," *Refocus* 4, 24–8.

American Chemistry Council (2003), "Frequently Asked Questions: The European Commission's Proposed Revisions to Europe's Chemical Regulatory System," unpublished paper, November 6.

American Chemistry Council (2007), "The Business of Chemistry: Essential to Our Quality of Life and the US Economy," <http://www.americanchemistry.com>.

Andersen, M.S. and Liefferink, D. (eds) (1997), *European Environmental Policy: The Pioneers* (Manchester, New York: Manchester University Press).

Andresen, S. and Hey, E. (eds) (2005), Special Issue of *International Environmental Agreements* 5:3, *The Effectiveness and Legitimacy of International Environmental Institutions*.

Andrews, D.M. (ed.) (2005), *The Atlantic Alliance under Stress: US-European Relations after Iraq* (Cambridge, UK: New York: Cambridge University Press).

Andrews, R.N.L. (1999), *Managing the Environment, Managing Ourselves: a History of American Environmental Policy* (New Haven, CT: Yale University Press).

Andrews, R.N.L. (2003), "Risk-Based Decision Making," in N.J. Vig and M.E. Kraft (eds), *Environmental Policy: New Directions for the Twenty-First Century* (Washington, DC: CQ Press).

Annan, K.A. (2002), "Toward a Sustainable Future," *Environment* 44.

Ansell, C.K. and Vogel, D. (eds) (2006), *What's the Beef? The Contested Governance of European Food Safety* (Cambridge, MA: MIT Press).

APHIS (2004), "Avian Influenza Trade Ban Status as of 12–27–04," <www.aphis.usda.gov/lpa/issues/ai_us/ai_trade_ban_status.html>.

Appleton, A.E. (1999), "Shrimp/Turtle: Untangling the Nets," *Journal of International Economic Law* 2, 477–96.

Arctic Monitoring and Assessment Programme (2002a), "Arctic Pollution 2002," Oslo: AMAP.

Arctic Monitoring and Assessment Programme (2002b), "Human Health in the Arctic," Oslo: AMAP.

Armstrong, K.A. and Bulmer, S. (1998), *The Governance of the Single European Market* (Manchester: Manchester University Press).

Arthur D. Little (2002), "Economic Effects of the EU Substances Policy: Summary of the BDI Research Project," October 31, Wiesbaden: Arthur D. Little GMbH.

Asmus, P. (2002), "Gone with the Wind: is California Losing its Clean Power Edge to Texas?," *Faultline* 16: August.

Audley, J.J. (1997), *Green Politics and Global Trade: NAFTA and the Future of Environmental Politics* (Washington, DC: Georgetown University Press).

Auld, G. (2001), "Explaining Certification Legitimacy: An Examination of Forest Sector Support for Forest Certification Programs in the United States Pacific Coast, the United Kingdom, and British Columbia, Canada," School of Forestry and Wildlife Sciences, Auburn University, Auburn, AL.

Austin, M.T. and Milner, H.V. (2001), "Strategies of European Standardization," *Journal of European Public Policy* 8, 411–31.

AWEA (2003), "Record Growth for Global Wind Power in 2002," *News Release*, March 3, Washington, DC: American Wind Energy Association.

Axelrod, R., Vig, N.J., and Schreurs, M.A. (2004), "The European Union as an Environmental Governance System," in N. Vig and R.S. Axelrod (eds), *The Global Environment: Institutions, Law and Policy* (Washington, DC: Congressional Quarterly Press).

Bache, I. and Flinders, M. (eds) (2004), *Multi-Level Governance* (Oxford: Oxford University Press).

Bailey, C.J. (1998), *Congress and Air Pollution: Environmental Policies in the USA* (Manchester: Manchester University Press).

Bailey, I. (2003), *New Environmental Policy Instruments in the European Union: Politics, Economics, and the Implementation of the Packaging Waste Directive* (Aldershot: Ashgate).

Baker, S. and McCormick (2004), "Sustainable Development: Comparative Understandings and Responses," in N.J. Vig and M. Faure (eds), *Green Giants? Environmental Policies of the United States and the European Union* (Cambridge, MA: MIT Press), 277–302.

Baker, S. (ed.) (1997), *The Politics of Sustainable Development: Theory, Policy and Practice within the European Union* (London and New York: Routledge).

Baker, S. (2006), *Sustainable Development* (Milton Park: Routledge).

Barker, M.L. and Soyez, D. (1994), "Think Locally Act Globally? The Transnationalization of Canadian Resource-Use Conflicts," *Environment* 36:5, 12–23.

Barlesz, M. and Loughlin, D. (2005), "Recycling Worldwide," *Waste Management Worldwide*.

Barringer, F. and Yardley, W. (2007), "Bush Splits on Greenhouse Gases with Congress and State Officials," *New York Times*, April 4.

Barschdorff, P. (2001), *Facilitating Transatlantic Cooperation after the Cold War: An Acquis Atlantique* (Münster: Lit Verlag).

Baumgartner, F.R. and Jones, B.D. (1993), *Agendas and Instability in American Politics* (Chicago: University of Chicago Press).

BBC News (2006), "Global Impact of Bird Flu," <http://news.bbc.co.uk/2/hi/health/4531500.stm>.

Bechberger, M. and Reiche, D. (2003), "RE in EU-28: Renewable Energy Policies in an Enlarged European Union," *Refocus* 4, 30–4.

Beck, U. (1995), *Ecological Politics in an Age of Risk* (Cambridge: Polity Press).

Benjelloun, H. (2000), "The European Union's Ban on Asbestos," *The Synergist/ American Industrial Hygene Association* 11:8.

Benz, A. and Eberlein, B. (1999), "The Europeanization of Regional Policies: Patterns of Multi-Level Governance," *Journal of European Public Policy* 6:2, 329–48.

Bergaud-Blackler, F. (2004), "Consumer Trust in Food—a European Study of the Social and Institutional Conditions for the Production of Trust," study commissioned by the European Commission.

Bernauer, T. (2003), *Genes, Trade, and Regulation: The Seeds of Conflict in Food Biotechnology* (Princeton, NJ: Princeton University Press).

Bernauer, T. and Aerni, P. (2006), "Competition for Public Trust: Causes and Consequences of Extending the Transatlantic Biotech Conflict to Developing Countries," in R. Falkner (ed.), *The International Politics of Genetically Modified Food* (New York: Palgrave Macmillan).

Bernauer, T. and Meins, E. (2003), "Technological Revolution Meets Policy and the Market: Explaining Cross-National Differences in Agricultural Biotechnology Regulation," *European Journal of Political Research* 42, 643–83.

Bernstein, S. and Cashore, B. (2000), "Globalization, Four Paths of Internationalization and Domestic Policy Change: The Case of Eco-Forestry in British Columbia, Canada," *Canadian Journal of Political Science* 33:1, 67–99.

Bernstein, S. and Cashore, B. (2007), "Can Non-State Global Governance be Legitimate? A Theoretical Framework," *Regulation and Governance* 1:1, 347–71.

Betsill, M. and Bulkeley, H. (2004), "Transnational Networks and Global Environmental Governance: The Cities for Climate Protection Program," *International Studies Quarterly* 48:2, 471–93.

Biermann, F. (2001), "The Rising Tide of Green Unilateralism in World Trade Law: Options for Reconciling the Emerging North-South Conflict," *Journal of World Trade* 35, 421–48.

"Biosafety Meeting Moves on Labeling, Compliance and Liability," (2004), *Bridges Weekly Trade News Digest* 4:2.

Bird, L., Bolinger, M., Gagliano, T., Wiser, R., Brown, M., and Parsons, B. (2005), "Policies and Market Factors Driving Wind Power Development in the United States," *Energy Policy* 33:11, 1397–407.

Blair, T. and Schwarzenegger, A. (2006), "United Kingdom and California Announcement on Climate Change and Clean Energy Collaboration, Mission Statement," July 31.

Bob, C. (2002), "Merchants of Morality," *Foreign Policy* 129 (January/ February), 36–45.

Bodansky, D. (1994), "The Precautionary Principle in US Environmental Law," in T. O'Riordan and J. Cameron (eds), *Interpreting the Precautionary Principle* (London: Earthscan), 201–11.

Bodansky, D. (2003), "Transatlantic Environmental Relations: The Growing Rift between US and European Climate Change Policies," in J. Peterson and M.A. Pollack (eds), *Europe, America, Bush: Transatlantic Relations in the Twenty-First Century* (London: Routledge), 59–68.

Bomberg, E. (2004), "Adapting Form to Function? From Economic to Sustainable Development Governance in the European Union," in W.M. Lafferty (ed.), *Governance for Sustainable Development. The Challenge of Adapting Form to Function* (Cheltenham: Edward Elgar), 61–94.

Bomberg, E. (2007), "Policy Learning in an Enlarged European Union: Environmental NGOs and New Policy Instruments," *Journal of European Public Policy* 14:2, 248–68.

Bomberg, E. and Schlosberg, D. (2008), "US Environmentalism in Comparative Perspective," *Environmental Politics* 17:2, 337–48.

Bomberg, E.E. (1998), *Green Parties and Politics in the European Union* (London and New York: Routledge).

Börzel, T.A. (2002), *States and Regions in the European Union: Institutional Adaptation in Germany and Spain* (Cambridge: Cambridge University Press).

Börzel, T.A. and Hösli, M. (2003), "Brussels between Bern and Berlin: Comparative Federalism Meets the European Union," *Governance: An International Journal of Policy and Administration* 16:2, 179–202.

Bourdieu, P. (1991), *Language and Symbolic Power* (Cambridge, MA: Harvard University Press).

BP (2004), "Statistical Review of World Energy 2004," London: British Petroleum.

Braden, J.B., Folmer, H., and Ulen, T.S. (eds) (1996), *Environmental Policy with Political and Economic Integration: The European Union and the United States* (Cheltenham: Edward Elgar).

Braun, D. (ed.) (2000), *Federalism and Public Policy* (Aldershot: Ashgate).

Bretherton, C. and Vogler, J. (1999), *The European Union as a Global Actor* (London and New York: Routledge).

Brickman, R., Jasanoff, S., and Ilgen, T. (1985), *Controlling Chemicals: The Politics of Regulation in Europe and the United States* (Ithaca, NY: Cornell University Press).

Brodeur, P. (1985), *Outrageous Conduct: The Complete New Yorker Reports* (New York: Pantheon Books).

Brown, P. (2005), "EU Votes to Continue Ban on GM Crops," *The Guardian*, June 25.

Brown, S.A. (1997), *Revolution at the Checkout Counter: The Explosion of the Bar Code* (Cambridge, MA: Harvard University Press).

Browne, J. (2004), "Beyond Kyoto," *Foreign Affairs* 83:4, 20–32.

Brunnée, J. (2004), "The United States and International Environmental Law: Living with an Elephant," *European Journal of International Law* 14:4, 617–49.

Bruton, J. (2007), "Letter to Barbara Boxer," February 22.

Bryner, G.C. (1993), *Blue Skies, Green Politics: The Clean Air Act of 1990* (Washington, DC: CQ Press).

Bryner, G.C. (2000), "The United States: 'Sorry, Not Our Problem'," in W.M. Lafferty and J. Meadowcroft (eds), *Implementing Sustainable Development* (Oxford: Oxford University Press), 273–302.

Bryner, G.C. (2008), "Failure and Opportunity: Environmental Groups in US Climate Change," *Environmental Politics* 17:2, 319–36.

Buccini, J. (2004), "The Global Pursuit of Sound Chemicals Management," Washington DC: The World Bank.

Buck, T. (2007), "Standard Barrier," *Financial Times*, July 10.

Buonanno, L., Zablotney, S., and Keefer, R. (2001), "Politics Versus Science in the Making of a New Regulatory Regime for Food in Europe," *European Integration Online Papers* 6:12.

Burchell, J. (2002), *The Evolution of Green Politics: Development and Change within European Green Parties* (London and Sterling, VA: Earthscan Publications).

Burgess, M. (1999), *Federalism and European Union: The Building of Europe 1950–2000* (London: Routledge).

Busby, J.W. (2003), "Climate Change Blues: Why the United States and Europe Just Can't Get Along," *Current History* 102:662, 113–18.

Busby, J.W. and Ochs, A. (2004), "From Mars and Venus Down to Earth: Understanding the Transatlantic Climate Divide," in D. Michel (ed.), *Climate Policy for the 21st Century: Meeting the Long-Term Challenge of Global Warming* (Washington DC: Center for Transatlantic Relations, Johns Hopkins University, School of Advanced International Studies), 35–76.

Busch, P.-O. and Jörgens, H. (2004), "Globale Ausbreitungsmuster Umweltpolitischer Innovationen," FFU-Report 02–2005, Berlin: Forschungsstelle für Umweltpolitik, Freie Universität Berlin.

Bzdera, A. (1993), "Comparative Analysis of Federal High Courts: A Political Theory of Judicial Review," *Canadian Journal of Political Science* 26:1, 3–29.

Caduff, L. (2005), "Vorsorge Oder Risiko? Umwelt- und Verbraucherschutz-politische Regulierung Im Europäisch-Amerikanischen Vergleich: Eine Politökonomische Analyse Des Hormonstreits und Der Elektronikschrott-Problematik," Center for Comparative and International Studies, ETH Zürich, Zürich.

Campbell, T. (2004), "San Luis Obispo County California will Likely Vote on GE Ban in November," Organics Consumer Association, June 24, 2004.

Cantley, M. (1995), "The Regulation of Modern Biotechnology: A Historical and European Perspective," in D. Bauer (ed.), *Biotechnology: Volume 12 Legal, Economic and Ethical Dimensions* (Weinheim: VCH), 503–681.

Cantley, M. (2004), "How Should Public Policy Respond to the Challenges of Modern Biotechnology?," *Current Opinion in Biotechnology* 15:3, 258–63.

Carmin, J. and VanDeveer, S.D. (eds) (2004), *EU Enlargement and the Environment: Institutional Change and Environmental Policy in Central and Eastern Europe* (London: Routledge).

Carson, M. (2004), *From Common Market to Social Europe?* (Stockholm: Almqvist and Wiksell International).

Carter, N. (2007), *The Politics of the Environment: Ideas, Activism, Policy* (Cambridge; New York: Cambridge University Press).

Cashore, B. (2002), "Legitimacy and the Privatization of Environmental Governance: How Non State Market-Driven (NSMD) Governance Systems Gain Rule Making Authority," *Governance* 15:4 (October), 503–29.

Cashore, B., Auld, G., and Newsom, D. (2004), *Governing through Markets: Forest Certification and the Emergence of Non-State Authority* (New Haven, CT: Yale University Press).

Cashore, B., Egan, B., Auld, G., and Newsom, D. (2007), "Revising Theories of Non-State Market Driven (NSMD) Governance: Lessons from the Finnish Forest Certification Experience," *Global Environmental Politics* 7:1, 1–44.

Cass, L.R. (2006), *The Failures of American and European Climate Policy: International Norms, Domestic Politics, and Unachievable Commitments* (Albany, NY: State University of New York Press).

Castelfranco, S. (2005), "US Rejection of Kyoto Protocol Brings Protest in Italy," *Voice of America*, February 15.

Castleman, B. (1999), "Global Corporate Policies and International 'Double Standards,' in Occupational and Environmental Health," *International Journal of Occupational and Environmental Health* 5, 61–4.

Castleman, B. (2001), *Asbestos: Medical and Legal Aspects* (New York: Aspen Law and Business).

Castleman, B. (2007), "Global Consensus That Asbestos Must Go: Statement to Senate Health, Education, Labor and Pensions Committee," March 1, Washington, DC: Senate Health, Education, Labor and Pensions Committee.

Castleman, B. and Lemen, R. (1998), "Corporate Junk Science in International Science Organizations," *Environment* 19:1, 1–2.

CEFIC (2006), "Facts and Figures: The European Chemical Industry in a Worldwide Perspective," December, Brussels: CEFIC.

Chen, C., Wiser, R., and Bolinger, M. (2007), "Weighing the Costs and Benefits of State Renewables Portfolio Standards: A Comparative Analysis of State-Level Policy Impact Projections," LBNL-61580, March: Ernest Orlando Lawrence Berkeley National Laboratory, Environmental Energy Technologies Division.

Cheng, E. (2003), "Genetically Modified Food: Bush Promotes a 'Biological Time Bomb'," *Green Left Weekly*, September 3.

Chertow, M.R. and Esty, D.C. (1997), *Thinking Ecologically: The Next Generation of Environmental Policy* (New Haven, CT: Yale University Press).

China Daily (2006), "EU Rules Dim Lights on Exports," June 14.

Christoforou, T. (2000), "Settlement of Science-based Trade Disputes in the WTO: A Critical Review of the Developing Case Law in the Face of Scientific Uncertainty," *Environmental Law Journal* 8:3, 622–48.

Christoforou, T. (2003), "The Precautionary Principle in European Community Law and Science," in J.A. Tickner (ed.), *Precaution, Environmental Science and Preventive Public Policy* (Washington, DC: Island Press).

Christoforou, T. (2004), "Conclusion: The Necessary Dialogue," in N.J. Vig and M. Faure (eds), *Green Giants? Environmental Policies of the United States and the European Union* (Cambridge, MA: MIT Press).

CIA (2008), "World Fact Book," Langley: Central Intelligence Agency.

Clapp, S. (2004), "EU Authorizes Center for Disease Prevention and Control to Begin Next Year in Stockholm," *Food Chemical News*, April 12, 23.

Clapp, S. and Lewis, S. (2004), "Flu-related Restrictions Eased on North American Poultry Imports," *Food Chemical News*, April 5, 25.

Clean Production Action (2005), "Moving Towards Sustainable Plastics: A Report Card on the Leading Six Automakers," Montreal: Clean Production Action.

CNN (2001), "Bush's Europe Tour Hit by Protests," June 18.

Cohen, J. and Paarlberg, R. (2004), "Unlocking Crop Biotechnology in Developing Countries—a Report from the Field," *World Development* 32:9, 1563–77.

Cohen-Tanugi, L. (2003), *An Alliance at Risk: The United States and Europe since September 11* (Baltimore, MD: Johns Hopkins University Press).

Coleman, W.D. (1988), *Business and Politics: A Study of Collective Action* (Kingston: McGill-Queen's University Press).

Collier, U. (2002), "EU Energy Policy in a Changing Climate," in A. Lenschow (ed.), *Environmental Policy Integration: Greening Sectoral Policies in Europe* (London: Earthscan), 175–92.

Committee for the Adaptation to Scientific and Technical Progress of EC Legislation on Waste (2005), "Summary Record of the Meeting on 28 April 2005," RVd/jh D (2005) 9144), Brussels: European Commission.

Cone, M. (2005), "Europe's Rules Forcing US Firms to Clean Up: Unwilling to Surrender Sales, Companies Struggle to Meet the EU's Tough Stand on Toxics," *Los Angeles Times*, May 16.

Conference Board Europe (2003), "Shifting Europe's Waste Burden," *Board Europe* 18.

Conlan, T. (1998), *From New Federalism to Devolution: Twenty-Five Years of Intergovernmental Reform* (Washington, DC: Brookings Institution Press).

Conrad, C.R. (2007), "The EC-Biotech Dispute and Applicability of the SPS Agreement," *World Trade Review* 6:2, 233–48.

Coon, C.E. (2001), "Why President Bush was Right to Abandon the Kyoto Protocol," *Issues: Energy and the Environment, Backgrounder*, 1437, May 11, Washington, DC: The Heritage Foundation.

Council of the European Union (1967), "Council Directive 67/548/EEC on the Approximation of Laws, Regulations and Administrative Provisions Relating to the Classification, Packaging and Labeling of Dangerous Substances," 67/548/EEC, August 16, Brussels: *Official Journal of the European Communities*, L 196.

Council of the European Union (1976), "Council Directive 76/769/EEC on the Approximation of Laws, Regulations and Administrative Provisions of the Member States Relating to Restrictions on the Marketing and Use of Certain Dangerous Substances and Preparations," 76/769/EEC, September 27, Brussels: *Official Journal of the European Communities*, L 262.

Council of the European Union (1983a), "Council Directive 83/477/EEC of 19 September 1983 on the Protection of Workers from the Risks Related to Exposure to Asbestos at Work," 83/477/EEC, September 24, Brussels: *Official Journal of the European Communities*, L 263.

Council of the European Union (1983b), "Council Directive 83/478/EEC of 19 September 1983 Amending for the Fifth Time (Asbestos) Directive 76/769/EEC on the Approximation of Laws, Regulations and Administrative Provisions of the Member States Relating to Restrictions on the Marketing and Use of Certain Dangerous Substances and Preparations," 83/478/EEC, September 24, Brussels: *Official Journal of the European Communities*, L 263.

Council of the European Union (1985), "Council Directive 85/610/EEC of 20 December 1985 Amending for the Seventh Time (Asbestos) Directive 76/769/EEC on the Approximation of Laws, Regulations and Administrative Provisions of the Member States Relating to Restrictions on the Marketing and Use of Certain Dangerous Substances and Preparations," 85/610/EEC, December 31, Brussels: *Official Journal of the European Communities*, L375.

Council of the European Union (1987), "Council Directive 87/217/EEC of 19 March 1987 on the Prevention and Reduction of Environmental Pollution by Asbestos," 87/217/EEC, March 28, Brussels: *Official Journal of the European Communities*, L85.

Council of the European Union (1991a), "Council Directive 91/382/EEC of 25 June 1991 Amending Directive 83/477/EEC on the Protection of Workers from the Risks Related to Exposure to Asbestos at Work," 91/382/EEC, July 29, Brussels: *Official Journal of the European Communities*, L 206.

Council of the European Union (1991b), "Council Directive 91/659/EEC of 3 December 1991 Adapting to Technical Progress Annex 1 to Council Directive 76/769/EEC on the Approximation of Laws, Regulations and Administrative Provisions of the Member States Relating to Restrictions on the Marketing and Use of Certain Dangerous Substances and Preparations

(Asbestos)," 91/659/EEC, December 31, Brussels: *Official Journal of the European Communities*, L 363.

Council of the European Union (2007a), "2785th Council Meeting," *Press Release*, 6272/07 (Presse 25), February 20, Brussels.

Council of the European Union (2007b), "Brussels European Council, 8/9 March 2007, Presidency Conclusions," 7224/07, March 9, Brussels.

Council of the European Union (2007c), "Revised Version Presidency Conclusions, Brussels, 8–9 March, 2007," 7224/1/07, Rev. 1, Concl 1, May 2, Brussels.

Cowell, A. (2007), "Britain Proposes Law to Curb Greenhouse Gases," *New York Times*, March 13.

Crepaz, M. (1995), "Explaining National Variation of Air Pollution Level: Political Institutions and their Impact on Environmental Policy-Making," *Environmental Politics* 4:1995, 391–414.

Crespi, J.M. and Marette, S. (2003), "'Does Contain' Vs. 'Does Not Contain': Does it Matter Which GMO Label is Used?," *European Journal of Law and Economics* 16:3, 327–44.

CSTEE (1998a), "Opinion on a Study Commissioned by Directorate General III (Industry) of the European Commission on Recent Assessments of the Hazards and Risks Posed by Asbestos and Substitute Fibres, and Recent Regulation of Fibres World-wide (Environmental Resources Management, Oxford)," February 9, Brussels: Committee on Toxicity, Ecotoxicity and the Environment.

CSTEE (1998b), "Opinion on Chrysotile Asbestos and Candidate Substitutes Expressed at the 5th CSTEE Plenary Meeting," September 15, Brussels: Committee on Toxicity, Ecotoxicity and the Environment.

CSTEE (1999), "Minutes of the 7th Plenary Meeting of the Scientific Committee on Toxicity, Ecotoxicity, and the Environment (CSTEE)," January 18, Brussels: Committee on Toxicity, Ecotoxicity and the Environment.

Cucic, S. (2000), "European Union Health Policy and its Implications for National Convergence," *International Journal for Quality in Health Care* 12:3, 217–25.

Davis, C. (2003), *Food Fights over Free Trade: How International Institutions Promote Agricultural Liberalization* (Princeton, NJ: Princeton University Press).

de Bruijn, T.J.N.M. and Norberg-Bohm, V. (eds) (2005), *Industrial Transformation: Environmental Policy Innovation in the United States and Europe* (Cambridge, MA: MIT Press).

De Grazia, V. (2005), *Irresistible Empire: America's Advance through Twentieth-Century Europe* (Cambridge, MA: Belknap Press of Harvard University Press).

de Schoutheete, P. (2006), "The European Council," in J. Peterson and M. Shackleton (eds), *The Institutions of the European Union* (Oxford: Oxford University Press).

Deere, C. and Esty, D.C. (eds) (2002), *Greening the Americas: NAFTA's Lessons for Hemispheric Trade* (Cambridge, MA: MIT Press).

DEFRA (2007a), "The Climate Change Levy," Department for Environment, Food, and Rural Affairs, <http://www.defra.gov.uk/environment/ccl/intro.htm>, accessed June 1, 2007.

DEFRA (2007b), "UK Emissions Trading Scheme," Department for Environment, Food, and Rural Affairs, <http://www.defra.gov.uk/environment/climatechange/trading/uk/index.htm>, accessed June 1, 2007.

Delmas, M.A. (2002), "The Diffusion of Environmental Management Standards in Europe and the United States: An Institutional Perspective," *Policy Sciences* 35:1, 91–119.

Demmke, C. (2004), "Implementation of Environmental Policy and Law in the United States and the European Union," in N.J. Vig and M.G. Faure (eds), *Green Giants? Environmental Policies of the United States and the European Union* (Cambridge, London: MIT Press), 135–57.

Desai, U. (ed.) (2002), *Environmental Politics and Policy in Industrialized Countries* (Cambridge, MA: MIT Press).

DeSombre, E.R. (2000), *Domestic Sources of International Environmental Policy: Industry, Environmentalists, and US Power* (Cambridge, MA: MIT Press).

DeSombre, E.R. (2005), "Understanding United States Unilateralism: Domestic Sources of US International Environmental Policy," in R. Axelrod, D. Downie, and N.J. Vig (eds), *The Global Environment: Institutions, Law, and Policy* (Washington, DC: CQ Press).

Deutscher Bundestag (1985), "Antwort der Bundesregierung auf die Große Anfrage der Abgeordneten Dr. Müller (Bremen), Vogel (München), Tatge und der Fraktion die Grünen: Haushaltspolitische, Ökologische und Entwicklungspolitische Risiken der Ausfuhrbürgschaften," Deutscher Bundestag, Bonn.

Deutscher Bundestag (2001a), "Für den Erhalt von Hermes als Instrument der Außenwirtschaftsförderung und eine Reform des Hermes-Instruments im Internationalen Rahmen—Antrag der CDU/CSU-Fraktion," Deutscher Bundestag, Berlin.

Deutscher Bundestag (2001b), "Für ein Effizientes und Transparentes Ausfuhrgewährleistungssystem—Antrag der F.D.P.-Fraktion," Deutscher Bundestag, Berlin.

Deutscher Bundestag (2001c), "Für ein Modernes Ausfuhrgewährleistungssystem—Antrag der Fraktionen von SPD und Bündnis 90/Die Grünen," Deutscher Bundestag, Berlin.

DG-IIII—DEN (1996), "Detailed Explanatory Note," *DG Enterprise*, <http://europa.eu.int/comm/dg03/directs/dg3c/asbdetainote.htm>, accessed February 22, 2001.

Dimitrov, R. (2006), *Science and International Environmental Policy: Regimes and Nonregimes in Global Governance* (Oxford: Rowman and Littlefield Publishers, Inc).

Dimitrov, R.S. (2005), "Hostage to Norms: States, Institutions and Global Forest Politics," *Global Environmental Politics* 5:4, 1–24.

DIY (1998), "Hodkinson Tells World Bank of B&Q's FSC Plans," *DIY* 5, January 23.

Dodd, C. (2007), "A Corporate Carbon Tax," *Boston Globe*, April 27.

Domask, J. (2003), "Chapter Eight: From Boycotts to Partnership: NGOs, the Private Sector, and the World's Forests," in J.P. Doh and H. Teegen (eds), *Globalization and NGOs: Transforming Business, Governments, and Society* (New York: Praeger).

Dowie, M. (1995), *Losing Ground: American Environmentalism at the Close of the Twentieth Century* (Cambridge, MA: MIT Press).

Downie, D.L. (2003), "Global Pops Policy: The 2001 Stockholm Convention on Persistent Organic Pollutants," in D.L. Downie and T. Fenge (eds), *Northern Lights against Pops: Combatting Toxic Threats in the Arctic* (Montreal: McGill-Queens University Press).

Downie, D.L., Krueger, J., and Selin, H. (2004), "Global Policy for Hazardous Chemicals," in R. Axelrod, D.L. Downie, and N. Vig (eds), *Global Environmental Policy: Institutions, Law and Policy* (Washington DC: CQ Press).

Drillisch, H., Unmüßig, B., and Zúñiga, C. (1998), "Hermes-Kampagne: Rundbrief 3," May 1998, Bonn: Weltwirtschaft, Ökologie & Entwicklung (WEED) and Urgewald.

Dryzek, J.S., Downes, D., Hunold, C., Schlosberg, D., and Hernes, H.-K. (2003), *Green States and Social Movements: Environmentalism in the United States, United Kingdom, Germany, and Norway* (Oxford and New York: Oxford University Press).

Dryzek, J.S., Hunold, C., Schlosberg, D., Downes, D., and Hernes, H.-K. (2002), "Environmental Transformation of the State: The USA, Norway, Germany and the UK," *Political Studies* 50:4, 659–82.

DSIRE (2005), "Database of State Incentives for Renewable Energy," Raleigh, NC: North Carolina Solar Center.

DTI (2004), "UK Energy in Brief," London: Department of Trade and Industry.

Dunlop, C. (2000), "GMOs and Regulatory Styles," *Environmental Politics* 9:2, 149–55.

Durant, R.F. (2004), "The Precautionary Principle," in R.F. Durant, D.J. Fiorino, and R. O'Leary (eds), *Environmental Governance Reconsidered: Challenges, Choices, and Opportunities* (Cambridge, MA: MIT Press).

Durant, R.F., Fiorino, D.J., and O'Leary, R. (eds) (2004), *Environmental Governance Reconsidered: Challenges, Choices, and Opportunities* (Cambridge, MA: MIT Press).

ECA Watch (2006a), "NGOs Urge OECD General Secretary to Intervene in Export Credit Environmental Standard Negotiations," *ECA Watch Press Advisory*, November 10: ECA Watch.

ECA Watch (2006b), "OECD Negotiations on Export Credit Environment Standards Break Down," *ECA Watch*, <http://www.eca-watch.org/problems/fora/oecd/CommonApproaches/ECAW_CA_talks_breakdown_28nov06.htm>, accessed July 23, 2007.

ECA Watch (2007), "German, Austrian and Swiss ECAs Refuse to Release Environmental and Social Conditions for Ilisu Financing," *ECA Watch*, <http://www.eca-watch.org/problems/mideast/turkey/ilisu/ECAW_Ilisu_Conditions_31jan07.htm>, accessed July 24, 2007.

Eckersley, R. (2004), "The Big Chill: The WTO and Multilateral Environmental Agreements," *Global Environmental Politics* 4:2, 24–50.

Eckhart, M. (2004), "Europe Galvanizes its Renewable Energy Ambitions," February 23: SolarAccess.com News.

Eckley, N. and Selin, H. (2004), "All Talk, Little Action: Precaution and European Chemicals Regulation," *Journal of European Public Policy* 11:1, 78–105.

Ecology Center (2005), "EPA Balks on Banning Major Automotive Lead Use," August 15, Ann Arbor, MI: Ecology Center.

Ecology Center, Great Lake United, and University of Tennessee Center for Clean Products and Clean Technologies (2001), "Toxics in Vehicles: Mercury—Implications for Recycling and Disposal," Ann Arbor, MI: Ecology Center.

EEB (2002a), "New EU Chemicals Policy, Key Elements of the New Legislation, the View of Environmental NGOs," July, Brussels: European Environmental Bureau.

EEB (2002b), "Sustainable Development: Making It Happen," *EEB Position Chapter*, March, Brussels: European Environmental Bureau.

EEB, and G-10 (2006), "A Programme for the Sustainable Development of the European Union," March, Brussels: European Environmental Bureau.

EIA (2004), "Electric Power Industry Restructuring Fact Sheet," Washington, DC: Energy Information Administration.

EIA (2005), "Policies to Promote Non-Hydro Renewable Energy in the United States and Selected Countries," Washington, DC: Energy Information Administration.

EIA (2007), "International Energy Price Information," Washington, DC: Energy Information Administration.

EIA (2008a), "Monthly Energy Review (MER)," May, Washington, DC: Energy Information Administration.

EIA (2008b), "Renewable Energy Annual 2006," Washington, DC: Energy Information Administration.

Eilperin, J. and Mufson, S. (2007), "Tax on Carbon Emissions Gains Support," *Washington Post*, April 1.

Eising, R. (2002), "Policy Learning in Embedded Negotiations: Explaining EU Electricity Liberalization," *International Organization* 56, 85–120.

Electronics Industry Association (2001), "AEE-EIA Principles on European Union WEEE and ROHS Directives," Washington DC: EIA.

Ellerman, A.D. and Joskow, P.L. (2008), "The European Union's Emissions's Trading System in Perspective," Washington, DC: Pew Center on Global Climate Change.

Elliott, D. (1994), "TQM-ing OMB: Why Regulatiory Review under Executive Order 12,291 Works Poorly and What President Clinton Should Do About It," *Law and Contemporary Problems* 57:2, 167–84.

Enderlein, H., Wälti, S., and Zürn, M. (eds) (2009), *Multilevel Governance Handbook* (Cheltenham: Edward Elgar).

ENDS "EC Ready to Move on End-of-Life Vehicles," *ENDS Report*: 246, 34.

ENDS (1996), "US Attacks European Plans for Electronics Take-Back Laws," *ENDS Report*: 253, 44.

ENDS (1997a), "Brussels Finalises Proposal on End-of-Life Vehicles," *ENDS Report*: 270.

ENDS (1997b), "Brussels Moves Ahead with Rules on Electronics Recycling," *ENDS Report*: 271.

ENDS (1999a), "Disputes over ELVs, GMOs Overshadow Environment Council," *ENDS Report*: 293, 41–2.

ENDS (1999b), "Producer Responsibility Retained for Household Electronic Waste," *ENDS Report*: 295, 46.

ENDS (2000), "Parliament Retains Free Take-Back of Existing Cars from 2006," *ENDS Report*: 301, 47.

ENDS (2001), "Parliament Champions Individual Producer Responsibility for WEEE," *ENDS Report*: 316, 48.

ENDS (2002a), "Agreement on WEEE Signals Revolution in Electronics Recycling," *ENDS Report*: 333, 46.

ENDS (2002b), "EC Study Calls for Light Touch on Extended Producer Responsibility," *ENDS Report*: 327, 52.

ENDS (2002c), "Individual Responsibility for WEEE," *ENDS Report*: 327, 52.

ENDS (2002d), "The WEEE Directive: A New Era for Electrical Goods Producers," *ENDS Report*: 334, 27.

ENDS (2004), "Chinese Firms Ignorant of New EU Eco-Design Rules," *ENDS Report*:351, 36.

ENS (2004), "Judge Upholds Maine's Regulation of Mercury from Cars," February 19.

EPA (2008), "Inventory of US Greenhouse Gas Emissions and Sinks: 1990–2006," USEPA #430–R-08–005, Washington, DC: US Environmental Protections Agency.

EPHA (2004), "European Public Health Association Briefing for Members on the European Centre for Disease Prevention and Control," September 30.

Esty, D.C. and Geradin, D. (eds) (2001), *Regulatory Competition and Economic Integration: Comparative Perspectives* (Oxford: Oxford University Press).

"EU Lifts British Beef Ban," (2006), *The Guardian*, March 8.

Euractiv (2005a), "Ministers Uphold National GMO Bans," *EurActiv.com*, <http://www.euractiv.com/>, accessed June 25, 2007.

Euractiv (2005b), "Permanent Phthalates Ban in Toys Approved," July 6.

Euractiv (2006), "Austria Finds Backing for for GMO Bans," *EurActiv.com*, <http://www.euractiv.com/>, accessed June 25, 2007.

Euractiv (2007a), "Commission Wants Binding Cuts on Car Emissions," *EurActiv.com*, <http://www.euractiv.com/en/climate-change/commission-wants-binding-cuts-car-emissions/article-161531>, accessed June 1, 2007.

Euractiv (2007b), "Parliament Sets up Climate Change Committee", <http://www.euractiv.com/en/climate-change/parliament-sets-climate-change-committee/article-162509>, accessed June 15, 2007.

Eurobarometer (2003), "A Report to the EC Directorate General for Research from the Project "Life Sciences and European Society" by G. Gaskell, N. Allum and S. Stares, 58:8.

Europa Press Release (2003), "EU Complies with WTO Ruling on Hormone Beef and Calls on USA and Canada to Life Trade Sanctions," IP/03/1393, October 15.

Europa Press Release (2004), "EU-US: EU Requests WTO to Confirm that there is no Justification for US/Canada to Continue to Apply Sanctions," IP/04/1345, November 8.

European Commission "Proposal for a Directive on Establishing a Framework for the Setting of Eco-Design Requirements for Energy-Using Products Com (2003) 453."

European Commission (1991), "A Community Strategy to Limit Carbon Dioxide Emissions and to Improve Energy Efficiency," *Communication from the Commission to the Council*, SEC (91) 1744 final, October 14, Brussels.

European Commission (1992), "Towards Sustainability," COM (1992) 624.

European Commission (1996a), "Energy for the Future: Renewable Sources of Energy," *Green Paper*, COM (96) 576, November 20, Brussels: European Commission.

European Commission (1996b), "Review of the Community Strategy for Waste Management," COM (96) 399.

European Commission (1997), "Energy for the Future: Renewable Sources of Energy," *White Paper*, COM (97) 599, 26 Novemer 1997, Brussels: European Commission.

European Commission (2000), "Proposal for a Directive of the European Parliament and of the Council on Waste Electrical and Electronic Equipment," COM 2000 (347), June 13.

European Commission (2001a), "European Governance: A White Chapter," COM (2001) 428.

European Commission (2001b), "White Paper on a Strategy for a Future Chemicals Policy," Brussels: European Commission.

European Commission (2002a), "A European Union Strategy for Sustainable Development," Brussels: European Commission.

European Commission (2002b), "EU Unanimously Ratifies Kyoto Protocol to Combat Climate Change," *Press Release*, EC02–108EN, May 30, Brussels.

European Commission (2002c), "European Union Ratifies the Kyoto Protocol," May 31, Brussels.

European Commission (2002d), "Towards a Global Partnership for Sustainable Development," *Commission Communication*, COM(2002)82.

European Commission (2003a), "Economic Impacts of Genetically Modified Crops on the Agri-Food Sector," *Working Document Rev. 2.*

European Commission (2003b), "Proposal for a Regulation of the European Parliament and of the Council Concerning the Registration, Evaluation, Authorisation and Restriction of Chemicals (REACH), Establishing a European Chemicals Agency and Amending Directive 1999/45/EC and Regulation (Ec) {on Persistent Organic Pollutants} Proposal for a Directive of the European Parliament and of the Council Amending Council Directive 67/548/EEC in Order to Adapt It to Regulation (EC) of the European Parliament and of the Council Concerning the Registration, Evaluation, Authorisation and Restriction of Chemicals," Brussels: European Commission.

European Commission (2004a), "Avian Influenza (AI)—Response to Outbreaks in Asia, Canada and the US in 2004, Chronology of Main Events and List of Decisions Adopted by the Commission," April 21.

European Commission (2004b), "Stakeholder Consultation on Adaptation to Scientific and Technical Progress under Directive 2002/95/EC—IBM Comments," July 1.

European Commission (2004c), "The Johannesburg Coalition on Renewable Energy, Information Note No. 1, Members and Objectives," Brussels: European Commission.

European Commission (2005a), "Avian Influenza," <http://ec.europa.eu/comm/food/animal/diseases/controlmeasures/avian/index_en.htm>.

European Commission (2005b), "Commission Confirms Quality of European GMO Legislative Framework," *Press Releases Rapid*, March 22: European Union On-Line.

European Commission (2005c), "European Union: Energy and Transport in Figures 2004," Brussels: European Commission, Directorate-General for Energy and Transport.

European Commission (2005d), "The TSE Roadmap," COM(2005) 322 final, July 15.

European Commission (2007a), "Energy for the Future: Renewable Sources of Energy," Directorate-General for Energy and Transport.

European Commission (2007b), "Energy, Transport and Environment Indicators," Luxembourg: EUROSTAT.

European Commission (2007c), "Limiting Global Climate Change to 2 Degrees Celsius: The Way Ahead for 2020 and Beyond," *Commission Communication*, COM(2007)0002, January 10.

European Commission (2007d), "Progress Report on the Sustainable Development Strategy 2007," *Communication from the Commission to the Council and the European Parliament*, COM(2007) 642 final, October 22.

European Commission (2008a), "20 20 by 2020, Europe's Climate Change Opportunity," *Communication from the Commission to the European Parliament, the Council, the European Economic and Social Committee and the Committee of the Regions*, COM(2008) 30 final, January 23, Brussels.

European Commission (2008b), "Climate Action," <http://ec.europa.eu/environment/climat/climate_action.htm>.

European Commission (2008c), "Energy for the Future: Renewable Sources of Energy."

European Council (2000), "Lisbon European Council Conclusions."

European Council (2001), "Gothenburg European Council Conclusions."

European Council (2005), "Guiding Principles for Sustainable Development," *Annex 1 to Brussels European Council Conclusions*, July 16/17.

European Environment Agency (2008a), "Annual European Community Greenhouse Gas Inventory 1990–2006," Technical Report No. 6/2008, Copenhagen.

European Environment Agency (2008b), "EN 27 Electricity Production by Fuel."

European Environment Agency and United Nations Environment Programme (1999), "Chemicals in the European Environment: Low Doses, High Stakes," *The EEA and UNEP Annual Message 2 on the State of Europe's Environment*, Copenhagen: European Environment Agency.

European Monitoring Center on Change (2004), "Trends and Drivers of Change in the European Automobile Industry: Mapping Report," Dublin.

European Parliament (2001), "Directive 2001/77/EC of the European Parliament and of the Council of 27 September 2001 on the Promotion of Electricity Produced from Renewable Energy Sources in the Internal Energy Market," 2001/77/EC, October 27: *Official Journal of the European Communities*, L 283/33–40.

European Parliament and Council of the European Union (1994), "European Parliament and Council Directive 94/62/EC of 20 December 1994 on Packaging and Packaging Waste," 94/62/EC, December 31: *Official Journal of the European Communities*, L365/10.

European Parliament and Council of the European Union (2000), "Directive 2000/53/EC of the European Parliament and of the Council of 18 September 2000 on End-of-Life Vehicles," 2000/53/EC, October 21: *Official Journal of the European Union*, L 269/34.

European Parliament and Council of the European Union (2003a), "Directive 2002/95/EC of the European Parliament and of the Council of 27 January 2003 on the Restriction of the Use of Certain Hazardous Substances in Electrical and Electronic Equipment," 2002/95/EC, February 13: *Official Journal of the European Union*, L37/19.

European Parliament and Council of the European Union (2003b), "Directive 2002/96/EC of the European Parliament and of the Council of 27 January 2003 on Waste Electrical and Electronic Equipment," 2002/96/EC, February 13: *Official Journal of the European Union*, L 37/24.

European Wind Energy Agency (2008), "The Progress of Wind Energy," <http://www.no-fuel.org/index.php?id=246>.

Eurostat "February 2001."

Evans, P.C. (2000), "Environment Time Line—Negotiating Common International Standards," Massachusetts Institute of Technology.

Evans, P.C. (2005), "International Regulation of Official Trade Finance: Competition and Collusion in Export Credits and Foreign Aid," Department of Political Science, Massachusetts Institute of Technology, Cambridge, MA.

Fabbrini, S. (2004), "Transatlantic Constitutionalism: Comparing the United States and European Union," *European Journal of Political Research* 43:4, 547–69.

Falkner, R. (2000), "Regulating Biotech Trade: The Cartagena Protocol on Biosafety," *International Affairs* 76:2, 299–314.

Falkner, R. (2002), "International Trade Conflicts over Agricultural Biotechnology," in A. Russell and J. Vogler (eds), *The International Politics of Biotechnology: Investigating Global Futures* (Manchester: Manchester University Press).

Farrell, A. and Sperling, D. (2007), "Getting the Carbon Out," *San Francisco Chronicle*, May 18.

Farrell, A.E. and Hanemann, W.M. (2009), "Field Notes on the Political Economy of California Climate Policy," in H. Selin and S.D. VanDeveer (eds), *Changing Climates in North American Politics: Institutions, Policymaking and Multilevel Governance* (Cambridge, MA: MIT Press).

Federal Government of Germany (2008), "The Case for an International Renewable Energy Agency (Irena)."

Finnish Forest Certification Council (1999), "Caring for Our Forests," Helsinki: Finnish Forest Certification Council.

Fischer, J. (2004), "California Energy Commission: Workshop on Renewable Development and Transmission Constraints in Southern California," May 10, Sacramento, CA: PPM Energy.

Fishbein, B. (1998), "EPR: What Does it Mean? Where is it Headed?," *P2: Pollution Prevention Review* 8, 43–55.

Foley, M. and Owens, J.E. (1996), *Congress and the Presidency: Institutional Politics in a Separated System* (Manchester: Manchester University Press).

Food Navigator (2005), "June 1, 2005 "US Argues Biotech Rules too Costly for Exporters" as Reported in the News," June 1: Pew Intitiative on Food and Biotechnology.

Forest Stewardship Council (1999), "FSC Principles and Criteria," Document 1.2, Revised January: Forest Stewardship Council.

Forsberg, B. (2005), "Getting the Lead Out: European Rules Force Electronics Companies to Clean Up," *San Francisco Chronicle*, January 20.

Foy, D. and Healy, R. (2007), "Cities Are the Answer," *Boston Globe*, April 4.

Fredriksson, P.G. and Millimet, D.L. (2002), "Strategic Interaction and the Determination of Environmental Policy across US States," *Journal of Urban Economics* 51:1, 101–22.

Freeman, H.B. (2002), "Trade Epidemic: The Impact of the Mad Cow Crisis on EU-US Relations," *Boston College International and Comparative Law Review* 25, 343–71.

Friends of the Earth Europe (2005), "Biotechnology Programme and European GMO Campaign."

Fuchs, D.A. (2004), "Channels and Dimensions of Business Power in Global Governance," paper presented at the *Annual Meeting of the International Studies Association*, Montreal, March 17–21.

Fues, T. (1994), *Reforming Export Guarantee Systems: Challenges Ahead for Northern NGOs* (Brussels: Eurodad).

Fuller, T. (2006), "Viceroys Long Gone, EU Grows in Asia," *International Herald Tribune*, March 16, 1 and 8.

Gagnon, L. (2003), "Comparing Power Generation Options: Greenhouse Gas Emissions," January 2003, Montreal, QC: Hydro Québec.

Gallagher, K. (2004), *Free Trade and the Environment: Mexico, NAFTA, and Beyond* (Stanford, CA: Stanford Law and Politics).

Garton Ash, T. (2004), *Free World: America, Europe, and the Surprising Future of the West* (New York: Random House).

Gaskell, G. (2004), "Science Policy and Society: The British Debate over GM Agriculture," *Current Opinion in Biotechnology* 15, 241–5.

Geddes, B. (1990), "How the Cases You Choose Affect the Answers You Get; Selection Bias in Comparative Politics," *Political Analysis* 2, 131–50.

Geddes, B. (2003), *Paradigms and Sand Castles: Theory Building and Research Design in Comparative Politics* (Ann Arbor, MI: University of Michigan Press).

Gee, D. and Greenberg, M. (2001), "Asbestos: From 'Magic' to Malevolent Mineral," in P. Harremoës, D. Gee, M. MacGarvin, A. Stirling, J. Keys, B. Bynne, and S. Guedes Vaz (eds), *Late Lessons from Early Warnings: The Precautionary Principle 1896–2000* (Copenhagen: European Environmental Agency), 52–63.

Geiser, K. and Tickner, J. (2003), "New Directions in European Chemicals Policies: Drivers, Scope and Status," October, Lowell, MA: Lowell Center for Sustainable Production.

GEN (2003), "Trade as an Environment Policy Tool? Environment as a Trade Policy Tool? GEN, Ecolabelling and Trade," June, Ottawa, ON: Global Ecolabelling Network.

Glim, M.C.S. (1994), "Environmental Policy and Management in the European Community and the United States: A Comparison," in R. Baker (ed.), *Comparative Public Management: Putting US Public Policy and Implementation in Context* (Westport, CN: Praeger), 219–31.

Global Wind Energy Council (2008), "US, China, and Spain Lead World Wind Power Market in 2007," <http://www.gwec.net/index.php?id=30&no_cache=1&tx_ttnews%5Btt_news%5D=139&tx_ttnews%5BbackPid%5D=4&cHash=6691aa654e>.

Goldberg, T. (2004), "Mercury Reduction and Education Legislation in the Imerc Member States," October: North-East Waste Management Officials Association.

Gordon, P. (2003), "Bridging the Atlantic Divide," *Foreign Affairs* January/February.

Gordon, P.H. and Shapiro, J. (2004), *Allies at War: America, Europe, and the Crisis over Iraq* (New York: McGraw-Hill).

Gore, C. and Robinson, P. (2009), "Local Government Response to Climate Change, Our Last, Best Hope?," in H. Selin and S.D. VanDeveer (eds), *Changing Climates in North American Politics: Institutions, Policymaking and Multilevel Governance* (Cambridge, MA: MIT Press).

Görlach, B., Knigge, M., and Schaper, M. (2007), "Transparency, Information Disclosure and Participation in Export Credit Agencies' Cover Decisions," in S. Thoyer and B. Martimort-Asso (eds), *Participation for Sustainability in Trade* (Aldershot: Ashgate).

Gottschlich, J. and Kreutzfeldt, M. (2007), "Land unter für Deutschen Expor—Trotz Kritik von Umwelt- und Menschenrechtsgruppen bewilligt die Bundesregierung Kreditgarantien für den umstrittenen Ilisu-Staudamm in der Türkei. Grund: Die vorgegebenen Kriterien seien erfüllt," *taz, die tageszeitung*, March 28, 10.

Government Offices of Sweden (2007), "Climate Policy," <http://www.sweden.gov.se/sb/d/5745/a/21787;jsessionid=adVFhogbxv_e>, accessed June 1, 2007.

Grande, E. (1996), "The State and Interest Groups in a Framework of Multi-Level Decision-Making: The Case of the European Union," *Journal of European Public Policy* 3:3, 318–38.

Grant, W., Matthews, D., and Newell, P. (2000), *The Effectiveness of European Union Environmental Policy* (New York: St Martin's Press).

Green-e (2003a), "Governance," <http://www.green-e.org/what_is/governance/governance_index.html>, accessed October 1, 2003.

Green-e (2003b), "Small Hydro, Low Impact Hydro," <http://www.green-e.org/what_is/dictionary/shydro.html>, accessed October 1, 2003.

Green-e (2003c), "Standard," <http://www.green-e.org/ipp/standard_for_marketers.html>, accessed October 1, 2003.

Greenpeace (2001), "Greenpeace Blocks US Oil Delivery to Protest G.W. Bush's Climate Stance, <http://archive.greenpeace.org/pressreleases/climate/2001jun10.html>, accessed June 1, 2007.

Guehlstorf, N. and Hallstrom, L. (2008), "Culture Wars over the Risks, Regulations, and Responsibilities in Genetic Agriculture: A Comparison of Food Biotechnology Policy in the United States and the European Union," paper presented at the *American Political Science Association annual meeting*, Boston, MA, August 28.

Gunningham, N., Grabosky, P.N., and Sinclair, D. (1998), *Smart Regulation: Designing Environmental Policy* (Oxford and New York: Oxford University Press).

Gupta, A. (2000), "Governing Trade in Genetically Modified Organisms: The Cartagena Protocol on Biosafety," *Environment:*May, 22–32.

Gupta, J. and Grubb, M. (eds) (2000), *Climate Change and European Leadership: A Sustainable Role for Europe?* (Dordrecht; Boston: Kluwer Academic).

Haas, P.M. (2003), "Environment Multilateralism and the United States," Amherst, MA: University of Massachusetts.

Haas, R. (ed.) (2001), *Review Report on Promotion Strategies for Electricity from Renewable Energy Sources in EU Countries* (Vienna: Institute of Energy Economics, Vienna University of Technology).

Haberman, A. (ed.) (2001), *Twenty-Five Years Behind Bars: The Proceedings of the Twenty-Fifth Anniversary of the U.P.C. At the Smithsonian Institution, Sept. 30, 1999* (Cambridge, MA: Harvard University Press).

Hagelüken, A. (2001), "Streit um Forderungen der Grünen: BDI gegen Umweltstandards bei Export-Krediten," *Süddeutsche Zeitung*, March 12, 2001, 24.

Hajer, M.A. (1995), *The Politics of Environmental Discourse: Ecological Modernization and the Policy Process* (Oxford: Clarendon Press).

Hall, P.A. (1986), *Governing the Economy: The Politics of State Intervention in Britain and France* (New York: Oxford University Press).

Hallman, W.K., Schilling, B.J., and Turvey, C.G. (2004), "Public Perceptions and Responses to Mad Cow Disease: A National Survey of Americans," New Brunswick: Food Policy Institute, Rutgers University.

Halluite, J., Linton, J.D., Yeomans, J.S., and Yoogalingam, R. (2005), "The Challenge of Hazardous Waste Management in a Sustainable Environment: Insights from Electronic Recovery Laws," *Corporate Social Responsibility and Environmental Management* 12:1, 31–7.

Hamilton, D.S. (2004), "Conflict and Cooperation in Transatlantic Relations," Washington, DC: The Center for Transatlantic Relations, Johns Hopkins University, School of Advanced International Studies.

Hanf, K. and Jansen, A.-I. (eds) (1998), *Governance and Environment in Western Europe: Politics, Policy and Administration* (Harlow: Addison-Wesley).

Hanisch, C. (2000), "Is Extended Producer Responsibility Effective?," *Environmental Science and Technology* 34:7, 170A-5A.

Hansen, E., Fletcher, R., Cashore, B., and McDermott, C. (2006), "Forest Certification in North America," EC 1518, January: Oregon State University Extension Service.

Hansen, E. and Juslin, H. (1999), "The Status of Forest Certification in the ECE Region," *Geneva Timber and Forest Discussion Papers*, ECE/TIM/DP/14, New York and Geneva: United Nations, Timber Section, Trade Division, UN-Economic Commission for Europe.

Hardin, R. (2002), *Trust and Trustworthiness* (New York: Russell Sage).

Harmelink, M., Voogt, M., and Cremer, C. (2006), "Analysing the Effectiveness of Renewable Energy Supporting Policies in the European Union," *Energy Policy* 34, 343–51.

Harrington, W. and Morgenstern, R.D. (2004), *Choosing Environmental Policy: Comparing Instruments and Outcomes in the United States and Europe* (Washington, DC: Resources for the Future).

Harris, P.G. (2000a), *Climate Change and American Foreign Policy* (New York: St Martin's Press).

Harris, P.G. (ed.) (2000b), *The Environment, International Relations, and US Foreign Policy* (Washington, DC: Georgetown University Press).

Harris, P.G. (2007), *Europe and Global Climate Change: Politics, Foreign Policy and Regional Cooperation* (Cheltenham: Edward Elgar).

Hart, L. (2004), "Bird Flu Threat Shuts Live Markets," *Los Angeles Times*, February 29, A20.

Harvard Center for Risk Analysis, and Harvard School of Public Health (2001), "Evaluation of the Potential for Bovine Spongiform Encephalopathy in the United States," November 26.

Hay, C. (1998), "The Tangled Web We Weave: The Discourse, Strategy and Practice of Networking," in D. Marsh (ed.), *Comparing Policy Networks* (Buckingham: Open University Press).

Hayes-Renshaw, F. (2006), "The Council of Ministers," in J. Peterson and M. Shackleton (eds), *The Institutions of the European Union* (Oxford: Oxford University Press).

Hebert, S. (1999), "HIV Blood Victims Demand Justice at Minister's Trial," *Electronic Telegraph*, February 10.

Heiman, M.K. and Solomon, B.D. (2004), "Power to the People: Electric Utility Restructuring and the Commitment to Renewable Energy," *Annals of the Association of American Geographers* 94, 94–116.

Held, A., Ragwitz, M., and Haas, R. (2006), "On the Success of Policy Strategies for the Promotion of Electricity from Renewable Energy Sources in the EU," *Energy and Environment* 17:6, 849–68.

Hellström, E. (2001), "Conflict Cultures—Qualitative Comparative Analysis of Environmental Conflicts in Forestry," *Silva Fennica Monographs* 2, 2–109.

Henley, J. (1999), "French Politicians on Trial for Mass AIDs Infection," *Guardian*, February 6.

Héritier, A., Knill, C., and Mingers, S. (1996), *Ringing the Changes in Europe: Regulatory Competition and the Transformation of the State: Britain, France, Germany* (Berlin and New York: Walter de Gruyter).

Hermes Kreditversicherungs-AG (1998), "Berücksichtigung von Umweltaspekten bei der Vergabe von Ausfuhrgewährleistungen," AGA-Report Nr. 72, June, Hamburg: Hermes Kreditversicherungs-AG.

Hix, S. (1999), *The Political System of the European Union* (Houndsmills, NY: Palgrave).

Hodge, C.C. (2005), *Atlanticism for a New Century: The Rise, Triumph, and Decline of NATO* (Upper Saddle River, NJ: Pearson/Prentice Hall).

Holdren, J.P. and Smith, K.R. (2000), "Energy, the Environment, and Health," in World Energy Assessment (ed.), *Energy and the Challenge of Sustainability* (New York: United Nations Development Programme), 61–110.

Holzinger, K., Knill, C., and Arts, B. (eds) (2008), *Environmental Policy Convergence in Europe: The Impact of International Institutions and Trade* (Cambridge and New York: Cambridge University Press).

Hooghe, L. and Marks, G. (2001), *Multi-Level Governance and European Integration* (Lanham, Boulder, New York, Oxford: Rowman and Littlefield).

Hooper, K. and She, J. (2003), "Lessons from the Polybrominated Diphenyl Ethers (Pbdes): The Precautionary Principle, Primary Prevention, and the Value of Community-Based Body-Burden Monitoring Using Breast Milk," *Environmental Health Perspectives* 111:1, 109–14.

Hoornbeek, J. (2000), "Information and Environmental Policy: A Tale of Two Agencies," *Journal of Comparative Policy Analysis* 2:2, 145–87.

Hoornbeek, J.A. (2004), "Policy-Making Institutions and Water Policy Outputs in the European Union and the United States: A Comparative Analysis," *Journal of European Public Policy* 11:3, 461–96.

Horlick, G., Schuchhardt, C., and Mann, H. (2002), "NAFTA Provisions and the Electricity Sector," *Environmental Challenges and Opportunities of the Evolving North American Electricity Market*, Background Paper 4, June 2002, Montreal, QC: North American Commission for Environmental Cooperation.

Hovden, E. (2004), "Norway: Top Down Europeanization by Fax," in A. Jordan and D. Liefferink (eds), *Environmental Policy in Europe: The Europeanization of National Environmental Policy* (London: Routledge), 154–71.

Hovi, J., Skodvin, T., and Andresen, S. (2004), "The Persistence of the Kyoto Protocol: Why Other Annex I Countries Move on without the United States," *Global Environmental Politics* 3:4, 1–23.

Howlett, M. and Rayner, J. (2006), "Globalization and Governance Capacity: Explaining Divergence in National Forest Programs as Instances of 'Next-Generation' Regulation in Canada and Europe," *Governance: An International Journal of Policy, Administration, and Institutions* 19:2, 251–75.

Humphreys, D. (2004), "Redefining the Issues: NGO Influence on International Forest Negotiations," *Global Environmental Politics* 4:2, 51–7.

Humphreys, D. (2006), " Forest for the Future: National Forest Programmes in Europe—Country and Regional Reports," in D. Humphreys. (ed.) (Brussels: European Science Foundation, COST Action E19).

Humphreys, D. (2007), *Logjam: Deforestation and the Crisis of Global Governance* (London: Earthscan).

Hunter, J.R. and Smith, Z.A. (2005), *Protecting Our Environment: Lessons from the European Union* (Albany, NY: State University of New York Press).

Hvelplund, F. (2001), "Political Prices or Political Quantities?," *New Energy* 5, 18–23.

Hvelplund, F. (2002), "Denmark," in D. Reiche (ed.), *Handbook of Renewable Energies in the European Union: Case Studies of All Member States* (Frankfurt am Main: Peter Lang), 63–76.

ICRE (2004), "Policy Recommendations for Renewable Energies," June 4, Bonn: International Conference for Renewable Energies.

IEA (1998), "Benign Energy? The Environmental Implications of Renewables," Paris: International Energy Agency.

IEA (2002), "Renewables Information 2002," Paris: International Energy Agency.

IEA (2004), "Renewable Energy: Market and Policy Trends in IEA Countries," Paris: International Energy Agency.

Iles, A. (2007), "Identifying Environmental Health Risks in Consumer Products: Non-Governmental Organizations and Civic Epistemology," *Public Understanding of Science* 19:4, 371–2.

ILO (1986), "C162—Convention Concerning Safety in the Use of Asbestos," International Labor Organization.

Innovest (2003), "Power Switch: Impacts of Climate Policy on the Power Sector," November, New York: Innovest Strategic Value Advisors.

International Chemical Secretariat (2004), "Cry Wolf: Predicted Cost by Industry in the Face of New Regulations," Göteborg: International Chemical Secretariat.

"International Roundup: Countries Respond to Avian Flu Outbreaks," (2004), *Food Chemical News*, March 1, 25.

IPCC (2007a), "Climate Change 2007: Impacts, Adaptation and Vulnerabilities," Geneva: Intergovernmental Panel on Climate Change.

IPCC (2007b), "Climate Change 2007: The Physical Science Basis," Geneva: Intergovernmental Panel on Climate Change.

Isaac, G. (2002), *Agricultural Biotechnology and Transatlantic Trade: Regulatory Barriers to GM Crops* (Wallingford and New York: CABI Pub.).

Jack, A. (2005), "Europe's New Disease Control Chief Faces Tough Challenge," *The Financial Times*, May 19.

Jacob, K., Beise, M., Blazejczak, J., Edler, D., Haum, R., Jänicke, M., Löw, T., Petschow, U., and Rennings, K. (2005), *Lead Markets for Environmental Innovations* (Heideberg; New York: Springer).

Jacobson, H.K. (2002), "Climate Change: Unilateralism, Realism and Two-Level Games," in S. Patrick and S. Forman (eds), *Multilateralism and US Foreign Policy: Ambivalent Engagement* (Boulder, CO: Lynne Rienner Publishers).

Jacques, P.j., Dunlap, R.E., and Freeman, M. (2008), "The Organization of Denial: Conservative Think Tanks and Environmental Skepticism," *Environmental Politics* 17:3, 349–85.

Jänicke, M. (2005), "Trend-Setters in Environmental Policy: The Character and Role of Pioneer Countries," *European Environment* 15:2, 129–42.

Jänicke, M. and Weidner, H. (eds) (1997), *National Environmental Policies: A Comparative Study of Capcity-Building* (Berlin: Springer).

Jansen, J.C. and Uyterlinde, M.A. (2004), "A Fragmented Market on the Way to Harmonisation? EU Policy-Making on Renewable Energy Promotion," *Energy for Sustainable Development* 8, 93–107.

Jasanoff, S. (1986), *Risk Management and Political Culture: a Comparative Study of Science in the Policy Context* (New York: Russell Sage Foundation).

Jasanoff, S. (1995), "Product, Process, or Programme: Three Cultures and the Regulation of Biotechnology," in M. Bauer (ed.), *Resistance to New Technology* (Cambridge: Cambridge University Press).

Jasanoff, S. (1997), "Civilization and Madness: The Great BSE Scare of 1996," *Public Understanding of Science* 6:3, 221–32.

Jasanoff, S. (2003), "A Living Legacy: The Precautionary Ideal in American Law," in J.A. Tickner (ed.), *Precaution, Environmental Science and Preventive Public Policy* (Washington DC: Island Press).

Jasanoff, S. (2005), *Designs on Nature: Science and Democracy in Europe and the United States* (Princeton, NJ: Princeton University Press).

Johansson, T.B. and Turkenburg, W. (2004), "Policies for Renewable Energy in the European Union and its Member States: An Overview," *Energy for Sustainable Development* 8, 5–24.

Jones, A. and Clark, J. (2001), *The Modalities of European Union Governance: New Institutionalist Explanations of Agri-environmental Policy* (Oxford: Oxford University Press).

Jones, C.A. and Levy, D.L. (2009), "Buisiness Strategies and Climate Change," in H. Selin and S.D. VanDeveer (eds), *Changing Climates in North American Politics: Institutions, Policymaking and Multilevel Governance* (Cambridge, MA: MIT Press).

Jordan, A. (ed.) (2002), *Environmental Policy in the European Union: Actors, Institutions and Processes* (London and Sterling, VA: Earthscan).

Jordan, A. (ed.) (2005), *Environmental Policy in the European Union* (London: Earthscan).

Jordan, A. and Liefferink, D. (eds) (2004), *Environmental Policy in Europe: The Europeanization of National Environmental Policy* (London: Routledge).

Jordan, A., Wurzel, R., and Zito, A.R. (2003), "'New' Instruments of Environmental Governance: Patterns and Pathways of Change," *Environmental Politics* 12:1, 1–24.

Jörgens, H.E. (2002), "Diffusion and Convergence of Environmental Policies in Europe," *European Environment. Special Issue* 15:2, 61–2.

Josling, T.E., Roberts, D., and Hassan, A. (1999), "The Beef-Hormone Dispute and its Implications for Trade Policy," *Working Paper*, September 1999, Stanford, CA: Freeman Spogli Insitute for International Studies, Stanford University.

Jozwiak, J.F., Jr and Crowley, P.M. (2004), "Comparing the EU and NAFTA Environmental Policies: Comparative Institutional Analysis and Case Studies," in G. Bouchard, P. Bowles, and W. Chandler (eds), *Crossing the Atlantic: Comparing the European Union and Canada* (Aldershot and Burlington, VT: Ashgate), 109–35.

Kagan, R. (2003), *Paradise and Power: America and Europe in the New World Order* (London: Atlantic Books).

Kantor, J. (2008), "Europeans Reconsider Biofuel Goal," *New York Times*, July 8.

Kastner, J. and Powell, D. (2002), "The SPS Agreement: Addressing Historical Factors in Trade Dispute Resolution," *Agriculture and Human Values* 19, 282–92.

Keck, M.E. and Sikkink, K. (1998), *Activists Beyond Borders: Advocacy Networks in International Politics* (Ithaca, NY: Cornell University Press).

Kelemen, R.D. (2000), "Regulatory Federalism: EU Environmental Regulation in Comparative Perspective," *Journal of Public Policy* 20:3, 133–67.

Kelemen, R.D. (2004a), "Environmental Federalism in the United States and the European Union," in N.J. Vig and M.G. Faure (eds), *Green Giants? Environmental Policies of the United States and the European Union* (Cambridge and New York: Cambridge University Press).

Kelemen, R.D. (2004b), *The Rules of Federalism: Institutions and Regulatory Politics in the EU and Beyond* (Cambridge, MA: Harvard University Press).

Kellerhals, M.D. (2001), "USAID Launches Biotechnology Initiatives with Africa: Programs Foster Improved Regulation, Research, Development," March 2, Washington, DC: Office of International Information Programs, US Department of State.

Kemikalieinspektionen (2004), "REACH—En Ny Kemikalielag För en Giftfri Miljö," *Fact Sheet*, September, Stockholm.

Kern, K. (2000), *Die Diffusion von Politikinnovationen: Umweltpolitische Innovationen im Mehrebenensystem der USA* (Opladen: Leske und Budrich).

King, G., Keohane, R.O., and Verba, S. (1994), *Designing Social Inquiry: Scientific Inference in Qualitative Research* (Princeton, NJ: Princeton University Press).

Kjær, A.M. (2004), *Governance* (Malden, MA: Polity Press).

Kleckner, D. (2006), "A Call to Green Revolution," *Agweb*, 10.

Klyza, C.M. and Sousa, D.J. (2008), *American Environmental Policy, 1990–2006: Beyond Gridlock* (Cambridge, MA: MIT Press).

Knigge, M. and Bausche, C. (2006), "Climate Change Policies at the US Subnational Level—Evidence and Implications," *Ecologic Discussion Chapter*, Berlin: Ecologic.

Knigge, M. and Collins, M. (2005), "An Ocean Apart? Subnational Transatlantic Environmental Cooperation," *Ecologic Discussion Chapter*, Berlin: Ecologic.

Knill, C. (2001), *The Europeanisation of National Administrations: Patterns of Institutional Change and Persistence* (Cambridge: Cambridge University Press).

Knill, C. and Lenschow, A. (1998), "Coping with Europe: The Impact of British and German Administrations on the Implementation of EU Environmental Policy," *Journal of European Public Policy* 5:4, 595–614.

Knill, C. and Lenschow, A. (eds) (2000), *Implementing EU Environmental Policy: New Directions and Old Problems* (Manchester and New York: Manchester University Press).

Koehn, P.H. (2008), "Underneath Kyoto: Emerging Subnational Government Initiatives and Incipient Issue-Bundling Opportunities in China and the United States," *Global Environmental Politics* 8:1, 53–77.

Koller, P. (1991), "Facetten der Macht," *Analyse und Kritik* 13, 107–33.

Koppen, I.J. (2002), "The Role of the European Court of Justice," in A. Jordan (ed.), *Environmental Policy in the European Union: Actors, Institutions and Processes* (London and Sterling, VA: Earthscan), 100–19.

Korenstein, S. (2005), "Managing Electronic Waste: The California Approach," *Journal of Environmental Health* 67:6, 36–7.

Korten, D. (2001), *When Corporations Rule the World* (Bloomfield, CT: Kumarian Press).

Kraak, M. and Pehle, H. (2001), "The Europeanisation of Environmental Policy: The German Perspective," paper prepared for the *Europeanization of Environmental Policy Workshop*, Cambridge, June 29–July 1.

Kraft, M.E. and Kamieniecki, S. (eds) (2007), *Business and Environmental Policy: Corporate Interests in the American Political System* (Cambridge, MA: MIT Press).

Kraft, M.E. and Scheberle, D. (1988), "Environmental Federalism at Decade's End: New Approaches and Strategies," *Publius: The Journal of Federalism* 28:1, 131–46.

Kraft, M.E. and Vig, N.J. (2006), "Environmental Policy from the 1970s to the Twenty-First Century," in N.J. Vig and M.E. Kraft (eds), *Environmental Policy: New Directions for the Twenty-First Century* (Washington, DC: CQ Press), 1–33.

Krämer, L. (2004), "The Roots of Divergence: A European Perspective," in N.J. Vig and M. Faure (eds), *Green Giants? Environmental Policies of the United States and the European Union* (Cambridge, MA: MIT Press), 53–72.

Kremer, M. and Zwane, A.P. (2005), "Encouraging Private Sector Research for Tropical Agriculture," *World Development* 33:1, 87–105.

Krueger, J. and Selin, H. (2002), "Governance for Sound Chemicals Management: The Need for a More Comprehensive Global Strategy," *Global Governance* 8:3, 323–42.

Lafferty, W.M. (2004), "Introduction: Form and Function in Governance for Sustainable Development," in W.M. Lafferty (ed.), *Governance for Sustainable Development. The Challenge of Adapting Form to Function* (Cheltenham: Edward Elgar).

Lafferty, W.M. and Meadowcroft, J. (2000a), "Concluding Perspectives," *Implementing Sustainable Development: Strategies and Initiatives in High Consumption Societies* (Oxford: Oxford University Press), 422–59.

Lafferty, W.M. and Meadowcroft, J. (eds) (2000b), *Implementing Sustainable Development: Strategies and Initiatives in High Consumption Societies* (Oxford: Oxford University Press).

Lafferty, W.M. and Meadowcroft, J. (2000c), "Introduction," *Implementing Sustainable Development: Strategies and Initiatives in High Consumption Societies* (Oxford: Oxford University Press), 1–22.

Lambrecht, B. (2001), *Dinner at the New Gene Café: How Genetic Engineering is Changing What We Eat, How We Live, and the Global Politics of Food* (New York: St. Martin's Griffin).

Lancet (2000), "Resisting Smoke and Spin," *The Lancet* 355:9211, 1197.

Lapointe, G. (1998), "Sustainable Forest Management Certification: The Canadian Programme," *The Forestry Chronicle* 74:2 (March/April), 227–30.

Lauber, V. (2002), "Renewable Energy at the EU Level," in D. Reiche (ed.), *Handbook of Renewable Energies in the European Union: Case Studies of All Member States* (Frankfurt am Main: Peter Lang), 25–36.

Lauber, V. and Mez, L. (2006), "Renewable Electricity Policy in Germany, 1974 to 2005," *Bulletin of Science, Technology, and Society* 26:2, 105–20.

Lautenberg, F.R. (2008), "Lautenberg, Solis, Waxman Introduce Legislation to Protect Americans from Hazardous Chemicals in Consumer Products," *Press Release*, May 20.

Layton, L. (2008), "Chemical Law has Global Impact," *Washington Post*, June 12.

Layzer, J.A. (2006), *The Environmental Case: Translating Values into Policy* (Washington, DC: CQ Press).

Lee, M. (2004), "Gene-Altered Rice is Hot Issue: Capital Biotech Firm Gets Static from Several Sides," *Sacramento Bee*, May 31.

Lee, Z. (2006), "EU "Green" Directives Cast Challenge to Chinese Electronics Industry," *Worldwatch Institute*, <http://www.worldwatch.org/node/3883>, accessed June 15, 2007.

Levi-Faur, D. and Jordana, J. (eds) (2005), *The Rise of Regulatory Capitalism: The Global Diffusion of a New Order*.

Levin, M.A. and Shapiro, M. (eds) (2004), *Transatlantic Policymaking in an Age of Austerity: Diversity and Drift* (Washington, DC: Georgetown University Press).

Levit, J.K. (2004), "The Dynamics of International Trade Finance Regulations: The Arrangement on Officially Supported Export Credits," *Harvard International Law Journal* 45, 65–182.

Levy, D., Pensky, M., and Torpey, J.C. (2005), *Old Europe, New Europe, Core Europe: Transatlantic Relations after the Iraq War* (London: Verso).

Lewis, S. (2004), "Latin American Exporters Seek to Profit from Bans on US Beef," *Food Chemical News*, January 12, 28–9.

Liefferink, D. and Andersen, M.S. (1998), "Strategies of the 'Green' Member States in EU Environmental Policy-Making," *Journal of European Public Policy* 5:2, 254–70.

Liefferink, D. and Andersen, M.S. (2002), "Strategies of the 'Green' Member States in EU Environmental Policy-Making," in A. Jordan (ed.), *Environmental Policy in the European Union* (London and Sterling, VA: Earthscan), 63–80.

Lindberg, T. (ed.) (2005), *Beyond Paradise and Power: Europe, America, and the Future of a Troubled Partnership* (New York: Routledge).

Loewenberg, S. (2003), "Precaution is for Europeans," *New York Times*, May 18.

Lohr, S. (2004), "Bar Code Détente: US Finally Adds One more Digit," *New York Times*, July 12.

Lopes, P. and Durfee, M. (eds) (1999), *The Social Diffusion of Ideas and Things* (Thousand Oaks, CA: Sage).

Low Carbon Vehicle Partnership (2007), "California Sets 'World First' Standard for Low Carbon Fuels," January 10.

Lowell Center for Sustainable Production (2003), "The Promise and Limits of the United States Toxic Substances Control Act," October 10, Lowell: University of Massachusetts Lowell.

Lucas, R. (2001), "End-of-Life Vehicle Regulation in Germany and Europe—Problems and Perspectives," Wuppertal: Wuppertal Institute.

Luhmann, N. (1993), *Risk: A Sociological Theory* (New York: A. de Gruyter).

Lukes, S. (1974), *Power: A Radical View* (London: Macmillan).

MacLehose, L., McKee, M., and Weinberg, J. (2002), "Responding to the Challenge of Communicable Disease in Europe," *Science* 295, 2047–50.

Macrory, R., Havercroft, I., and Purdy, R. (eds) (2004), *Principles of European Environmental Law* (Groningen: Europa Law Publishers).

Mahoney, C. (2005), "Killing or Compromising? How the Potential for Policy Change Effects Lobbying Positions in the US and the EU," paper presented at the *Annual Meeting of the American Political Science Association*, Washington, DC, September 1–4.

Mahoney, C. (2007), "Networking Vs. Allying: The Decision of Interest Groups to Join Coalitions in the US and the EU," *Journal of European Public Policy* 14:3, 366–83.

Mahoney, C. (2008), *Brussels Versus the Beltway: Advocacy in the United States and the European Union* (Washington, DC: Georgetown University Press).

Majone, G. (2002), "What Price Safety? The Precautionary Principle and its Policy Implications," *Journal of Common Market Studies* 40:1, 89–110.

Maniates, M. (2002), "Individualization: Plant a Tree, Buy a Bike, Save the World?," in T. Princen, M. Maniates, and K. Conca (eds), *Confronting Consumption* (Cambridge, MA: MIT Press), 43–67.

March, J.G. and Olsen, J.P. (1989), *Rediscovering Institutions: The Organizational Basis of Politics* (New York: Free Press).

Margolis, M. (2006), "Flying South," *Newsweek Magazine*: January, 44–6.

Markell, D.L. and Knox, J.H. (eds) (2003), *Greening NAFTA: The North American Commission for Environmental Cooperation* (Stanford, CA: Stanford Law and Politics).

Marris, C. (2000), "Swings and Roundabouts: French Public Policy on Agricultural GMOs 1996–1999," Center d'Economie et d'Ethique pour l'Environnement et le Développment (C3ED).

Marsh, D. (ed.) (1998), *Comparing Policy Networks* (Buckingham: Open University Press).

Marsh, S. and Mackenstein, H. (2005), *The International Relations of the European Union* (Harlow: Pearson/Longman).

Martinelli, A. (ed.) (2007), *Transatlantic Divide: Comparing American and European Society* (Oxford and New York: Oxford University Press).

Mazey, S. and Richardson, J. (2002), "Environmental Groups and the EC: Challenges and Opportunities," in A. Jordan (ed.), *Environmental Policy in the European Union: Actors, Institutions and Processes* (London and Sterling, VA: Earthscan), 106–21.

Mazmanian, D.A. and Kraft, M.E. (eds) (1999), *Toward Sustainable Communities: Transition and Transformations in Environmental Policy* (Cambridge, MA: MIT Press).

McCormick, J. (2001), *Environmental Policy in the European Union* (Houndmills, Basingstoke, New York: Palgrave).

McCormick, J. (2008), *The European Union: Politics and Policies* (Boulder, CO: Westview Press).

McCright, A.M. and Dunlap, R.E. (2003), "Defeating Kyoto: The Conservative Movement's Impact on US Climate Change Policy," *Social Problems* 50:3, 348–73.

McCully, P. (2002), "Flooding the Land, Warming the Earth: Greenhouse Gas Emissions from Dams," Berkeley, CA: International Rivers Network.

McGrath, C. (2002), "Comparative Lobbying Practices: Washington, London, Brussels," paper presented at the *Annual Conference of the Political Studies Association*, Aberdeen.

Medearis, D. and Swett, B. (2003), "International Best Practices and Innovation—Strategically Harvesting Environmental Lessons from Abroad," *Ecologic Working Paper*, Berlin: Ecologic.

Menz, F.C. (2005), "Green Electricity Policies in the United States: Case Study," *Energy Policy* 33:18, 2398–410.

Menz, F.C. and Vachon, S. (2006), "The Effectiveness of Different Policy Regimes for Promoting Wind Power: Experiences from the States," *Energy Policy* 34, 1786–96.

Mercer Management Consulting (2003), "The Likely Impact of the Future European Legislation in the Area of Chemical Substances," April 1: Mercer Management Consulting.

Meyer, N.I. (2003), "European Schemes for Promoting Renewables in Liberalised Markets," *Energy Policy* 31, 665–76.

Mikkelä, H., Sampo, S., and Kaipainen, J. (eds) (2001a), *Final Report: Working Group Steering the Further Development of Criteria and Indicators for Sustainable Forest Management* (Helsinki: Ministry of Agriculture and Forestry).

Mikkelä, H., Sampo, S., and Kaipainen, J. (2001b), "The State of Forestry in Finland 2000: Criteria and Indicators for Sustainable Forest Management in Finland," in H. Mikkelä, S. Sampo, and J. Kaipainen (eds), *Final Report: Working Group Steering the Further Development of Criteria and Indicators for Sustainable Forest Management* (Helsinki: Ministry of Agriculture and Forestry), 16.

Miljödepartementet (2008), "Regeringens Klimatpolitik: Insatser Och Initiativ I Klimatarbetet," *informationsblad från Miljödepartementet*, April.

Mishra, R. (2007), "Mayor Aims to Cut City's Greenhouse Gas Emissions," *Boston Globe*, April 13.

Moffat, A.C. (1998), "Forest Certification: An Examination of the Comptability of the Canadian Standards Association and Forest Stewardship Council Systems in the Maritime Region," *Environmental Studies*, Dalhousie University, Halifax, Nova Scotia, 53.

Mol, A.P.J. and Sonnenfeld, D.A. (eds) (2000), *Ecological Modernization around the World. Perspectives and Critical Debates Special Issue of Environmental Politics* 9:1. Spring.

Montfort, J.-P. (2002), "The EU Chemicals Review from a Legal Perspective: For a Progressive, Coherent and Integrated Approach that Preserves the Internal Market," *European Environmental Law Review* 11:10, 270–86.

Moore, C. and Ihle, J. (1999), "Renewable Energy Policy Outside the United States," Issue Brief No. 14, October, Washington, DC: Renewable Energy Policy Project.

Moravcsik, A.M. (1989), "Disciplining Trade Finance: The OECD Export Credit Arrangement," *International Organization* 43:1, 173–205.

Morel-ál "Huissier, P. (1995), "Chrysotile," *Canadian Minerals Yearbook*: Natural Resources Canada.

Morgenstern, R.D. and Pizer, W.A. (eds) (2007), *Reality Check: The Nature and Performance of Voluntary Environmental Programs in the United States, Europe, and Japan* (Washington, DC: Resources for the Future).

Mossman, M. (2007), "Turks Fear Dam Will Drown out Nation's History— International Movement Mobilizes to Save Community Atop Archeological Gold Mine," *The Globe and Mail*, January 30, A16.

Motavalli, J. (2007), "The Can-Do Congress?," *E Magazine*: May/June.

Motolla, K. (ed.) (2006), *Transatlantic Relations and Global Governance* (Washington, DC: Center for Transatlantic Relations, Johns Hopkins University, School of Advanced International Studies).

MTK (2001), "Forest Management Associations—Protecting at Forest Owners Interests (30.5.2001)," *MTK*, <http://www.mtk.fi/sivu.asp?path=2918;2935 ;3183;4497;6984>, accessed May 5, 2003.

Müller-Rommel, F. and Poguntke, T. (eds) (2002), *Green Parties in National Governments* (London and Portland, OR: F. Cass).

Murphy, J. (2004), "Avian Flu Hits Texas; Live Bird Markets Likely Culprit," *Food Chemical News*, March 1, 24–5.

National Home Center News (1998), "B&Q Pledges to Work Towards Forest Certification," *National Home Center News* 24:2, 9.

Newsom, D. (2001), Achieving Legitimacy? Exploring Competing Forest Certification Programs" Actions to Gain Forest Manager Support in the US Southeast, Germany, and British Columbia, Canada, *School of Forestry and Wildlife Sciences*, Auburn University, Auburn, Alabama.

Newsom, D., Cashore, B., Auld, G., and Granskog, J. (2002), "Forest Certification in the Heart of Dixie: A Survey of Alabama Landowners," in L. Teeter, B. Cashore, and D. Zhang (eds), *Forest Policy for Private Forestry: Global and Regional Challenges* (Wallingford: CABI Publishing), 291–300.

Nicolaïdis, K. and Howse, R. (eds) (2001), *The Federal Vision: Legitimacy and Levels of Governance in the United States and the European Union* (Oxford: Oxford University Press).

NNC (2005), "The Medium-Term Philippine Plan of Action for Nutrition 2005–2010," National Nutrition Council of the Philippines.

Nordbeck, R. and Faust, M. (2003), "European Chemicals Regulation and its Effect on Innovation: An Assessment of the EU's White Paper on the Strategy for a Future Chemicals Policy," *European Environment* 13:1, 79–99.

NRC (2001), "Chrysotile Asbestos: Fact Sheet," Natural Resources Canada.

O'Leary, R. (2003), "Environmental Policy in the Courts," in N.J. Vig and M.E. Kraft (eds), *Environmental Policy: New Directions for the Twenty-First Century* (Washington DC: CQ Press), 151–73.

O'Neill, K. (2003), "A Vital Fluid: Risk, Controversy and the Politics of Blood Donation in the Era of 'Mad Cow Disease'," *Public Understanding of Science* 12:4, 359–80.

O'Neill, K. (2005), "US Beef Industry Faces New Policies and Testing for Mad Cow Disease," *California Agriculture* 59:4, 203–11.

O'Neill, K., Balsiger, J., and VanDeveer, S.D. (2004), "Actors, Norms, and Impacts: Recent International Cooperation Theory and the Influence of the Agent-Structure Debate," *Annual Review of Political Science* 7, 149–75.

O'Riordan, T. and Cameron, J. (eds) (1994), *Interpreting the Precautionary Principle* (London: Earthscan Publications).

O'Rourke, D. (2005), "Market Movements: Nongovernmental Organization Strategies to Influence Global Production and Consumption," *Journal of Industrial Ecology* 9, 115–28.

Oates, W.E. (1998), "Environmental Policy in the European Community: Harmonization or National Standards?," *Empirica* 25:1, 1–13.

OECD (2001a), "An Overview of Green Tax Reform and Environmentally Related Taxes in OECD Countries," *Environmentally Related Taxes in OECD Countries: Issues and Strategies*, September, Paris: Organisation for Economic Cooperation and Development.

OECD (2001b), "Policies to Enhance Sustainable Development," Paris: Organisation for Economic Cooperation and Development.

OECD (2005a), "Economic Survey of the United States 2005: Energy and Environmental Issues," *Report*, Paris: Organisation for Economic Cooperation and Development.

OECD (2005b), "Environmental Data Compendium 2004," Paris: Organisation for Economic Cooperation and Development.

OECD (2005c), "Updated Recommendation on Common Approaches on Environment and Officially Supported Export Credits," *TD/ECG(2005)3*, February 25, Paris: OECD.

OECD (2007), "Revised Council Recommendation on Common Approaches on the Environment and Officially Supported Export Credits," TAD/ECG(2007)9, June 12, Paris: OECD.

OIE (2004), "The OIE Standards on BSE: A Guide for Understanding and Proper Implementation," <http://www.oie.int/eng/press/en_040109.htm>.

OIE (2008a), "Number of Cases of BSE Reported in the United Kingdom," *International Organization for Animal Health*, <http://www.oie.int/eng/info/en_esbmonde.htm >, accessed May 20, 2008.

OIE (2008b), "Number of Reported Cases of BSE Worldwide (Excluding UK)," *International Organization for Animal Health*, <http://www.oie.int/eng/info/ en_esbmonde.htm >, accessed May 20, 2008.

Oliver, C. (1991), "Strategic Responses to Institutional Processes," *Academy of Management Review* 16:1, 145–79.

Olson, E. (2000), "Health Tops Free Trade in WTO Ruling," *International Herald Tribune*, June 16, 16.

Olson, S. (2004), "Banning GMOs in Mendocino County," *SF Bay Area IndyMedia*, Wednesday, March 3.

"Open Sesame. Editorial," *Nature Biotechnology* 23:6, 633.

Ortiz-García, S., et al. (2005), "Absence of Detectable Transgenes in Local Landraces of Maize in Oaxaca, Mexico (2003–2004)," *Proceedings of the National Academy of Sciences* 102:35, 12338–43.

Ott, H. (1998), "The Kyoto Protocol: Unfinished Business," *Environment* July-August, 16.

Paarlberg, R. (2001), *The Politics of Precaution: Genetically Modified Crops in Developing Countries* (Baltimore, MD: Johns Hopkins University Press).

Paarlberg, R. (2003), "Reinvigorating Genetically Modified Crops," *Issues in Science and Technology*:Spring.

Paarlberg, R. (2008), *Starved for Science: How Biotechnology is Being Kept out of Africa* (Cambridge, MA: Harvard University Press).

Palmer, T.C., Jr (2007), "Cambridge Sets $70m Energy Initiative," *Boston Globe*, March 29.

Papadakis, E. and Schreurs, M.A. (2007), *Historical Dictionary of the Green Movement* (Lanham, MD: Scarecrow Press).

Parry, J. (2004), "Mortality from Avian Flu is Higher than in Previous Outbreak," *British Medical Journal*, 368.

Parsons, T. (1967), *Sociological Theory and Modern Society* (New York: The Free Press).

Patlitzianas, K.D., Kagiannas, A.G., Askounis, D.T., and Psarras, J. (2005), "The Policy Perspective for RES Development in the New Member States of the EU," *Renewable Energy* 30, 477–92.

Pavolka, M. (2004), "Vermont Progressive Counter GMOs," *In These Times*, June 7.

PCSD (1999), "Towards a Sustainable America," Washington, DC: President's Council on Sustainable Development.

Pelosi, N. (2007), "We Will Work Together to Tackle Global Warming, One of Humanity's Greatest Challenges," <http://www.speaker.gov/newsroom/ speeches?id=0013>, accessed June 1, 2007.

Perron, L. (1999), "Chrysotile," *Canadian Minerals Yearbook*.

Petersmann, E.-U. and Pollack, M.A. (eds) (2003), *Transatlantic Economic Disputes: The EU, the US, and the WTO* (Oxford: Oxford University Press).

Peterson, J. (2004), "Policy Networks," in A. Wiener and T. Diez (eds), *European Integration Theory* (Oxford: Oxford University Press), 117–36.

Peterson, J. and Bomberg, E.E. (1999), *Decision-Making in the European Union* (New York: St Martin's Press).

Peterson, J. and Pollack, M.A. (eds) (2003), *Europe, America, Bush: Transatlantic Relations in the Twenty-First Century* (London: Routledge).

Pew Center on Global Climate Change (2003), "Summary of the McCain-Lieberman Climate Stewardship Act," October 30.

Pew Initiative on Food and Biotechnology (2004), "Results of Several Surveys of Americans on Questions Pertaining to Agricultural Biotechnology," Washington, DC.

Pew Initiative on Food and Biotechnology (2006), "Survey: Public Sentiments About Genetically Modified Food," Washington, DC.

Pfander, J.E. (1996), "Environmental Federalism in Europe and the United States: A Comparative Assessment of Regulation through the Agency of Member States," in J.B. Braden, H. Folmer, and T.S. Ulen (eds), *Environmental Policy with Political and Economic Integration* (Cheltenham and Brookfield, VT: Edward Elgar), 59–131.

Pianin, E. (2001), "US Aims to Pull out of Warming Treaty: 'No Interest' in Implementing Kyoto Pact, Whitman Says," *Washington Post*, March 28, A01.

Pickelsimer, C. and Wahl, T.I. (2002), "Mad Cow Disease: Implications for World Beef Trade," *IMPACT Center Information Series #96*.

Pinstrup-Andersen, P. and Schiaoler, E. (2000), *Seeds of Contention: World Hunger and Global Controversy over GM Crops* (Baltimore, MD: Johns Hopkins University Press).

Pohl, O. (2004), "European Environmental Rules Propel Change in US," *New York Times*, July 6.

Point Carbon (2007a), "Legislation," May 9.

Point Carbon (2007b), "Top US Energy Advisors Ramp up Plan to Slash Greenhouse Gas Emissions," April 25.

Pollack, M.A. (1997), "Representing Diffuse Interests in EC Policy-Making," *Journal of European Public Policy* 4:4, 572–90.

Pollack, M.A. (2003), "The Political Economy of Transatlantic Trade Disputes," in E.-U. Petersmann and M.A. Pollack (eds), *Transatlantic Economic Disputes: The EU, the US, and the WTO* (Oxford: Oxford University Press), 65–118.

Pond, E. (2004), *Friendly Fire: The Near-Death of the Transatlantic Alliance* (Pittsburgh, PA: European Union Studies Association).

Posner, P.L. (1998), *The Politics of Unfunded Mandates: Whither Federalism?* (Washington DC: Georgetown University Press).

Possani, L.D. (2003), "The Past, Present and Future of Biotechnology in Mexico," *Nature Biotechnology* 21:5, 582–3.

Powell, C. (2002), "Making Sustainable Development Work: Governance, Finance, and Public-Private Cooperation," remarks at State Department Conference, Meridian International Center, July 12, Washington, DC.

Powell, C. (2003), "Cable to US Posts in EU Member States and the US Mission to the EU," April 29.

Powell, D. and Leiss, W. (1997), *Mad Cows and Mother's Milk: The Perils of Poor Risk Communication* (Montreal: McGill-Queen's University Press).

Powell, M.R. (1999), *Science at EPA: Information in the Regulatory Process* (Washington DC: Resources for the Future).

Priest, S.H., Bonfadelli, H., and Rusanen, M. (2003), "The 'Trust Gap' Hypothesis: Predicting Support for Biotechnology across National Cultures as a Function of Trust in Actors," *Risk Analysis* 23:4, 751–66.

Princen, S. (2002), *EU Regulation and Transatlantic Trade* (The Hague: Kluwer).

Purvis, N. (2003), "Greening US Foreign Aid through the Millennium Challenge Account," *Policy Brief*, #119, June, Washington, DC: Brookings.

Putnam, R.D. (1988), "Diplomacy and Domestic Politics: The Logic of Two-Level Games," *International Organization* 42:3, 427–60.

Putnam, R.D. (1995), "Bowling Alone: America's Declining Social Capital," *Journal of Democracy* 6:1, 65–78.

Quist, D. and Chapela, I.H. (2001), "Transgenic DNA Introgressed into Traditional Maize Landraces in Oaxaca, Mexico," *Nature* 414, 541–3.

Rabe, B.G. (2002), "Statehouse and Greenhouse: The States Are Taking the Lead on Climate Change," *The Brookings Review* 20:2.

Rabe, B.G. (2004), *Statehouse and Greenhouse: The Emerging Politics of American Climate Change Policy* (Washington, DC: Brookings Institution Press).

Rabe, B.G. (2006a), "Power to the States: The Promise and Pitfalls of Decentralization," in N. Vig and M.E. Kraft (eds), *Environmental Policy: New Directions for the Twenty-First Century* (Washington, DC: CQ Press), 34–56.

Rabe, B.G. (2006b), "Race to the Top: The Expanding Role of US State Renewable Portfolio Standards," Washington, DC: Pew Center on Global Climate Change.

Rabe, B.G. (2008), "States on Steroids: The Intergovernmental Odyssey of American Climate Policy," *Review of Policy Research* 25:2, 105–28.

Rabe, B.G. (2009), "Second Generation Climate Policies in the States: Proliferation, Diffusion and Regionalization," in H. Selin and S.D. VanDeveer (eds), *Changing Climates in North American Politics: Institutions, Policymaking and Multilevel Governance* (Cambridge, MA: MIT Press).

Rametsteiner, E. (1999), "The Attitude of European Consumers toward Forests and Forestry," *Unasylva* 50:196.

Rampton, S. and Stauber, J. (1997), *Mad Cow USA: Could the Nightmare Happen Here?* (Monroe, ME: Common Courage Press).

Randall, E. (2000), "European Union Health Policy with and without Design: Serendipity, Tragedy and the Future of EU Health Policy," *Policy Studies* 21:2, 133.

Recycling Today (2004), "Maine Governor Signs Law to Boost Electronics Recycling," April 23.

Reiche, D. and Bechberger, M. (2004), "Policy Differences in the Promotion of Renewable Energies in the EU Member States," *Energy Policy* 32, 843–49.

Reid, T.R. (2004), *The United States of Europe: The New Superpower and the End of American Supremacy* (New York: Penguin Press).

"Renewed Controversy over the International Reach of NEPA," (1977), *Environmental Law Reporter* 7, 10205–9.

Reuters UK (2005), "EU States Get Final Warning on Electronic Waste," July 11.

Revkin, A.C. and Healy, P. (2007), "Global Coalition Make Buildings Energy-Efficient," *New York Times*, May 17.

Rickenbach, M., Fletcher, R., and Hansen, E. (2000), "An Introduction to Forest Certification," EC 1518, July, Corvallis, OR: Oregon State University Extension Service.

Rickerson, W., Bennhold, F., and Bradbury, J. (2008), "Feed-in Tariffs and Renewable Energy in the USA: A Policy Update," Washington, DC: Heinrich Böll Foundation.

Rifkin, J. (2004), *The European Dream: How Europe's Vision of the Future is Quietly Eclipsing the American Dream* (New York: Jeremy P. Tarcher/Penguin).

Ringquist, E.J. (1993), *Environmental Protection at the State Level: Politics and Progress in Controlling Pollution* (Armonk NY and London: M.E. Sharpe).

Root, R. (2000), "Lead Loading of Urban Streets by Motor Vehicle Wheel Weights," *Environmental Health Perspectives* 108:10.

Rose-Ackerman, S. (1995), *Controlling Environmental Policy: the Limits of Public Law in Germany and the United States* (New Haven, CT: Yale University Press).

Rosenbaum, W.A. (2002), *Environmental Politics and Policy* (Washington, DC: CQ Press).

Rosenbaum, W.A. (2003), "Still Reforming after All These Years: George W. Bush's 'New Era' at the EPA," in N.J. Vig and M.E. Kraft (eds), *Environmental Policy: New Directions for the Twenty-First Century* (Washington, DC: CQ Press).

Rosenbaum, W.A. (2005), *Environmental Politics and Policy* (Washington, DC: CQ Press).

Rosendal, K. (2001), "Impact of Overlapping Regimes: The Case of Biodiversity," *Global Governance* 7:2, 95–117.

Rowlands, I.H. (2003), "Renewable Electricity and Transatlantic Relations: exploring the Issues," *Working Paper*, 2003/17, San Domenico, IT: Robert Schuman Centre for Advanced Studies.

Rowlands, I.H. (2005a), "Renewable Energy and International Politics," in P. Dauvergne (ed.), *The International Handbook of Environmental Politics* (Cheltenham: Edward Elgar), 78–94.

Rowlands, I.H. (2005b), "The European Directive on Renewable Electricity: Conflicts and Compromises," *Energy Policy* 33, 965–74.

Rowlands, I.H. and Patterson, M.J. (2002), "A North American Definition for 'Green Electricity,' Implications for Sustainability," *International Journal of Environment and Sustainable Development* 1, 249–64.

Royal Commission on Environmental Pollution (2003), "Chemicals in Products: Safeguarding the Environment and Human Health," June, London: Royal Commission on Environmental Pollution.

Ruggie, J.G. (2004), "Reconstituting the Global Public Domain—Issues, Actors, and Practices," *European Journal of International Relations* 10, 499–531.

Runge, C.F., Senauer, B., Pardey, P.G., and Rosegrant, M.W. (2003), *Ending Hunger in Our Lifetime: Food Security and Globalization* (Baltimore, MD: Johns Hopkins University Press).

Sadeleer, N.D. (ed.) (2007), *Implementing the Precautionary Principle: Approaches from the Nordic Countries, EU and USA* (London and Sterling, VA: Earthscan).

Sampson, G.P. and Chambers, W.B. (eds) (2002), *Trade, Environment, and the Millennium* (Tokyo: United Nations University Press).

Sands, P. (2000), "'Unilateralism,' Values, and International Law," *European Journal of International Law* 11, 291–302.

Sanger, D.E. (2003), "Bush Links Europe's Ban on Bio-Crops with Hunger," *New York Times*, May 22.

Sasser, E.N. (2002), "The Certification Solution: NGO Promotion of Private, Voluntary Self-Regulation," paper presented at the *74th Annual Meeting of the Canadian Political Science Association*, Toronto, Ontario, May 29–31.

Sbragia, A. (2000), "The European Union as Coxswain: Governance by Steering," in J. Pierre (ed.), *Debating Governance: Authority, Steering, and Democracy* (Oxford: Oxford University Press), 219–40.

Schaper, M. (2005), "Environmental Politics across the Atlantic," *Global Environmental Politics* 5:1, 131–5.

Schaper, M. (2007), "Leveraging Green Power: Environmental Rules for Project Finance," *Business and Politics* 9:3.

Scharpf, F.W. (1997), "Introduction: The Problem Solving Capacity of Multi/Level Governance," *Journal of European Public Policy* 4:4, 520–38.

Scheberle, D. (1997), *Federalism and Environmental Policy: Trust and the Politics of Implementation* (Washington, DC: Georgetown University Press).

Scheberle, D. (2004), "Devolution," in R.F. Durant, D.J. Fiorino, and R. O'Leary (eds), *Environmental Governance Reconsidered: Challenges, Choices, and Opportunities* (Cambridge, MA: MIT Press), 361–92.

Schiffer, L.J. (2004), "The National Environmental Policy Act Today, with an Emphasis on its Application across US Borders," *Duke Environmental Law and Policy Forum* 14:2, 325–45.

Schlosberg, D. and Bomberg, E. (2008), "Perspectives on American Environmentalism," *Environmental Politics* 17:2, 187–99.

Schlosberg, D. and Rinfret, S. (2008), "Ecological Modernization, American Style," *Environmental Politics* 17:2, 254–75.

Schmid, K.-P. (2001), "Ökonom—Für Welche Exporte Soll der Staat Bürgen?," *Die Zeit*, 22 March, 25.

Schmidt, C. (2002), "E-Junk Explosion," *Environmental Health Perspectives* 110:4, A188–94.

Schmidt, C.W. (2007), "Environment: California out in Front," *Environmental Health Perspectives* 115:3, A145–7.

Schmitter, P.C. and Streeck, W. (1981), "The Organization of Business Interests," *Discussion Paper*, IIM/LMP: 81–83, Berlin: Wissenschaftszentrum Berlin.

Schneider, A. (2000), "Asbestos Industry has Shown its Might," *Seattle Post-Intelligencer*, November 16.

Schörling, I. and Lund, G. (2004), "REACH—the Only Planet Guide to the Secrets of Chemicals Policy in the EU. What Happened and Why?," Brussels.

Schreurs, M.A. (2002), *Environmental Politics in Japan, Germany, and the United States* (Cambridge: Cambridge University Press).

Schreurs, M.A. (2004a), "Environmental Protection in an Expanding European Community: Lessons from Past Accessions," *Environmental Politics* 13, Special Edition EU Enlargement and the Environment: Institutional Change and Environmental Policy in Central and Eastern Europe: Supplement March 1, 27–51.

Schreurs, M.A. (2004b), "National Energy Policy: United States," in C.J. Cleveland (ed.), *Encyclopedia of Energy* (San Diego, CA: Elsevier), 173–79.

Schreurs, M.A. (2004c), "The Climate Change Divide: The European Union, the United States, and the Future of the Kyoto Protocol," in N.J. Vig and M. Faure (eds), *Green Giants? Environmental Policy of the United States and the European Union* (Cambridge, MA: MIT Press), 207–30.

Schreurs, M.A. (2005a), "Environmental Policy-Making in the Advanced Industrialized Countries: Japan, the European Union and the United States of America Compared," in H. Imura and M.A. Schreurs (eds), *Environmental Policy in Japan* (Cheltenham and Northampton, MA: Edward Elgar), 315–41.

Schreurs, M.A. (2005b), "Global Environment Threats and a Divided Northern Community," *International Environmental Agreements: Politics, Law, and Economics* 5:3, 349–76.

Schreurs, M.A. and Epstein, M.W. (2007), "Environmental Federalism in the United States and the Swing of the Pendulum from the Federal to the State

Level," *Paper prepared for the American Center for Contemporary German Studies*, June 13, Berlin.

Schreurs, M.A. and Tiberghien, Y. (2007), "Multi-Level Reinforcement: Explaining EU Leadership in Climate Change Mitigation," *Global Environmental Politics* 7:4, 19–46.

Schulz, W. (2000), "Promotion of Renewable Energy in Germany," in P. Jasinski and W. Pfaffenberger (eds), *Energy and Environment: Multiregulation in Europe* (Aldershot: Ashgate Publishing Ltd.), 128–49.

Schwarzenegger, A. and Rell, J. (2007), "Lead or Step Aside, EPA: States Can't Wait on Global Warming," *Washington Post*, May 21.

Schweiger, T. and Ritsema, G. (2003), "The Introduction of GMOs by the Back Door of EU Accession," Friends of the Earth Europe and the Northern Alliance for Sustainability.

Scruggs, L.A. (2003), *Sustaining Abundance: Environmental Performance in Industrial Democracies* (Cambridge: Cambridge University Press).

SCS (2003), "Environmentally Preferable Electricity Sources," Emeryville, CA: Scientific Certification Systems.

SEDA (2003), "The Green Power Customer Logo," Grosvenor Place: Sustainable Energy Development Authority.

Seifert, W. (2006), "Synchronised National Publics as Functional Equivalent of an Integrated European Public: The Case of Biotechnology," *European Integration online Papers (EIoP)* 10:8.

Selin, H. (2003), "Regional Pops Policy: The UNECE/CLRTAP Pops Agreement," in D.L. Downie and T. Fenge (eds), *Northern Lights against Pops: Combatting Toxic Threats in the Arctic* (Montreal: McGill-Queens University Press), 111–32.

Selin, H. (2007), "Coalition Politics and Chemicals Management in a Regulatory Ambitious Europe," *Global Environmental Politics* 7:3, 63–93.

Selin, H. and Eckley, N. (2003), "Science, Politics, and Persistent Organic Pollutants: Scientific Assessments and their Role in International Environmental Negotiations," *International Environmental Agreements: Politics, Law and Economics* 3:1, 17–42.

Selin, H. and VanDeveer, S.D. (2003), "Mapping Institutional Linkages in European Air Pollution Politics," *Global Environmental Politics* 3:3, 14–46.

Selin, H. and VanDeveer, S.D. (2005), "Canadian-US Environmental Cooperation: Climate Change Networks and Regional Action," *American Review of Canadian Studies* 35:2, 353–78.

Selin, H. and VanDeveer, S.D. (2006a), "Canadian-US Cooperation: Regional Climate Change Action in the Northeast," in P.G. Le Prestre and P.J. Stoett (eds), *Bilateral Ecopolitics: Continuity and Change in Canadian-American Environmental Relations* (Aldershot: Ashgate).

Selin, H. and VanDeveer, S.D. (2006b), "Raising Global Standards: Hazardous Substances and E-Waste Management in the European Union," *Environment* 48:10, 6–18.

Selin, H. and VanDeveer, S.D. (2007), "Political Science and Prediction: What's Next for US Climate Change Policy?," *Review of Policy Research* 24:1, 1–27.

Selin, H. and VanDeveer, S.D. (eds) (2009a), *Changing Climates in North American Politics: Institutions, Policymaking and Multilevel Governance* (Cambridge, MA: MIT Press).

Selin, H. and VanDeveer, S.D. (2009b), "Climate Leadership in Northeast North America," in H. Selin and S.D. VanDeveer (eds), *Changing Climates in North American Politics: Institutions, Policymaking and Multilevel Governance* (Cambridge, MA: MIT Press).

Shabecoff, P. (2000), *Earth Rising: American Environmentalism in the 21st Century* (Washington, DC: Island Press).

Shears, P., Zollers, F., and Hurd, S. (2001), "Food for Thought: What Mad Cows have Wrought with Respect to Food Safety Regulation in the EU and UK," *British Food Journal* 103:1, 63–87.

Sheppard, H.E. (2002), "Revamping the Export-Import Bank in 2002: The Impact of this Interim Solution on the United States and Latin America," *Legislation and Public Policy* 6:89, 89–130.

Shiroyama, H. (2007), "The Harmonization of Automobile Environmental Standards between Japan, the United States and Europe: The 'Depoliticizing Strategy' by Industry and the Dynamics between Firms and Governments in a Transnational Context," *The Pacific Review* 20:3, 351–70.

Shiva, V. (2000), *Stolen Harvest: The Hijacking of the Global Food Supply* (Cambridge, MA: South End Press).

Shiva, V. (2003), "Towards a People Centred Fair Trade Agreement on Agriculture," *Z-Magazine* (December 24).

Silicon Valley Toxics Coalition and Basel Action Network (2002), "Exporting Harm: The High-Tech Trashing of Asia," San Jose, CA: SVTC.

Sissel, K. (2003), "California Imposes Bans on Two Brominated Flame Retardants," *Chemical Week* 165:31, 42.

Sissine, F., Cunningham, L.J., and Gurevitz, M. (2008), "Energy Efficiency and Renewable Energy Legislation in the 110th Congress," *Congressional Research Service Report for Congress*, RL33831, June 10, Washington, DC: Congressional Research Service.

Slaughter, A.-M. (2004), *A New World Order* (Princeton, NJ: Princeton University Press).

Smith, M. and Steffenson, R. (2005), "The European Union and the United States," in C. Hill and M. Smith (eds), *International Relations and the European Union* (Oxford: Oxford University Press).

Sommestad, L. and Trittin, J. (2004), "Eine Sichere Chemikalienpolitik für Europa," *Frankfurter Rundschau*, June 26.

Speth, J.G. (2004), *Red Sky at Morning: America and the Crisis of the Global Environment* (New Haven, CT: Yale University Press).

SSC (2001), "Opinion on Requirements for Statistically Authoritative BSE/Tse Surveys, Adopted 29/30 November 2001," *Scientific Steering Committee of the European Food Safety Authority*, <http://europa.eu.int/comm/food/fs/sc/ssc/outcome_en.html >, accessed June 17, 2004.

Stanbury, W.T. (2000), *Environmental Groups and the International Conflict over the Forests of British Columbia, 1990 to 2000* (Vancouver: SFU-UBC Centre for the Study of Government and Business).

Stanbury, W.T. and Vertinsky, I. (1997), "Boycotts in Conflicts over Forestry Issues: The Case of Clayoquot Sound," *Commonwealth Forestry Review* 76:1.

Steinberg, P.F. and VanDeveer, S.D. (eds) (forthcoming), *Comparative Environmental Politics* (Cambridge, MA: MIT Press).

Steinberg, R.H. (ed.) (2002), *The Greening of Trade Law: International Trade Organizations and Environmental Issues* (Lanham, MD: Rowman and Littlefield).

Steiner, A., Wälde, T., and Bradbrook, A. (2004), "International Institutional Arrangements: Bundling the Forces—but How?," Bonn: International Conference for Renewable Energies.

Steinmo, S., Thelen, K.A., and Longstreth, F. (1992), *Structuring Politics: Historical Institutionalism in Comparative Analysis* (Cambridge: Cambridge University Press).

Stewart, R.B. (1977), "Pyramids of Sacrifice? Problems of Federalism in Mandating State Implementation of National Environmental Policy," *Yale Law Journal* 86:1977, 1196–272.

Stolberg, S.G. (2007), "Bush Proposes Goals on Greenhouse Gas Emissions," *New York Times*, June 1.

Strittmatter, K. (2007a), "Bau des ersten Staudamms am Tigris verzögert sich—Türkische Regierung Sträubt sich offenbar gegen vereinbarte Umweltauflagen," *Süddeutsche Zeitung*, May 26, 8.

Strittmatter, K. (2007b), "Kulturkampf am Tigris—Berlin fördert trotz Protesten Staudamm-Projekt in der Türkei," *Süddeutsche Zeitung*, March, 28, 1.

Sucharipa-Behrmann, L. (1994), "Eco-Labelling Approaches for Tropical Timber: The Austrian Experience," in OECD (ed.), *Life-Cycle Management and Trade* (Paris: Organisation for Economic Co-operation and Development).

Suchman, M.C. (1995), "Managing Legitimacy: Strategic and Institutional Approaches," *Academy of Management Review* 20:3, 571–610.

Sugarman, C. (2004), "Agribusiness Dominates Usda, Study Charges," *Food Chemical News*, August 2.

"Summary of the World Summit on Sustainable Development: 26 August–4 September 2002," (2002), *Earth Negotiations Bulletin* 22.

Sundelius, B. and Grönvall, J. (2004), "Strategic Dilemmas of Biosecurity in the European Union," *Biosecurity and Bioterrorism* 2:4, 17–23.

Sussman, G., Daynes, B.W., and West, J.P. (2002), *American Politics and the Environment* (New York: Longman).

Sutherland, P. (2001), "Preface," *Resolving and Preventing US-EU Trade Disputes: Six Prize-Winning Essays from the BP/EUI Transatlantic Essay Contest* (San Domenico di Fiesole: European University Institute), 1–2.

Svensson, L. (2008), *Combating Climate Change: A Transatlantic Approach to Common Solutions* (Washington, DC: Brookings Institution Press).

Switzer, J.V. (2001), *Environmental Politics: Domestic and Global Dimensions* (Bedford, MA: St Martin's Press).

TAED (2000), "Statement by the Climate Change, Clean Air and Energy Working Group of the Transatlantic Environment Dialogue," May, Brussels: Transatlantic Environment Dialogue.

Taylor, M.R. (2004), "Lead or React? A Game Plan for Modernizing the Food Safety System in the United States," *Food Drug Law Journal* 59, 399–403.

Tews, K., Busch, P.-O., and Jörgens, H. (2003), "The Diffusion of New Environmental Policy Instruments," *European Journal of Political Research* 42:4, 569–600.

The Economist (2006), "Interpreting Smoke Signals," July 22, 39–40.

The White House (2001a), "Press Briefing by Ari Fleischer," March 28: The White House, Office of the Press Secretary.

The White House (2001b), "Text of a Letter to Senators Hagel, Helms, Craig, and Roberts," March 13, Washington: The White House, Office of the Press Secretary.

The White House (2002), "President Announces Clear Skies and Global Climate Change Initiatives," February 14, Washington: The White House, Office of the Press Secretary.

Thieme, D. and Rudolf, B. (2002), "Preussenelektra Ag V. Schleswag Ag. Case C-379/98," *The American Journal of International Law* 96, 225–30.

Thorpe, B. and Kruszewska, I. (1999), "Strategies to Promote Clean Production—Extended Producer Responsibility," January, Montreal: Clean Production Action.

Tickell, O. (2000), "Why the UK's Ancient Woodland is Still under Threat," Grantham, UK: The Woodland Trust.

Tojo, N. (2000), "Analysis of EPR Policies and Legislation through Comparative Study of Selected EPR Programmes for EEE," IIEEE Communications 10.

Toke, D. (2004), *The Politics of GM Food: A Comparative Study of the UK, USA, and EU* (New York: Routledge).

US Environmental Protection Agency (2003), "Overview: Office of Pollution Prevention and Toxics Programs," December 24.

"US Faces Uphill Battle on Tough OECD Rules on Export Credit Loans," (2001), *Inside Trade* (July 27), 10.

UK CJD Surveillance Unit (2008), "CJD Statistics," <http://www.cjd.ed.ac. uk/figures.htm>, accessed May 20, 2008.

UNCED (1992), "Agenda 21," New York: United Nations Conference of Environment and Development.

UNEP (2000), "The Cartagena Protocol on Biosafety," United Nations Environment Program.

Union of Concerned Scientists (2002), "Clean Energy Backgrounder: Public Utility Regulatory Policy Act (PURPA)," Cambridge, MA: Union of Concerned Scientists.

Union of Concerned Scientists (2007), "State Minimum Renewable Electricity Requirements (as of January 2007)," Cambridge, MA: Union of Concerned Scientists.

United Kingdom Parliament (2001), "Export Credits and the Environment: US Decision," Written Answers, Column WA 77, December 20.

United Nations (2002), "Report of the World Summit on Sustainable Development," August 26–September 4, Johannesburg, South Africa.

United States (2004), "Notification G/TBT/N/EEC/52 Regarding European Commission Regulation Com(2003)644," June 21.

United States Congress (1978), S. 3077: A Bill to Amend and Extend the Export-Import Bank Act of 1945, and for Other Purposes *Hearing before the Subcommittee on Resource Protection of the Committee on Environment and Public Works*, US Government Printing Office, Washington, DC.

United States Congress (1992a), Export Enhancement Act of 1992—Bill Summary.

United States Congress (1992b), *Export Enhancement Act of 1992. Senate Report 102–320* (Washington: US Government Printing Office).

United States Congress (1994), "Tied Aid Practices of US Competitors," Hearing before the Subcommittee on Economic Policy, Trade and Environment of the Committee on Foreign Affairs, US Government Printing Office, Washington, DC.

United States Department of Energy (2008), "20% Wind Energy by 2030: Increasing Wind Energy's Contribution to US Electricity Supply," May.

United States General Accounting Office (1994), "Toxic Substances Control Act: Legislative Changes Could Make the Act More Effective," GAO/ RCED-94-103, September, Washington DC.

United States General Accounting Office (2002), "Mad Cow Disease: Improvements in the Animal Feed Ban and Other Areas Would Strengthen US Prevention Efforts," *Report to Congressional Requesters*, GAO-02-183, January, Washington, DC.

United States General Accounting Office (2005), "Chemical Regulation: Options Exists to Improve EPA's Ability to Assess Health Risks and Management its Chemical Review Program," GAO-05-458, June, Washington, DC.

United States General Accounting Office (2006), "Chemical Regulation: Actions Are Needed to Improve the Effectiveness of EPA's Chemical Review Program," GAO-06-1032T, August, Washington, DC.

United States House of Representatives (2004), "A Special Interest Case Study: The Chemical Industry, the Bush Administration, and European Efforts to Regulate Chemicals," Committee on Government Reform—Minority Staff, Special Investigations Division, April 1, Washington DC.

United States Senate (1997), "Byrd-Hagel Resolution," 105th Congress, 1st Session, S. Res. 98, Report No. 105–54.

USAID (2003a), "Cabio: Mobilizing Science and Technology to Reduce Poverty and Hunger," Press Release, 2003–063, Washington, DC.

USAID (2003b), "Educating Scientists for Management," Press Release, 2003–070, Washington, DC.

USDA (2002), "Low Pathogenic Avian Influenza Virus," <http://www.aphis. usda.gov/lpa/pubs/fsheet_faq_notice/fs_ahlpai.html>.

USDA (2003), "USDA Actions to Prevent Bovine Spongiform Encephalopathy (BSE)," <www.aphis.usda.gov/lpa/issues/bse/bsechron.html>.

USDA (2004), "Highly Pathogenic Avian Influenza," <http://www.aphis.usda. gov/lpa/pubs/fsheet_faq_notice/fs_ahavianflu.html>.

USDA (2005), "Chronology of the European Union's Hormone Ban," <http:// www.fas.usda.gov/itp/policy/chronology.html>, accessed June 25, 2007.

USDA (2008), "USDA Beef Import and Export Figures," <http://www.fas. usda.gov/currwmt.asp>.

USGS (2007), "United States Geological Survey," <http://ww.usgs.gov>.

Victor, D. (2006), "Recovering Sustainable Development," *Foreign Affairs* 85:1.

Victor, D., House, J., and Joy, S. (2005), "A Madisonian Aproach to Climate Policy," *Science* 309:16, 1820–21.

Vig, N.J. and Faure, M. (2004a), "Conclusion: The Necessary Dialogue," in N.J. Vig and M. Faure (eds), *Green Giants? Environmental Policies of the United States and the European Union* (Cambridge, MA: MIT Press), 347–75.

Vig, N.J. and Faure, M. (eds) (2004b), *Green Giants? Environmental Policies of the United States and the European Union* (Cambridge, MA: MIT Press).

Vig, N.J. and Faure, M. (2004c), "Introduction," in N.J. Vig and M. Faure (eds), *Green Giants? Environmental Policies of the United States and the European Union* (Cambridge, MA: MIT Press), 1–14.

Vig, N.J. and Kraft, M.E. (eds) (2006), *Environmental Policy: New Directions for the Twenty-First Century* (Washington, DC: CQ Press).

Vincent, K. (2004), "'Mad Cows' and Eurocrats—Community Responses to the BSE Crisis," *European Law Journal* 10:5, 499–517.

Vlosky, R.P. (2000), "Certification: Perceptions of Non-Industrial Private Forestland Owners in Louisiana," *Working Paper* 41, March 20, Baton

Rouge, Louisiana: Louisiana Forest Products Laboratory, Louisiana State University Agricultural Center.

Vogel, D. (1986), *National Styles of Regulation: Environmental Policy in Great Britain and the United States* (Ithaca, NY: Cornell University Press).

Vogel, D. (1995), *Trading Up: Consumer and Environmental Regulation in a Global Economy* (Cambridge, MA: Harvard University Press).

Vogel, D. (1997a), *Barriers or Benefits? Regulation in Transatlantic Trade* (Washington, DC: Brookings Institution Press).

Vogel, D. (1997b), "Trading Up and Governing Across: Transnational Governance and Environmental Protection," *Journal of European Public Policy* 4:4, 556–71.

Vogel, D. (2003a), *National Styles of Business Regulation: A Case Study of Environmental Protection* (Washington, DC: Beard Books).

Vogel, D. (2003b), "The Hare and the Tortoise Revisited: The New Politics of Consumer and Environmental Regulation in Europe," *British Journal of Political Science* 33:4, 557–80.

Vogel, D. (2004), "The Hare and the Tortoise Revisited: The New Politics of Risk Regulation in Europe and the United States," in M. Levin and M. Shapiro (eds), *TransAtlantic Policymaking in an Era of Austerity* (Washington DC: Georgetown University Press), 177–202.

Vogel, D. (2005), *The Market for Virtue: The Potential and Limits of Corporate Social Responsibility* (Washington, DC: Brookings Institution Press).

Vogel, L. (1999), "The WTO Asbestos Dispute: Workplace Health Dictated by Trade Rules?," *Lannee Sociale*, Brussels: Institute of Sociology.

Vos, E. (2000), "EU Food Safety Regulation in the Aftermath of the BSE Crisis," *Journal of Consumer Policy* 23, 227–55.

Wallace, W. (2001), "Europe, the Necessary Partner," *Foreign Affairs* 80:1.

Wälti, S. (2004), "How Multilevel Structures Affect Environmental Policy in Industrialized Countries," *European Journal of Political Research* 43:4, 597–632.

Walzenbach, G.P.E. (1999), "Export Promotion: Gaining Advantage through Credit and Insurance," in T.C. Lawton (ed.), *European Industrial Policy and Competitiveness: Concepts and Instruments* (New York: St. Martin's Press), 93–113.

Waterton, C. and Wynne, B. (1996), "Building the European Union: Science and the Cultural Dimensions of Environmental Policy," *Journal of European Environmental Policy* 3:3, 421–40.

Watson, R. and Shackleton, M. (2003), "Non-Institutional Actors and Lobbying in the EU," in E.E. Bomberg and A.C.G. Stubb (eds), *The European Union: How Does It Work?* (Oxford: Oxford University Press), 88–107.

Watts, R.L. (1999), *Comparing Federal Systems* (Kingston, Ontario: McGill/ Queen's University Press).

Weale, A., Pridham, G., Cini, M., Konstadakopulos, D., Porter, M., and Flynn, B. (2000), *Environmental Governance in Europe: An Ever Closer Ecological Union?* (Oxford; New York: Oxford University Press).

West, D. and Loomis, B.A. (1999), *The Sound of Money: How Political Interests Get what they Want* (New York: Norton).

Wettestad, J. (2005), "The Making of the 2003 Emissions Trading Directive: An Ultra-Quick Process Due to Entrepreneurial Proficiency?," *Global Environmental Politics* 5:1, 1–23.

Wiener, J.B. (2004), "Convergence, Divergence, and Complexity in US and European Risk and Regulation," in N.J. Vig and M. Faure (eds), *Green Giants? Environmental Policies of the United States and the European Union* (Cambridge, MA: MIT Press), 73–109.

Wiener, J.B. and Rogers, M.D. (2002), "Comparing Precaution in the United States and Europe," *Journal of Risk Research* 5:4, 317–49.

Wigzell, H. (2005), "A European CDC?," *Science* 307, 1691.

Wijnholds, H.D.B. (2000), "Competitive Electricity Markets in the US And the EU: Can they Learn from Each Other?," *Journal of Euromarketing* 9, 37–55.

Williamson, H. and Luce, E. (2007), "Tensions Mount over G8 Climate Discussions," *Financial Times Weekend*, May 26–27, 1.

Wilson, M.P., Chia, D.A., and Ehlers, B.C. (2006), "Green Chemistry in California: A Framework for Leadership in Chemicals Policy and Innovation," report prepared for the California Senate Environmental Quality Committee and the California Assembly Committee on Environmental Safety and Toxic Materials, Berkeley, CA: California Policy Research Center.

Wirth, D.A. (2002), "European Communities—Measures Affecting Asbestos and Asbestos-Containing Products," *American Journal of International Law* 96, 435–9.

Wiser, R., Pickle, S., and Goldman, C. (1998), "Renewable Energy Policy and Electricity Restructuring: A California Case Study," *Energy Policy* 26, 465–75.

Woll, C. (2006), "Lobbying in the European Union: From Sui Generis to a Comparative Perspective," *Journal of European Public Policy* 13:3, 456–69.

World Commission on Environment and Development (1987), *Our Common Future* (New York: Oxford University Press).

World Health Organization (2002), "WHO Avian Influenza Fact Sheet," <http://www.who.int/mediacentre/factsheets/avian_influenza/en/>.

World Health Organization (2004), "Avian Influenza and Human Health: Report by the Secretariat for the 114th Session of the Executive Board," April 8.

World Wide Fund for Nature (2002), "A New Regulatory System for Chemicals in Europe: A Step Towards a Cleaner, Safer World?," September, Brussels.

World Wide Fund for Nature (2004), "Reach and 'Proportionality' under WTO Rules," June, Surrey.

Wright, D.S. (1998), *Understanding Intergovernmental Relations* (Pacific Grove, CA: Brooks/Cole).

WTO (2003), "Report to the 5th Session of the WTO Ministerial Conference in Cancún," WT/CTE/8, July 11, Geneva: World Trade Organisation Committee on Trade and Environment.

WTO (2008), "Chemicals and Toys Main Focus of Members" Trade Concerns," *WTO 2008 News Items*, March 20.

Wurzel, R.K.W. (2004), "Germany: From Environmental Leadership to Partial Mismatch," in A. Jordan and D. Liefferink (eds), *Environmental Policy in Europe: The Europeanization of National Environmental Policy* (London: Routledge), 99–117.

Wyman, R.A. (1980), "Control of Toxic Substances: The Attempt to Harmonize the Notification Requirements of the US Toxic Substances Control Act and the European Community Sixth Amendment," *Virginia Journal of International Law* 20:2, 417–58.

Yandle, B. (1983), "Bootleggers and Baptists: The Education of a Regulatory Economist," *Regulation*:May/June, 12–6.

Yorobe, C., et al. (2004), "Economic Impact of BT Corn in the Philippines," paper presented at the *45th National PAEDA Convention*, BSWM, Quezon City, October 12–13.

Zito, A.R. (1999), "Task Expansion: A Theoretical Overview," *Environment and Planning C: Government and Policy* 17:1, 19–35.

Zito, A.R. (2000), *Creating Environmental Policy in the European Union* (Houndmills: Palgrave Macmillan).

Index

Milton Keynes UK
Ingram Content Group UK Ltd.
UKHW031143141024
449569UK00024B/1118